1988

University of St. Francis
GEN 541.28 Y322
Yates, Keith.
Hückel molecular orbital theo

3 0301 00075728 2

D1566599

# HÜCKEL MOLECULAR ORBITAL THEORY

# Hückel Molecular Orbital Theory

**KEITH YATES**
*Department of Chemistry*
*Lash Miller Chemical Laboratories*
*University of Toronto*
*Toronto, Canada*

 **ACADEMIC PRESS, INC.**
(Harcourt Brace Jovanovich, Publishers)
**Orlando  San Diego  New York  London**
**Toronto  Montreal  Sydney  Tokyo**

COPYRIGHT © 1978, BY ACADEMIC PRESS, INC.
ALL RIGHTS RESERVED.
NO PART OF THIS PUBLICATION MAY BE REPRODUCED OR
TRANSMITTED IN ANY FORM OR BY ANY MEANS, ELECTRONIC
OR MECHANICAL, INCLUDING PHOTOCOPY, RECORDING, OR ANY
INFORMATION STORAGE AND RETRIEVAL SYSTEM, WITHOUT
PERMISSION IN WRITING FROM THE PUBLISHER.


ACADEMIC PRESS, INC.
Orlando, Florida 32887



*United Kingdom Edition published by*
ACADEMIC PRESS, INC. (LONDON) LTD.
24/28 Oval Road, London NW1 7DX




Library of Congress Cataloging in Publication Data

Yates, Keith.
   Hückel molecular orbital theory.

   Includes bibliographical references and index.
   1. Molecular orbitals.  I. Title.
QD461.Y37     541'.28   78-8409
ISBN 0-12-768850-1



PRINTED IN THE UNITED STATES OF AMERICA

84 85 86 87    9 8 7 6 5 4 3 2

# CONTENTS

PREFACE     x

## I INTRODUCTION     1

    I.1 The Basic Postulates of Quantum Mechanics     2
    I.2 Hamiltonian Operators     8
    I.3 Theories of Chemical Bonding     10
    Problems     25
    References     26
    Supplementary Reading     26

## II HÜCKEL MOLECULAR ORBITAL THEORY     27

    II.1 Basic Assumptions     27
    II.2 The Variation Principle     30
    II.3 The Basic Hückel Method     37
    II.4 Application of the HMO Method to Simple $\Pi$-Systems     44

II.5 Calculation of the MO
 Coefficients 60
II.6 Bond Orders and Electron
 Densities 69
II.7 Alternant and Nonalternant
 Hydrocarbons 78
 Problems 85
 References 87
 Supplementary Reading 87

## III THE USE OF SYMMETRY PROPERTIES IN SIMPLIFYING HMO CALCULATIONS 88

III.1 Application of Elementary
 Group Theory 88
III.2 Use of Symmetry to Simplify
 Secular Determinants 97
 Problems 124
 References 126
 Supplementary Reading 126

## IV POLYENE STABILITIES, HÜCKEL'S RULE, AND AROMATIC CHARACTER 127

IV.1 Acyclic and Cyclic Polyenes 129
IV.2 Even- and Odd-Numbered
 Linear Polyenes 131
IV.3 Linear and Branched Chain
 Polyenes 133
IV.4 Cyclic Systems Containing
 $(4n + 2)$ $\pi$-Electrons 134
IV.5 Hückel's Rule and the
 Annulenes 143
IV.6 Polycyclic Systems 150
 Problems 152

Contents                                                                    vii

|              | References                  | 153 |
|              | Supplementary Reading       | 154 |

## V  EXTENSIONS AND IMPROVEMENTS OF THE SIMPLE HÜCKEL METHOD    156

| V.1 | Systems Involving Heteroatoms | 156 |
| V.2 | Inclusion of Differential Overlap | 170 |
| V.3 | Self-Consistent HMO Methods | 180 |
| V.4 | The Extended Hückel (EHMO) Method | 190 |
|     | Problems | 201 |
|     | References | 205 |
|     | Supplementary Reading | 205 |

## VI  THE QUANTITATIVE SIGNIFICANCE OF HMO RESULTS    206

| VI.1 | The Relationship of HMO Results to Molecular Properties | 206 |
| VI.2 | Conclusion | 233 |
|      | Problems | 234 |
|      | References | 237 |
|      | Supplementary Reading | 238 |

## VII  THE PRINCIPLE OF CONSERVATION OF ORBITAL SYMMETRY    239

| VII.1 | Selection Rules for Intramolecular Cycloadditions | 242 |

|  |  |  |
|---|---|---|
| VII.2 | The Woodward–Hoffmann Rules | 245 |
| VII.3 | Energy Level Correlation Diagrams | 250 |
| VII.4 | State-Correlation Diagrams | 256 |
| VII.5 | Intermolecular Cycloadditions | 263 |
| VII.6 | Sigmatropic Reactions | 276 |
| VII.7 | Generalized Selection Rules for Pericyclic Reactions | 284 |
|  | Problems | 288 |
|  | References | 290 |
|  | Supplementary Reading | 290 |

# VIII  THE MÖBIUS–HÜCKEL CONCEPT  291

|  |  |  |
|---|---|---|
| VIII.1 | Hückel Systems | 291 |
| VIII.2 | Möbius Systems | 293 |
| VIII.3 | Application of the Möbius–Hückel Differentiation to Concerted Reactions | 301 |
|  | Problems | 311 |
|  | References | 313 |
|  | Supplementary Reading | 313 |

# IX  SYMMETRY, TOPOLOGY, AND AROMATICITY  314

|  |  |  |
|---|---|---|
| IX.1 | Aromaticity for Pericyclic and Other Topologies | 314 |
| IX.2 | HMO Calculations on Nonplanar Systems | 334 |

| | | | |
|---|---|---|---|
| | IX.3 | Orbital Interaction Diagrams | 345 |
| | IX.4 | MO Following | 357 |
| | | Problems | 364 |
| | | References | 365 |
| | | Supplementary Reading | 366 |

*INDEX*                                               367

# PREFACE

As pointed out more than a decade ago by J. D. Roberts in his concise and valuable introduction to the simple LCAO method,[1] any organic chemist with no more than high school algebra can make useful calculations of semiempirical electronic energies and electron distributions for typical organic systems of interest. At that time Roberts stated, "there is no excuse for a modern organic chemist not to be able to use the LCAO method." With the passage of time, which has led to significant advances in our understanding of structure and reactivity, and particularly because of the development of powerful methods based on the principle of conservation of orbital symmetry, there is even less excuse today for such an inability on the part of organic chemists. The original development and continued use of the simple Hückel Molecular Orbital (HMO) method has provided organic chemists with a great many results that, although crude and approximate, have provided many significant insights into the properties and reactivities of organic molecules containing $\pi$-electron systems and increased our theoretical understanding of the nature of chemical bonding. Simple molecular orbital approaches of the HMO type have probably been more successful than many more advanced and sophisticated theoretical approaches in providing organic chemists with real predictions (and not *ex post facto* explanations or rationalizations of well-known properties) concerning the stabilities and reactivities of organic systems. One has only to consider the outstanding successes of Hückel's $(4n + 2)$ rule and of the Woodward–Hoffman rules in this regard. It is also

# Preface

clear that the simple HMO technique has given us a great deal of information and results, both of a qualitative and a semiquantitative nature, that could not have been obtained by use of valence-bond (resonance) theory. It is probably fair to say that the simple Hückel approach seldom, if ever, gives results or suggestions that are clearly at variance with experimental evidence.

This book is intended to be as simple, descriptive, and nonmathematical an introduction as possible to Hückel molecular orbital theory and its application to organic chemistry. It is suggested that the text could provide the basis for a one-semester or one-term course in theoretical organic chemistry, suitable either for juniors or seniors. The book is not intended to give an exhaustive or comprehensive treatment of the subject, but is meant to provide a simpler and more basic text that would complement available and more advanced works such as Streitwieser's "Molecular Orbital Theory for Organic Chemists"[2] and Woodward and Hoffmanns' "The Conservation of Orbital Symmetry."[3] Each of these is an excellent and comprehensive treatment, replete with examples and illustrations from the current literature. Thus, specific examples have been included in the present work only to illustrate particular points. The reader is referred to these and other more comprehensive treatments for examples of the widespread application of simple molecular orbital concepts.

The emphasis in the present text is on basic concepts and methods. It is the author's feeling that many organic chemistry students, both undergraduate and graduate, frequently apply important ideas and approaches such as the Woodward–Hoffmann rules without a sufficiently sound understanding of their theoretical basis. Similarly, many students go on to make calculations of more sophisticated types such as the various CNDO methods or even the LCAO-SCF type in a very mechanical way, without ever having developed a feeling for the quantum mechanical concepts involved, or equally importantly, the nature of the approximations and limitations inherent in these methods.

One advantage of the simple LCAO approaches described in this book is that they can at the same time provide practicing organic students with a basic understanding of quantum mechanical ideas that they can profitably apply in many areas of experimental organic chemistry, and also provide a basis for further theoretical calculations if they become more deeply interested in the theoretical aspects of the subject. The mathematics involved in these simple approaches need deter no organic student; in fact the mathematical simplicity of the HMO approach is one of its great advantages, in that basic quantum mechanical ideas are not obscured in a welter of complex equations and integrals. Numerous problems are included at the end of most chapters. These should be solved as they are approached in the text. (In many cases the answers are very easy to obtain and verify, and in other cases detailed numerical solutions are readily available in standard reference works.) In addition, supplementary texts and references are included at the end of every chapter. (No attempt has been made to include all pertinent references.) Those works indicated with an asterisk are strongly recommended as supplementary reading.

I hope that any organic students who work their way through this text and the problems will derive as much enjoyment from it as I have in writing it.

Finally I would like to thank Miss Helen Ohorodnyk, Professor I. G. Csizmadia, and Professor W. Forst for reading and commenting on various sections of the text and Mrs. Sue McClelland for her patience with the typing.

## REFERENCES

1. J. D. Roberts, "Molecular Orbital Calculations." Benjamin, Reading, Massachusetts, 1962.
2. A. Streitwieser, Jr., "Molecular Orbital Theory for Organic Chemists." Wiley, New York, 1961.
3. R. B. Woodward and R. Hoffmann, "The Conservation of Orbital Symmetry." Verlag Chemie, Weinheim, 1970; and references therein.

# I | INTRODUCTION

The problem of what constitutes the electronic structure of complex molecules is of fundamental importance to most of chemistry, and particularly to organic chemistry, since all of the experimental work in this field involves molecules and not atoms, and usually fairly large molecules. To obtain a satisfactory theoretical understanding of the molecular structure and properties of organic systems we must ultimately turn to quantum mechanics to answer questions concerning thermodynamic stabilities, spectroscopic properties, and chemical reactivities. These questions all resolve themselves finally into the question of the nature of the chemical bonding or electron distribution in the organic molecules. However, the fundamental wave equations that could be applied to the types of covalent bonding normally found in these molecules are impossible to solve exactly for any multibody problem such as that involved in a real molecular system. Even for the simplest molecular species $H_2^+$, which has only two nuclei and one electron, it is necessary to make a simplifying assumption (the Born–Oppenheimer approximation)[1] to arrive at any quantum mechanical solution. For more complex molecules, the situation becomes increasingly more difficult, and further simplifying assumptions or approximations must necessarily be made if we are to obtain even very approximate answers to questions about molecular electronic structures and energies. This has led to the development of various quantum mechanical methods, whose results, albeit crude and approximate, have shed significant light on the nature of chemical bonding and reactivity, particularly in organic chemistry.

The simplest and most approximate of these approaches has come to be known as the *Hückel molecular orbital* (HMO) method,[2] on which most of this book is based. This approach, despite its theoretical naiveté, has several important advantages, particularly for practicing organic chemists. First, because of its simplicity, it is easy to understand and apply, even to fairly complex systems. Second, it does not differ fundamentally from other more sophisticated and less inexact methods, and hence is very useful for developing a basic understanding of and feeling for quantum mechanical concepts, treatments, and results in a way that would not be possible using mathematically more sophisticated methods. Third, and perhaps most important, the results of this approximate theoretical treatment of organic molecules have contributed very significantly to the understanding of organic chemists, particularly in the last ten or fifteen years. As will be shown later, the results of HMO calculations have helped to correlate and explain a wide range of both physical and chemical properties, and have also in recent years helped to provide very significant predictive rules concerning the course of many organic reactions.

In order to provide a satisfactory conceptual basis for such approximate quantum mechanical methods, and to give some understanding of the necessity for and the nature of the simplifying assumptions and approximations that must be made, the remainder of this chapter will deal with basic concepts and definitions that underly all quantum chemical calculations.

## I.1  THE BASIC POSTULATES OF QUANTUM MECHANICS

The quantum mechanical equations that are used to calculate molecular properties are based on a set of fundamental statements or postulates. These concern atomic and molecular properties and hence are outside everyday experience. They cannot be derived a priori or justified in any absolute sense, and the reader is thus asked at first to take these statements purely on faith. However, the main point is that these postulates *are* justified if and only if they are able to explain and

## I.1 The Basic Postulates of Quantum Mechanics

correlate experimentally observed data, to make predictions, and to be generally applicable to the chemical systems in which we are interested.

This set of postulates will serve to introduce the basic concepts and definitions we make frequent use of, either explicitly or implicitly, in all of the calculations and discussions in the rest of this book.

**Postulate I** (a) Any state of a system of $n$ particles (such as a molecule) is described as fully as possible by a function $\Psi$, which is a function only of the spatial coordinates of the particles and the time, that is,

$$\Psi(q_1 q_2 q_3, q_1' q_2' q_3', \ldots, q_1^n q_2^n q_3^n, t)$$

where $q_1 q_2 q_3$ are the coordinates of the first particle, and so forth for each of the $n$ particles.

(b) If we know that such a state is described by a particular $\Psi$, then the quantity $\Psi\Psi^* d\tau$,[†] where $d\tau$ is a volume element based on generalized spatial coordinates, gives the probability of finding $q_1$ for the first particle between $q_1$ and $(q_1 + dq_1)$, $q_2$ between $q_2$ and $(q_2 + dq_2)$, and so on for each of the $n$ particles at a specific time $t$.

Part (a) of this postulate tells us that all the information we need about the properties of any molecular system is contained in some mathematical function $\Psi$, a wave function, which is a function only of the spatial coordinates of the system and the time. If this $\Psi$ includes $t$ explicitly, it is called a time-dependent wave function. However, if the observable properties we are interested in do not change with time, the system is said to be in a stationary state. In this case, the time dependence can be separated out, and we are left with a time-independent or stationary-state wave function $\Psi(q_1 q_2 q_3, \ldots, q_1^n q_2^n q_3^n)$. Since most properties of interest to organic chemists—such as energies, electron densities, dipole moments, bond orders, and bond lengths—are time independent (or time averaged, as far as experimental data are concerned), we will need to deal only with stationary states and their wave functions.

Part (b) of the postulate gives us a physical interpretation of $\Psi$ in terms of its square, or more properly in terms of its $\Psi\Psi^*$ product. However, in order that the $\Psi$ are in accord with physical reality, these

---

[†] If $\Psi$ is the wave function, $\Psi^*$ is its complex conjugate. For example, if $\Psi = f + ig$, $\Psi^* = f - ig$. The approximate wave functions used in this book do not contain complex parts, thus in general $\psi\psi^*$ reduces to $\Psi^2$.

functions are subject to certain reasonable restrictions:

(i) $\Psi$ is everywhere finite, that is, the particles whose behavior $\Psi$ describes must be bound to a nucleus or an assemblage of nuclei.

(ii) $\Psi$ is single-valued; that is, each particle in the system must be in one place at a time only.

(iii) $\Psi^2$ or $\Psi\Psi^*$ must be an integrable function.

These restrictions arise mainly because $\Psi^2\,d\tau$ is a probability distribution function, and therefore its integral over a given region of space can have only one value under a given set of conditions, and this value must be finite. A special case of (iii) is when

$$\int_0^\infty \Psi^2\,d\tau = 1$$

where $\int_0^\infty$ is taken to mean integration over the limits of all coordinates over all space. When this is true, the function $\Psi$ is said to be normalized. Thus the probability of finding each particle in *some* region of space, or the total probability of finding the system in some configuration or other, must be unity. All wave functions dealt with in this book will be normalized by multiplying them where necessary by an appropriate normalization factor.

**Postulate II** For every observable property of a system that is described by some $\Psi$, there exists a linear Hermitian operator. The physical properties we are interested in can be obtained from this operator and the wave function that describes the system.

An operator $\hat{O}$ is linear if

$$\hat{O}(f+g) = \hat{O}f + \hat{O}g \quad \text{and} \quad \hat{O}(af) = a\hat{O}f$$

where $f$ and $g$ are functions and $a$ is a constant. Frequent use will be made of the linearity property of certain operators in handling quantum mechanical equations.

The Hermitian property of the operators we will use ensures that we always obtain real solutions, since Hermitian operators are defined by the relation

$$\int_{\substack{\text{all}\\\text{space}}} \Psi^*\hat{O}\Phi\,d\tau = \int_{\substack{\text{all}\\\text{space}}} \Phi\hat{O}^*\Psi^*\,d\tau$$

where $\Psi$ and $\Phi$ are any two functions that satisfy the above conditions of acceptability and $\hat{O}$ is the operator of interest (usually the Hamilto-

## 1.1 The Basic Postulates of Quantum Mechanics

nian or total energy operator). Since we will only use real $\Psi$ (or $\Phi$), this means that in any quantum mechanical integral of the form

$$\int \Psi_i \hat{O} \Psi_j \, d\tau$$

the order of multiplication and operation with respect to $\Psi_i$ and $\Psi_j$ is immaterial, and the value of the integral will always be real. We will also make frequent use of this property of the operators of interest. It is sufficient for the purposes of this book to know that such a class of operators exists and that all operators to be used (particularly Hamiltonian operators) belong to this class.

**Postulate III** To obtain the operator associated with a given observable, simply take the classical expression for that observable in terms of the coordinates, momenta, and time and make the following replacements:

(i) Each component of momentum $p_q$ is replaced by the differential operator $-i\hbar \, \partial/\partial q$, where $q$ is a generalized coordinate, $i = \sqrt{-1}$, and $\hbar = h/2\pi$. (Note that these operators are both linear and Hermitian, whereas $\partial/\partial q$ alone is not Hermitian.)

(ii) Time $t$ and all spatial coordinates $q_i$ are left formally unchanged, and the corresponding operators are simply multiplication operators.

Although other operators will be considered later on, the operator that is of principal interest is the one associated with the energy values for the system under consideration. For example, the classical expression for the kinetic energy of a single particle, such as an electron, using Cartesian coordinates is

$$T = \frac{1}{2m}(p_x^2 + p_y^2 + p_z^2)$$

where $p_q$ is the momentum component along the $q$ direction. Using step (i) this becomes the kinetic energy operator

$$\hat{T} = \frac{1}{2m}\left[\left(-i\hbar \frac{\partial}{\partial x}\right)^2 + \left(-i\hbar \frac{\partial}{\partial y}\right)^2 + \left(-i\hbar \frac{\partial}{\partial z}\right)^2\right]$$

$$= -\frac{\hbar^2}{2m}\left(\frac{\partial^2}{\partial x^2} + \frac{\partial^2}{\partial y^2} + \frac{\partial^2}{\partial z^2}\right) = -\frac{\hbar^2}{2m}\nabla^2$$

where $\nabla^2$ is the Laplacian operator and $m$ is the mass of the particle at rest.

The potential energy $V$ is a function only of the spatial coordinates of the particle, and contains physical constants such as the electronic and nuclear charge. For example, the classical expression for the potential energy of a single electron $i$ of charge $e$ in the field of a nucleus $n$ of charge $Ze$ is given by

$$V = -Ze^2/r_{ni}$$

where $r_{ni}$ is the distance between the electron and the nucleus, which can be expressed in terms of coordinates $x, y, z$. This expression remains unchanged in quantum mechanics and hence $V(x, y, z) \to \hat{V}(x, y, z)$. Hence the operation involved under $\hat{V}$ is simply multiplication of some function by $\hat{V}$.

The classical expression for the total energy $E$ is Hamilton's function

$$E = T + V$$

and thus the associated quantum mechanical operator is

$$\hat{\mathscr{H}} = \hat{T} + \hat{V} = -\frac{\hbar^2}{2m}\nabla^2 + \hat{V}(x, y, z)$$

This is the Hamiltonian or total energy operator $\hat{\mathscr{H}}$ for the system. (Examples of other Hamiltonian operators are given in Section I.2.)

**Postulate IV** If $\hat{P}$ is an operator corresponding to some physical observable $p$, and there is a set of states each described by a function $\Psi_s$, which is an eigenfunction of $\hat{P}$, then a series of measurements of the physical quantity corresponding to $p$ would always give the same result $p_s$ for a given state $s$, which would be the eigenvalue of the operator $P_s$, where

$$\hat{P}\Psi_s = p_s\Psi_s$$

such that $p_s$ is a real number.

For example, measurements of the energy of a series of identical systems in a state described by some wave function $\Psi_n$ that is an eigenfunction of the total energy operator $\hat{\mathscr{H}}$ will always give the same result $E_n$. Thus

$$\hat{\mathscr{H}}\Psi_n = E_n\Psi_n$$

and the eigenvalue is always exactly $E_n$. Therefore the problem of computing allowed energy states of a given system is reduced to the problem of finding the functions $\Psi_n$ and energy values $E_n$ that satisfy this

## I.1 The Basic Postulates of Quantum Mechanics

equation, since one can write down the expression for $\hat{\mathcal{H}}$ for any molecular system (at least in principle). For example, for a one-particle system in a stationary state we obtain (dropping the operator subscripts)

$$\hat{\mathcal{H}}\psi = E\psi$$

or

$$\left(-\frac{\hbar^2}{2m}\nabla^2 + V\right)\psi = E\psi$$

or

$$\frac{\hbar^2}{2m}\nabla^2\psi + (E - V)\psi = 0$$

which is the well-known Schrödinger wave equation for a single particle. We can at least write down similar expressions for molecular many-particle systems using the appropriate Hamiltonian operators (as Section I.2).

Frequently, however, we may wish to know something about a property of a real molecular system that cannot be characterized by an eigenfunction strictly appropriate to that property (i.e., we may not be able to obtain the correct eigenfunction of an operator, or we may not be able to write down, or handle mathematically, the correct operator). For this reason, one more postulate is necessary.

**Postulate V**  Given an operator $\hat{P}$ and a set of identical systems characterized by some function $\Phi_s$, which is not a true eigenfunction of $\hat{P}$, measurement of the property associated with $\hat{P}$ would not give the same answer every time. Instead, a distribution of results would be obtained, the average of which would be given by

$$\frac{\int \Phi_s \hat{P} \Phi_s \, d\tau}{\int \Phi_s^2 \, d\tau} = \langle p_s \rangle$$

where all integrals are taken over all space.

The quantity $\langle p_s \rangle$ is the average or expectation value of the physical quantity associated with $\hat{P}$. This expectation value may or may not be close to the true or exact eigenvalue $p_s$, but there is frequently nothing we can do about this. This is generally the best we can do and we must therefore try to obtain the best possible expectation values of the property of interest, usually the energy. This last postulate is needed quite generally, since we will not use exact Hamiltonians and therefore can never get exact eigenfunctions or eigenvalues (although these terms are used nonetheless for the approximate wave functions and energies derived from them).

## I.2  HAMILTONIAN OPERATORS

To assess the nature of the problem of solving the equation

$$\mathscr{H}\Psi = \mathscr{E}\Psi$$

to obtain the wave functions and energies for any real atomic or molecular system, we consider the form of the Hamiltonian operator for several typical systems of interest to chemists. The simplest chemical system is the hydrogen atom, with one electron and one nucleus, for which

$$\mathscr{H} = \left( -\frac{\hbar^2}{2m_e}\nabla^2 - \frac{e^2}{r_{ni}} \right)$$

In this case the operator contains only three kinetic energy terms and one potential energy term. The wave equation can be solved exactly to yield the well-known hydrogen atomic orbital wave functions (1s, 2s, 2p, 3s, 3p, 3d, and so forth),[3] and the energies of these allowed states of an electron in a hydrogen atom are thus known exactly. Similarly, exact solutions can be obtained for any other one electron atomic system such as $He^+$ or $Li^{2+}$.

The Hamiltonian for any other atom (with $n$ electrons and nuclear charge $Z_\mu$) is

$$\mathscr{H} = \sum_{i=1}^{n} \left( -\frac{\hbar^2}{2m_i}\nabla_i^2 - \frac{Z_\mu e^2}{r_{i\mu}} + \sum_{\substack{j \\ (i<j)}} \frac{e^2}{r_{ij}} \right)$$

where $r_{i\mu}$ is the electron–nucleus distance and $r_{ij}$ the interelectronic distance. This expression contains $3n$ electronic kinetic energy terms, $n$ electron–nucleus attraction terms, and $n(n-1)/2$ interelectronic repulsion terms. The resulting wave equations cannot be solved exactly

## I.2  Hamiltonian Operators

for any polyelectronic atom, thus approximations must be made.[†] However, by using approximate methods, reasonably exact energies can be obtained for simple atoms, although the resulting wave functions cannot be represented in closed analytical form.

If we now turn to molecules (with $n$ electrons and $\eta$ nuclei of charge $Z_\mu$), the expression becomes even more complex, and more approximations and simplifying assumptions must be made in order to handle the resulting wave equations

$$\mathcal{H} = \sum_{i=1}^{n} \left( -\frac{\hbar^2}{2m_i} \nabla_i^2 - \sum_{\mu=1}^{n} \frac{Z_\mu e^2}{r_{i\mu}} + \sum_{\substack{j \\ (i<j)}} \frac{e^2}{r_{ij}} \right)$$

Even this expression has been simplified by making the Born–Oppenheimer approximation and omitting all purely nuclear terms. An example of a molecule of interest to organic chemists is ethylene with 16 electrons and six nuclei. For this molecule it is easy to write down the Hamiltonian, which is

$$\mathcal{H} = \sum_{i=1}^{16} \left( -\frac{\hbar^2}{2m_i} \nabla_i^2 - \sum_{\mu=1}^{6} \frac{Z_\mu e^2}{r_{i\mu}} + \sum_{\substack{j=1 \\ (i<j)}}^{16} \frac{e^2}{r_{ij}} \right)$$

Again, this operator has been simplified by omitting purely nuclear terms, yet still contains 48 kinetic energy terms for the electrons, 96 potential energy terms for the electron–nucleus attractions, and 120 potential energy terms for the electron–electron repulsions. It should thus be clear that even by making few approximations and using high-speed computers, solution of the wave equation for ethylene is a very difficult and time-consuming problem; yet this is the simplest "exact" Hamiltonian for any organic molecule containing a $\pi$-electron system.

The difficulty or impossibility of obtaining solutions to the wave equations for molecular systems has led to the development of the two main approximate approaches to chemical bonding. These are the well-known *valence-bond* (VB) theory, due mainly to Heitler and

---

[†] The basic difficulty is that any polyelectronic atom or real molecular system contains at least three particles, and the three-body problem, which even the simplest of these systems constitutes, cannot be solved exactly; although approximate solutions can be obtained to any desired degree of accuracy for very simple systems, this becomes less and less easy the more particles there are in the system (or the more terms there are in the Hamiltonian).

London, Slater, and Pauling,[4] and the *molecular orbital* (MO) theory, due mainly to Hund and Mulliken.[5] These two approaches will be illustrated in the next section, mainly by considering the bonding in a simple case—the hydrogen molecule. A comparison of the two methods will be used to show that although each has its advantages, MO theory is mathematically simpler and easier to apply in approximate form to organic systems.

## I.3  THEORIES OF CHEMICAL BONDING

### I.3.1  VALENCE-BOND THEORY

According to VB theory, molecules are composed of atoms that are bonded together by the valence electrons through individual orbital overlaps. Each overlap (or bond) contains two electrons of opposite spin and bonds two atoms together at a time, such as X and Y in X—Y, no matter what else X and Y might also be bonded to. Each overlap is described by a wave function, which can be called a bond eigenfunction. The properties of the total molecule are then considered to be the total of the properties of the constituent atoms (nuclei and inner shell or nonbonding electrons) and of the individual bonds that bind them together. The total *electronic* structure of the molecule is described by some molecular wave function that is made up of a product of bond eigenfunctions. Thus

$$\psi_{mol}^{VB} = \prod_i \psi_{(x-y)} \prod_j \phi_j$$

where the $\psi_{(x-y)}$ are individual bond functions and the $\phi_j$ represent the functions for the nonbonding electrons. In this case the total electronic energy is given by the sum

$$E_{mol}^{VB} = \sum_i \varepsilon_i + \sum_j \varepsilon_j$$

where the $\varepsilon_i$ are the individual bond energies and the $\varepsilon_j$ are the energies of the nonbonding electrons.

This is a very pictorial approach since each molecular wave function can be associated with a physical picture of what the molecule is like, in terms of a familiar valence bond structure. For example, if we take a simple case such as methane and consider only the bonding electrons,

## I.3 Theories of Chemical Bonding

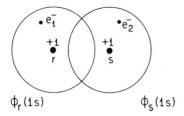

$\phi_r(1s)$  $\phi_s(1s)$

**FIGURE I.1**

the bonding part of the wave function can be represented by

$$\psi_{CH_4} = \prod_{i=1}^{4} \psi_{(C-H)_i}$$

which corresponds to the familiar chemical structure (I).

$$\require{mhchem} H-\underset{H}{\overset{H}{C}}-H$$

I

From this function we would obtain

$$E_{CH_4} = \sum_{i=1}^{4} \varepsilon_{(C-H)_i}$$

To illustrate how the individual bond functions are constructed and how their energies are derived, it is simpler to consider the hydrogen molecule, since there are only two atoms and one bond. Starting from two hydrogen atoms labeled $r$ and $s$, each containing an electron in a hydrogen 1s atomic orbital labeled 1 and 2, these can be brought together so that the 1s functions overlap. Thus one suitable approximate bond function would be $\phi_r(1)\phi_s(2)$, which could be represented pictorially as shown in Fig. I.1. In these representations each electron is mainly associated with only one nucleus at a time. This is a possible bond function and at the same time a possible wave function for $H_2$. An equally possible function would be $\phi_r(2)\phi_s(1)$, in which we have simply exchanged the electrons. Since there is no reason to prefer one over the other, a more acceptable function is a linear combination of the two

$$\psi_{H_2}^{VB} = (\phi_r(1)\phi_s(2) + \phi_r(2)\phi_s(1))$$

where $\phi_r$ is a 1s AO on nucleus $r$ and $\phi_s$ is a 1s AO on nucleus $s$. This VB function is not of the linear combination of atomic orbitals (LCAO) type, which will be described later, since it is a sum of product functions in which each electron is mainly associated with one nucleus at a time, even though the two electrons can exchange nuclei. It is rather a linear combination of two possible molecular wave functions. In simple VB theory, the two electrons are never both associated with one nucleus in any one of the terms; that is, their motion is completely "correlated." A bond is formed between atoms $r$ and $s$ by the overlapping of two AOs (these can be pure AOs or hybrid orbitals) of suitable symmetry, and the bond is represented by the above function $\psi_{H_2}^{VB}$, which is a sum function since there are two possible and equally probable ways of doing this.

The energy of this interaction, represented by $\psi_{H_2}^{VB}$, can be calculated from the wave equation

$$H\Psi_{H_2}^{VB} = E_{H_2}\psi_{H_2}^{VB}$$

(It will be assumed that the full Hamiltonian for this system can be handled mathematically, and that the individual integrals can be evaluated numerically; thus it is not necessary to represent this operator explicitly. In fact, for this simple system the mathematical operations can be performed.) Multiplying each side of the above equation by $\psi_{H_2}^{VB}$ and integrating over the limits of each coordinate gives

$$\psi_{H_2}^{VB}\mathscr{H}\psi_{H_2}^{VB} = E_{H_2}(\psi_{H_2}^{VB})^2$$

and

$$\int_0^\infty \psi_{H_2}^{VB}\mathscr{H}\psi_{H_2}^{VB}\, d\tau = E_{H_2} \int_0^\infty (\psi_{H_2}^{VB})^2\, d\tau$$

which can be rearranged to give

$$E_{H_2} = \frac{\int_0^\infty \psi_{H_2}^{VB}\mathscr{H}\psi_{H_2}^{VB}\, d\tau}{\int_0^\infty (\psi_{H_2}^{VB})^2\, d\tau}$$

Representing this expression as one possible electronic energy ($E_1$) of $H_2$ in abbreviated form (dropping subscripts and superscripts)[†] and

---

[†] From here on, all quantum mechanical integrals will be assumed to be taken over the limits of all coordinates.

## I.3 Theories of Chemical Bonding

substituting for $\psi$ in terms of the expanded function, we obtain

$$E_1 = \frac{\int \psi \mathcal{H} \psi \, d\tau}{\int \psi^2 \, d\tau}$$

$$= \frac{\int [\phi_r(1)\phi_s(2) + \phi_r(2)\phi_s(1)]\mathcal{H}[\phi_r(1)\phi_s(2) + \phi_r(2)\phi_s(1)] \, d\tau}{\int [\phi_r(1)\phi_s(2) + \phi_r(2)\phi_s(1)]^2 \, d\tau}$$

$$= \frac{N}{D}$$

Expansion of this expression (recalling Postulate II) gives for the numerator

$$N = \int \phi_r(1)\phi_s(2)\mathcal{H}\phi_r(1)\phi_s(2) \, d\tau + \int \phi_r(2)\phi_s(1)\mathcal{H}\phi_r(2)\phi_s(1) \, d\tau$$
$$+ 2\int \phi_r(1)\phi_s(2)\mathcal{H}\phi_r(2)\phi_s(1) \, d\tau$$

Thus we can set

$$N = 2Q + 2J$$

where

$$Q = \int \phi_r(1)\phi_s(2)\mathcal{H}\phi_r(1)\phi_s(2) \, d\tau = \int \phi_r(2)\phi_s(1)\mathcal{H}\phi_r(2)\phi_s(1) \, d\tau$$

since the energy terms are obviously unaffected by how we choose to label the two electrons, and

$$J = \int \phi_r(1)\phi_s(2)\mathcal{H}\phi_r(2)\phi_s(1) \, d\tau$$

Terms such as $Q$ are called *Coulomb integrals* and the terms $J$ are called *exchange integrals*.

In addition to these terms we obtain on expanding the denominator

$$D = \int [\phi_r(1)\phi_s(2) + \phi_r(2)\phi_s(1)]^2 \, d\tau$$
$$= \int (\phi_r(1)\phi_s(2))^2 \, d\tau_1 \, d\tau_2 + [\phi_r(2)\phi_s(1)]^2 \, d\tau_1 \, d\tau_2$$
$$+ 2\int \phi_r(1)\phi_s(2)\phi_r(2)\phi_s(1) \, d\tau_1 \, d\tau_2$$
$$= \int \phi_r^2(1) \, d\tau_1 \int \phi_s^2(2) \, d\tau_2 + \int \phi_r^2(2) \, d\tau_2 \int \phi_s^2(1) \, d\tau_1$$
$$+ 2\int \phi_r(1)\phi_s(1) \, d\tau_1 \int \phi_s(2)\phi_r(2) \, d\tau_2$$

Because we are dealing with 1s AOs, which are normalized,

$$\int \phi_i(k)\phi_j(k)\,d\tau_k = 1 \quad \text{if } i = j$$
$$= S \quad \text{if } i \neq j$$

Then

$$D = 1 \cdot 1 + 1 \cdot 1 + 2S \cdot S = 2 + 2S^2$$

where $S$ is called an *overlap integral*. Thus the overall energy of $H_2$ using this VB formulation is given by

$$E_1 = \frac{2Q + 2J}{2 + 2S^2} = \frac{Q + J}{1 + S^2}$$

The significance of the various terms is as follows. The Coulomb integral $Q$ represents mainly the coulombic interaction energy between each electron and its own nucleus, *in the molecule*, and also the attraction of each electron for the other nucleus, since it is partly in the field of this nucleus also. Such terms are obviously negative, since each electron is attracted not only by its own nucleus but by the other as well. Numerical calculation of $Q$ for $H_2$ shows that this term corresponds to about 15% of the total binding energy of the molecule. The exchange integral $J$ is not as easy to associate with a physical description. Quantum mechanical integrals of this type represent the energy associated with the process of exchanging electrons between the two nuclei, and in the case of $H_2$, this term corresponds to about 85% of the total binding energy. Terms such as $J$ are also negative, providing $S > 0$, and in general the larger $S$ is the more negative is $J$. The overlap integral $S$ gives a measure of the effectiveness of interpenetration or overlap of the two 1s AOs at the equilibrium internuclear distance of $H_2$. This relationship between the sign of $J$ and magnitude of $S$ provides a basis for the criterion of maximum overlap in chemical bonding. Since both $Q$ and $J$ are negative and the numerator is positive, $E_1$ represents an attractive or bonding state of $H_2$.

At this point, it should be noted that we could equally well have chosen as an alternative linear combination of functions for $H_2$, the negative combination,

$$\psi = \phi_r(1)\phi_s(2) - \phi_r(2)\phi_s(1)$$

## I.3 Theories of Chemical Bonding

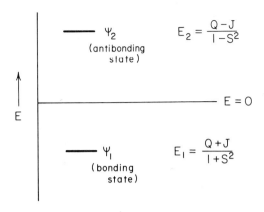

FIGURE I.2

Using the same equation and procedure, this would have yielded as a second possible energy for $H_2$,

$$E_2 = (Q - J)/(1 - S^2)$$

Since $J$ is a negative energy quantity and is greater in magnitude than $Q$, and $(1 - S^2)$ can never be negative, this corresponds to a positive energy. Thus the above wave function for $H_2$ represents a repulsive or antibonding state of the molecule.

The VB picture of the $H_2$ system in schematic form is shown in Fig. I.2, where the spatial parts of the wave function for each state[†] are

$$\psi_1 = \phi_r(1)\phi_s(2) + \phi_r(2)\phi_s(1)$$
$$\psi_2 = \phi_r(1)\phi_s(2) - \phi_r(2)\phi_s(1)$$

Since the two electrons will naturally occupy the bonding state in the stable $H_2$ molecule, the ground-state VB wave function for $H_2$ can be represented more fully as

$$\Psi_{H_2}^{VB} = N[\phi_r(1)\phi_s(2) + \phi_r(2)\phi_s(1)] \cdot \frac{1}{\sqrt{2}} [\alpha(1)\beta(2) - \beta(1)\alpha(2)]$$

where $N$ is a normalization constant, and an electronic spin function

---

[†] Note that these two states are not necessarily symmetrically disposed about zero, which is taken to be the energy of an isolated hydrogen atom.

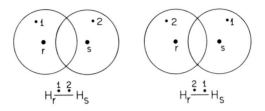

*FIGURE I.3*

has been introduced.† Since the spatial part of $\Psi_{H_2}^{VB}$ is symmetric with respect to interchange of the two electrons, the spin function must be antisymmetric in order to make the total wave function obey the Pauli principle.[6]

The spatial part of this valence bond function for the $H_2$ molecule can be represented pictorially in either of the ways shown in Fig. I.3. In this picture of the $H_2$ molecule bonding arises from two equivalent covalent contributions or terms.

It may be noted that a calculation using the above function and the full Hamiltonian gives a result of $-3.14$ eV or 72 kcal for the binding energy of the $H_2$ molecule.[7] This corresponds to about 70% of the experimental value, which is quite good in view of the approximations made.

This kind of treatment can be applied to most ordinary single bonds in organic molecules, and values of $Q$, $J$, and $S$ can be calculated numerically for one bond at a time, albeit with somewhat more difficulty than for $H_2$. However, for simple systems the calculated energies give quite a reasonable account of the observed binding energies. For example, if methane is treated as a system of four isolated or localized

---

† For two electrons, either of which can have $\alpha(\uparrow)$ or $\beta(\downarrow)$ spin, possible combinations are

$$\alpha(1)\alpha(2)$$

$$\beta(1)\beta(2)$$

$$\frac{1}{\sqrt{2}}[\alpha(1)\beta(2) + \alpha(2)\beta(1)]$$

- - - - - - - - - - - - - - - - - - - -

$$\frac{1}{\sqrt{2}}[\alpha(1)\beta(2) - \alpha(2)\beta(1)]$$

Of these four, only the last is antisymmetric with respect to interchange of the two electrons, and corresponds to a spin-paired ground, or singlet state of the molecule.

## I.3 Theories of Chemical Bonding

single bond interactions (analogous to the H—H bond in $H_2$) to obtain the total binding energy, this does not differ greatly from the result obtained by treating the molecule as a whole (using a product of four bond eigenfunctions). It should be noted, however, that the best values of the energy in methane are obtained by using combinations of pure 1s AOs for hydrogen and $sp^3$ hybrid orbitals for carbon.

The major difficulty with the VB approach arises when treating delocalized bond systems of the types that are common in organic chemistry. This problem can be illustrated in a qualitative way by considering the VB approach to a calculation of the bond energy in a molecule like benzene. First of all, a trial molecular VB function could be set up as a product of bond functions for the various C—C and C—H σ-bond interactions of the benzene framework, and of the C—C π-bond interactions, as represented by the valence bond formula II.

II

This molecular wave function $\Psi_{II}^{VB}$ could then be used to calculate an approximate energy for benzene. The calculation would be considerably more difficult than for $H_2$ or $CH_4$. The approximate energy $E_{II}$ would not give a very good account of the binding energy in benzene, particularly because the π-contribution would be grossly underestimated. The trial VB wave function could be improved by considering a second possible (and equally probable) representation III, corresponding to the

III

valence bond formula and by combining $\Psi_{II}^{VB}$ and $\Psi_{III}^{VB}$ in the form of a linear combination, with weights $a$ and $b$ (which would be equal in magnitude in this simple case). Thus an improved VB function

would be

$$\Psi_{imp}^{VB} = a\Psi_{II} + b\Psi_{III}$$

Calculations using this function would give a better account of the binding energy of benzene, but would be extremely difficult, since the function $\Psi_{imp}^{VB}$ is now a linear combination of two separate and complex *molecular* wave functions. Further improvement could be achieved by including additional functions in the linear combination, such as those represented by the formulas (IV–VI). However, by this time the

      IV            V            VI

calculations would have become mathematically completely intractable, using a starting wave function such as

$$\Psi_{benz}^{VB} = a\Psi_{II} + b\Psi_{III} + c\Psi_{IV} + d\Psi_{V} + \cdots$$

Yet in order to give a reasonable account of the properties and bonding of delocalized systems like benzene, such an approach would be necessary using the VB method.

For this reason, use of the VB method is restricted to fairly simple, localized bonding systems and this approach is not generally applicable to the delocalized systems that are of considerable interest to organic chemists. The main use of the VB approach in organic chemistry has come through application of resonance theory, which is based on VB arguments. It is of considerable interest to compare the predictions of qualitative valence bond–resonance theory[8] with those based on molecular orbital calculations. This type of comparison will be attempted wherever possible throughout the remainder of this book.

## I.3.2 MOLECULAR ORBITAL THEORY

This approach differs from the VB method in that approximate wave functions are constructed to encompass the entire molecule, and the electrons occupying these functions are simultaneously associated with all nuclei in the molecule, and not just two at a time, as in the VB theory. The electronic motion is not restricted in any way, thus the MO approach gives a completely delocalized picture of the bonding

## I.3 Theories of Chemical Bonding

in any molecule. These approximate electronic functions are called molecular orbitals (MOs) and are similar to the more familiar atomic orbitals (AOs) in that each electron is assigned a wave function $\Psi$ that determines its motion, energy, and other properties. The MOs differ from AOs only in that each $\psi_{MO}$ extends over the whole molecular framework, but otherwise are taken to have similar properties to AO wave functions for an electron in a single atom, namely:

(i) $\psi_{MO}^2 \, d\tau$ is a measure of the probability of finding the electron in a particular region of space $d\tau$ at a given time.

(ii) The Pauli principle is assumed to apply to MOs in the same way as to AOs, that is, no two electrons can occupy the same MO with the same spin, thus each MO can hold two electrons at most.

(iii) The size, shape, and energy of the MOs and hence the size, shape, and energy of the molecules themselves are determined by the quantum numbers of the AOs that combine to form the MOs and also by the way in which they combine.

The approximation most commonly used to set up these MO functions (and hence obtain information about molecular structure) is the LCAO method. In this approach, which will be used extensively in this book, each MO wave function is taken to be a *linear combination of atomic orbitals* that describes the energy and motion of one electron at a time (unlike VB functions) and not all electrons in the system. Since electronic spin functions will not be treated explicitly, these one electron MO functions are the same for an electron with either $\alpha$ or $\beta$ spin. Thus each MO can hold two electrons providing their spins are paired.

For a diatomic molecule AB a possible MO function would be the linear combination

$$\Psi_j = C_{jA}\phi_A + C_{jB}\phi_B$$

where $\phi_A$ is an AO on atom A, $\phi_B$ is an AO on atom B, $C_{jA}$ is the coefficient of the Ath AO in the jth MO, and $C_{jB}$ is the coefficient of the Bth AO in the jth MO. We will assume for now that each of the constituent atoms has only one valence shell AO. These AO functions are said to form the *basis set* for the construction of the MO functions and can be obtained (at least in principle) by solving the appropriate wave equations for atoms A and B.† This is much simpler than trying to

---

† The AO functions are most commonly used in the LCAO method are hydrogenlike AOs of the type 1s, 2s, $2p_x$, $2p_y$, $2p_z$, etc., which can be approximated by various methods[9] for nuclei other than hydrogen.

obtain the molecular wave functions that form the linear combinations in the VB approach.

Note that if there are $j$ AOs altogether on the $N$ atoms that form the molecule, there will necessarily be $j$ such MOs (or $j$ linearly independent combinations) that differ from each other only in the values of the coefficients $C_{jN}$ which appear in the LCAO functions. Although each $\psi_{MO}$ is considered separately from all others, the method of obtaining them is the same for each one. Also the method of treating each $\psi_{MO}$ in the calculations based on the wave equation for the molecule is the same as for any other $\psi_{MO}$. It is these two points that make MO calculations mathematically much simpler than VB calculations.

In the above simple case, there are two possible MO functions:

$$\psi_1 = C_{1A}\phi_A + C_{1B}\phi_B$$
$$\psi_2 = C_{2A}\phi_A + C_{2B}\phi_B$$

The coefficients $C_{jN}$ are constants whose values are to be determined in such a way as to give the best possible approximate solutions to the wave equation for the molecule being considered. Once a complete set (in this simple case, two) of these $\psi_i$ is arrived at and their energies determined, the valence electrons are then fed into them, starting with the lowest energy MO, following Pauli's principle and Hund's rule[10] (i.e., degenerate MOs are filled singly as far as possible, before doubly occupying them with paired spins). It can then be determined if a stable molecule, ion, or radical will result (relative to the constituent free atoms), how stable it will be relative to other molecular systems, and what kind of electron distribution it will have.

For more complex molecules containing $m$ atoms, which together have a total of $n$ valence orbitals (AOs), we will obtain $n$ MO wave functions of the form

$$\psi_j = C_{ji}\phi_1 + C_{j2}\phi_2 + C_{j3}\phi_3 + \cdots + C_{jn}\phi_n$$

or

$$\psi_j = \sum_{i=1}^{n} C_{ji}\phi_i$$

where $j = 1, 2, 3, \ldots, n$. In general we will deal with cases where $m = n$ (i.e., only one valence AO will be considered on each atom), although this is in no way necessary to the treatment.

It is important to stress at this point that the LCAO approach to constructing acceptable functions *is* an approximation, although it is

## I.3 Theories of Chemical Bonding

widely used in all molecular orbital calculations. There are also certain conditions for the LCAO approach to be a reasonably valid approximation:

(i) All $\phi_n$ in the basis set of functions used for each MO should be comparable in energy.

(ii) The electron distributions of adjacent $\phi_i$ and $\phi_j$ in the MOs should overlap appreciably, that is, the total extent of overlap of adjacent AOs in the MO constructed from them is one measure of the strength of bonding interactions (which may be either favorable or unfavorable) between them.

(iii) Any $\phi_i$ and $\phi_j$ that are combined in an MO must have the same symmetry with respect to molecular or bond axes.

Applying this treatment to the hydrogen molecule, we can set up MOs using the LCAO approach, which will be of the form:

$$\psi_j^{MO} = C_{jr}\phi_r + C_{js}\phi_s$$

where $\phi_r$ is a 1s AO on hydrogen atom $r$, $\phi_s$ is a 1s AO on hydrogen atom $s$, $C_{jr}$ is the weight of the $r$th AO in the $j$th MO, and $C_{js}$ is the weight of the $s$th AO in the $j$th MO. Since there is only one AO on each atom, there are only two functions in the basis set. Thus there will be two MOs of the above form, and from symmetry considerations each $C_{jr}$ will be equal to $C_{js}$ in absolute magnitude. In simple form,[†] the only two linearly independent MOs for the $H_2$ molecule are

$$\psi_1^{MO} = (\phi_r + \phi_s)$$
$$\psi_2^{MO} = (\phi_r - \phi_s)$$

Using the appropriate wave equation as before, we can obtain expressions for the energy of the two MOs of the $H_2$ molecule. Thus for $\psi_1^{MO}$

$$\mathcal{H}\psi_1^{MO} = E_1\psi_1^{MO}$$

---

[†] Note that a normalization factor $N$ has been omitted from these functions for simplicity. This can be reinserted later. Note also that the two MOs are orthogonal to each other, that is, $\int \psi_1^{MO}\psi_2^{MO} \, d\tau = 0$. This is true for all sets of MOs for any molecule, namely, that any two MOs must be mutually orthogonal. What this means is that the probability of any one electron simultaneously having the spatial characteristics of two MO wave functions is zero, or, in other words, an electron cannot occupy two different MOs at the same time.

and multiplying both sides by $\psi_1^{MO}$ integrating over all space, and rearranging, gives

$$E_1 = \frac{\int \psi_1^{MO} \mathcal{H} \psi_1^{MO} \, d\tau}{\int (\psi_1^{MO})^2 \, d\tau}$$

Inserting the full expression for $\psi_1^{MO}$ and using the basic postulates and operator properties, this can be expanded to give

$$E_1 = \frac{\int (\phi_r + \phi_s) \mathcal{H} (\phi_r + \phi_s) \, d\tau}{\int (\phi_r + \phi_s)^2 \, d\tau}$$

$$= \frac{\int \phi_r \mathcal{H} \phi_r \, d\tau + \int \phi_s \mathcal{H} \phi_s \, d\tau + 2 \int \phi_r \mathcal{H} \phi_s \, d\tau}{\int \phi_r^2 \, d\tau + \int \phi_s^2 \, d\tau + 2 \int \phi_r \phi_s \, d\tau}$$

If we set

$$\alpha = \int \phi_r \mathcal{H} \phi_r \, d\tau = \int \phi_s \mathcal{H} \phi_s \, d\tau$$

$$\beta = \int \phi_r \mathcal{H} \phi_s \, d\tau$$

$$S = \int \phi_r \phi_s \, d\tau$$

and recall that $\phi_r$ and $\phi_s$ are normalized 1s AOs, the above expression becomes

$$E_1 = \frac{2\alpha + 2\beta}{2 + 2S} = \frac{\alpha + \beta}{1 + S}$$

Terms such as $\alpha$ are called *Coulomb integrals* and represent negative energy quantities corresponding to the energy of a hydrogen 1s electron in the field of its own nucleus (note that despite the same name, these Coulombic integrals are of different form from those that arise in the VB treatment). Terms such as $\beta$ are called *bond* or *resonance integrals* and are also negative energy quantities, but correspond to the energy of an electron in the field of both nuclei simultaneously. It is these terms that give rise (in the MO approach) to the net bonding energy and stability of the $H_2$ molecule. (Note that these integrals do not correspond directly to the exchange integrals in the VB theory.) Both types of integral $\alpha$ and $\beta$ are soluble numerically for this simple molecular system.

## I.3 Theories of Chemical Bonding

The terms $S$ are as before the overlap integrals that measure the extent of overlap of the two 1s AOs when the two nuclei are brought together to the equilibrium distance in $H_2$. Again, mathematically, $\beta < 0$ only if $S \neq 0$, and the larger $S$ is the more negative is $\beta$.

Thus for $\psi_1^{MO}$ the calculated energy $E_1 = (\alpha + \beta)/(1 + S)$ corresponds to a negative energy level, that is, the function $\psi_1^{MO}$ is a *bonding molecular orbital*, or more simply a bonding orbital.

For the second MO

$$\psi_2^{MO} = (\phi_r - \phi_s)$$

it is clear that a similar treatment would give

$$E_2 = \frac{\alpha - \beta}{1 - S}$$

and since $\beta$ is generally greater in absolute magnitude than $\alpha$ (and both are negative energy terms), this corresponds to a positive energy level or an *antibonding molecular orbital*.

The two allowed energy levels for electrons in the $H_2$ molecule can be represented diagramatically as shown in Fig. I.4 and the two electrons are placed in the lower MO with spins paired. (Note that these energies always correspond to allowed *levels* for a given electron in the MO approach, whereas the corresponding energies in the VB approach always correspond to allowed total molecular *states*. In this simple case the two correspond directly since there are only two electrons and one bonding interaction in the molecule, but this is not the case in general.)

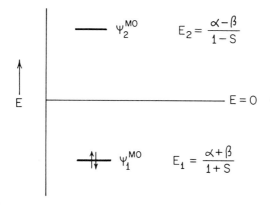

FIGURE I.4

$$\psi_{H_2}^{MO} = \left[ \phi_r(1)\,\phi_r(2) + \phi_s(1)\,\phi_s(2) + \phi_r(1)\,\phi_s(2) + \phi_r(2)\,\phi_s(1) \right]$$

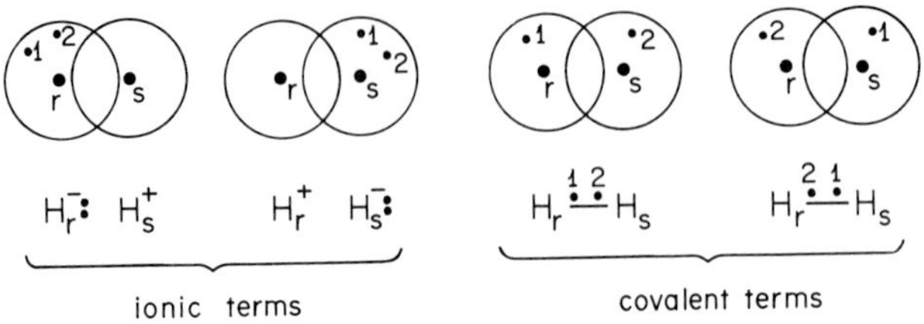

$$\underbrace{H_r^{-}\!:\;H_s^{+} \qquad H_r^{+}\;H_s^{-}\!:}_{\text{ionic terms}} \qquad \underbrace{H_r\overset{1\;2}{\cdot\cdot}H_s \qquad H_r\overset{2\;1}{\cdot\cdot}H_s}_{\text{covalent terms}}$$

**FIGURE I.5**

With the two electrons [designated by (1) and (2)] placed in the bonding MO, the complete MO wave function for $H_2$ is thus

$$\psi_{H_2}^{MO} = N[(\phi_r + \phi_s)(1)][(\phi_r + \phi_s)(2)]\sigma_{\text{spin}}$$

where $N$ is the appropriate normalization factor, and $\sigma_{\text{spin}}$ is a similar spin function to that designated previously. It is instructive to expand the spatial part of this function and compare it with the VB representation of $H_2$, in pictorial form (Fig. I.5). This approach gives a reasonable account of the bonding energy of $H_2$, but its deficiency is that it gives too much weight to the ionic terms and not enough to the covalent terms.

On the other hand, the VB approach gives the wave function (spatial part only)

$$\psi_{H_2}^{VB} = [\phi_r(1)\phi_s(2) + \phi_r(2)\phi_s(1)]$$

which contains only the two covalent terms and no contribution from ionic terms. Thus both approaches are deficient in the sense that the MO function allows the electronic motion to be completely unrestricted, whereas the VB function causes the motion of each electron to be completely correlated with respect to the other. Each function can be improved; in the case of the MO function by introducing a measure of electron correlation by including some antibonding character,[†] and for the VB function by including some contribution from ionic terms. Thus, neglecting spin, these improved wave functions

---

[†] This is a simple form of configuration interaction (Ref. 11).

would have the form

$$\Psi_{H_2}^{MO}(\text{imp}) = [(\phi_r + \phi_s)(1)(\phi_r + \phi_s)(2)]$$
$$+ k[(\phi_r - \phi_s)(1)(\phi_r - \phi_s)(2)]$$
$$\Psi_{H_2}^{VB}(\text{imp}) = [\phi_r(1)\phi_s(2) + \phi_r(2)\phi_s(1)]$$
$$+ \lambda[\phi_r(1)\phi_s(2) + \phi_s(1)\phi_s(2)]$$

Expansion of these expressions shows that apart from a normalizing factor, they are entirely equivalent providing

$$\lambda = -\frac{1+k}{1-k}$$

If we were to improve either the VB or MO function for $H_2$ until no further change in calculated energy were obtained, we would get precisely the same energy and wave function using either approach. Thus both approaches become equivalent if we take them beyond their simple forms by correcting for their deficiencies. This can be shown to be true in general for complex molecules as well as $H_2$.

Thus inherently neither theory, MO or VB, is superior to the other if carried far enough. However, the MO theory and its approach to more complex molecules (particularly those containing delocalized $\pi$-systems) is much simpler mathematically, especially in the approximate forms that are necessary for calculations on most molecules of interest to organic chemists. It is for this reason that we will essentially consider only the MO approach for the remainder of this book.

## PROBLEMS

1. Show that the operator $id/dx$ is Hermitian whereas the operator $d/dx$ is not. (In expressions such as $\int \psi^* \hat{O} \Phi \, d\tau$ let $\psi^*$ and $\Phi$ be functions of $x$ only, such as $\psi^* = A - iB$ and $\Phi = E + iF$, where $A$, $B$, $E$, and $F$ are functions of $x$.)
2. Write down full expressions for the Hamiltonians for the helium and lithium atoms.
3. Write down simplified Hamiltonians (omitting purely nuclear terms) for the molecules carbon monoxide, carbon dioxide, and benzene. Count the total number of kinetic and potential energy terms each contains.
4. Using the modified VB and MO formulations for the hydrogen molecule (see p. 25) calculate the total energy in each case in terms

of the appropriate integrals. Show that these energies are equivalent when $\lambda = -[(1 + k)/(1 - k)]$.

5. Set up reasonable LCAOs for the He$_2$ molecule and use the same procedure as for H$_2$ to obtain the MO energies. Write down a final expression for $\psi_{He_2}^{MO}$ excluding spin. Calculate the total energy of the molecule (in terms of $\alpha$ and $\beta$ units). Would the resulting molecule be stable relative to two isolated He atoms if $\alpha$ and $\beta$ were approximately equal in magnitude? To what extent is the stability of the He$_2$ molecule dependent on the magnitude of the overlap integral $S$?

6. How would you go about trying to construct the corresponding VB wave function for He$_2$? What step would be necessary initially?

## REFERENCES

1. M. Born and R. Oppenheimer, *Ann. Phys. (Leipzig)* **84**, 457 (1927).
2. E. Hückel, *Z. Phys.* **70**, 204 (1931).
3. C. A. Coulson, "Valence," p. 22. Oxford Univ. Press, London and New York, 1952. H. E. White, *Phys. Rev.* **37**, 1416 (1931).
4. W. Heitler and F. London, *Z. Phys.* **44**, 455 (1927); J. C. Slater, *Phys. Rev.* **38**, 1109 (1921); L. Pauling, *J. Chem. Phys.* **1**, 280 (1933).
5. F. Hund, *Z. Phys.* **40**, 742 (1927); R. S. Mulliken, *Phys. Rev.* **32**, 186 (1928); R. S. Mulliken, *J. Chem. Phys.* **3**, 375 (1935); and references therein. See also R. S. Mulliken, *Int. J. Quantum Chem* **1**, 103 (1967).
6. See Coulson, Ref. 3, p. 136.
7. R. G. Parr, "Quantum Theory of Molecular Electronic Structure," p. 14. Benjamin, New York, 1963.
8. G. W. Wheland, "Resonance in Organic Chemistry." Wiley, New York, 1955.
9. J. C. Slater, *Phys. Rev.* **36**, 57 (1930). See also Ref. 3, p. 40.
10. F. Hund, *Z. Phys.* **33**, 345 (1925); **34**, 296 (1925).
11. M. J. S. Dewar, "The Molecular Orbital Theory of Organic Chemistry," p. 102. McGraw-Hill, New York, 1969. R. L. Flurry, Jr., "Molecular Orbital Theories of Bonding in Organic Molecules," p. 248. Dekker, 1968, New York.

## SUPPLEMENTARY READING

Coulson, C. A., "Valence." Oxford Univ. Press, London and New York, 1952.
*Hanna,[†] M. W., "Quantum Mechanics in Chemistry." Benjamin, New York, 1969.
Flurry, R. L., "Molecular Orbital Theories of Bonding in Organic Molecules." Dekker, New York, 1968.

[†] Those references marked with an asterisk in each chapter are especially recommended.

# II | *HÜCKEL MOLECULAR ORBITAL THEORY*

## II.1 BASIC ASSUMPTIONS

The HMO method is a simple but powerful approach toward explaining the stabilities, physical properties, and chemical reactivities of organic $\pi$-systems. The $\sigma$-electronic frameworks of these systems are relatively uninteresting in a theoretical sense since the properties of $\sigma$-bonds do not vary significantly from structure to structure. The HMO results that one can obtain easily are admittedly crude and approximate, being based on a number of gross assumptions or approximations. However, these results are capable of explaining and predicting a large amount of interesting chemistry and seldom if ever give answers that are clearly in disagreement with either chemical experience or chemical intuition.

If we start with the equation

$$\mathscr{H}\Psi + E\Psi$$

in which $\mathscr{H}$ is the electronic Hamiltonian operator, including all electronic and nuclear interaction terms, this is completely insoluble for any organic molecule containing $\pi$-electrons (recall the form of this operator for ethylene). Therefore, it is necessary to make several basic assumptions and to use Postulate V to obtain approximate "eigenvalues" and approximate "eigenfunctions" for these $\pi$-systems.

The first assumption is that the total wave function for a polyelectronic system can be factored into sets of independent, noninteracting electronic systems $\Phi$, each of which describes the behavior of a particular set of electrons. If we consider molecules in which the electrons can reasonably be described as being either $\sigma$- or $\pi$-electrons, then the assumption is that

$$\Psi_{\text{polyelect}} = \Phi_\sigma \Phi_\pi$$

and further that

$$E_{\text{tot}} = E_\sigma + E_\pi$$

This is not a bad approximation for many organic molecules since $\sigma$-systems seem to be reasonably independent of the $\pi$-system in the same molecule, and to a lesser extent vice versa. (For example, inductive and resonance effects seem to be approximately additive.)

It is further assumed that $\Phi_\sigma$ is divisible into a set of relatively localized two-center $\sigma$-bonds as in saturated molecules and that these can be adequately treated, if so desired, by localized (two-center) VB or MO methods. Thus

$$\Phi_\sigma = \prod_i \theta_{\sigma_i}$$

and

$$E_\sigma^{\text{tot}} = \sum_i \varepsilon_{\sigma_i}$$

where the $\theta_{\sigma_i}$ are localized bond functions (as in the $H_2$ molecule) and the $\varepsilon_{\sigma_i}$ are the $\sigma$-bond energies. This again is not an unreasonable assumption since $\sigma$-bond energies seem to be fairly constant from molecule to molecule and to a good approximation additive in saturated systems.

Clearly no corresponding assumption can be made in general about $\pi$-electron systems, since these frequently have properties that reflect extensive delocalization. Thus $\Phi_\pi$ is treated as a product of individual MOs, each of which is delocalized over the entire framework of the molecule. Each $\psi_{\text{MO}}$ is obtained by the LCAO approximation, as a linear combination of $2p_z$-type AOs all sharing the same nodal plane (if this is geometrically possible; if it is not, each $\pi$-system can be treated separately).

## II.1 Basic Assumptions

Thus

$$\Phi_\pi = \prod_j \psi_j$$

where each $\psi_j$ is of the form

$$\psi_j = C_{j1}\phi_1 + C_{j2}\phi_2 + C_{j3}\phi_3 + \cdots + C_{jn}\phi_n = \sum_{i=1}^{n} C_{ji}\phi_i$$

and there are $n$ such MO wave functions ($j = 1, 2, \ldots, n$). In these expressions $\phi_i$ is a $2p_z$ AO on the $i$th carbon atom[†] in a $\pi$-system containing $n$ sp²-hybridized carbon atoms and $C_{ji}$ is the coefficient of the $i$th AO in the $j$th MO.

These $\psi_j$ are "eigenfunctions" of a Hamiltonian operator ($\mathcal{H}_{\text{eff}}$) that is considered to be effective for the energy and behavior of the $\pi$-system only. Thus

$$\mathcal{H}_{\text{eff}}\psi_j = E_j\psi_j$$

The operator $\mathcal{H}_{\text{eff}}$ is an approximate "core" Hamiltonian that is considered for the motion of the $\pi$-electrons about a "core" consisting of the nuclei, the nonbonding and inner-shell electrons, and the $\sigma$-electronic framework of the molecule. The $\pi$-electrons are taken to move in the total potential field of this "core." The function $\psi_j$ is an approximate wave function that describes *one* possible distribution of up to two electrons, with opposite spin but with identical other properties. For this reason, these $\psi_j$ are called "one-electron" wave functions. Similarly, $E_j$ represents the approximate energy of *one* allowed state of up to two electrons.

It is because the above approximations are needed for complex organic systems that a fifth Postulate was necessary. Therefore, we can only deal in terms of expectation values for the allowed energies. Fortunately, we do not need to consider the explicit form of the "core" Hamiltonian. The problem is to find a set of one-electron $\pi$-wave functions ($\psi_j$) such that the expectation values of their energies are a minimum for every member of the set. This is the best we can possibly do within the present theoretical framework and in view of the complexity of the molecules to be considered.

---

[†] Only hydrocarbon $\pi$-systems will be considered for the present. The question of heteroatoms will be introduced later.

Using Postulate V, we have for each $\psi_j$

$$\langle E_j \rangle = \frac{\int \psi_j \mathcal{H}_{\text{eff}} \psi_j \, d\tau}{\int \psi_j^2 \, d\tau}$$

and it is required to minimize $\langle E_j \rangle$ for each $j$ in the set.

The problem is thus to find sets of coefficients $C_{ji}$ for each MO $\psi_j$ that give the lowest possible values for each energy $E_j$, and therefore the lowest value for $E_\pi^{\text{tot}}$, where

$$E_\pi^{\text{tot}} = 2 \sum_{j=1}^{n} (E_j)_{\text{occ}}$$

To do this, we make use of the *variation method* based on the principle described in the next section.

## II.2 THE VARIATION PRINCIPLE

This principle can be stated as follows: Given any approximate wave function $\psi_j$ satisfying the boundary conditions of the problem, the expectation value of the energy $\langle E_j \rangle$, calculated from this function, will always be higher than the true energy $E_0$ of the ground-state configuration that is described approximately by $\psi_j$. Thus, for any approximate $\psi_j$,

$$\langle E_j \rangle = \frac{\int \psi_j \mathcal{H} \psi_j \, d\tau}{\int \psi_j^2 \, d\tau} > E_0$$

An intuitive proof of this principle is the following: The ground-state configuration of any set of electrons must be the most stable distribution possible for that set, and must be describable by *some* function $\psi_0$ with energy $E_0$. Any other distribution of these electrons, and hence any other $\psi$, must represent a less probable arrangement and thus must correspond to a higher energy.[1]

If this is so, the best we can do is to minimize $\langle E_j \rangle$ with respect to $\psi_j$ and in this way approach $E_0$ and $\psi_0$ as closely as possible under the present LCAO–MO treatment. The energies obtained may not be close to the exact energy, and the wave functions may not give a very close description of the real electron distribution, but there is nothing

## II.2 The Variation Principle

we can do about this. Thus the function

$$\langle E_j \rangle = \varepsilon_j = \frac{\int \psi_j \mathcal{H} \psi_j \, d\tau}{\int \psi_j^2 \, d\tau}, \quad \text{where} \quad \psi_j = \sum_{i=1}^{n} C_{ji} \phi_i$$

is to be minimized with respect to each coefficient $C_{ji}$ for a given value of $j$ (i.e., for one particular MO), and this is to be repeated for each value of $j$ (i.e., for all possible MOs).

Thus for the $j$th MO $\psi_j$, the problem is to find sets of $C_{ji}$ such that

$$\partial \varepsilon_j / \partial C_{ji} = 0$$

for each value of $i$ up to $n$. This is then repeated for each of the $j$ MOs up to $\psi_n$. (Note that the minimization treatment is exactly the same for each MO, because they are all of the same form.) Since this treatment is the basis of most approximate MO methods, it is instructive to go through the procedure for a simple $\pi$-system, ethylene ($j = 1, 2$, $i = 1, 2$), then illustrate the final results in the more general case (Fig. II.1).

*Ethylene $\pi$-system*

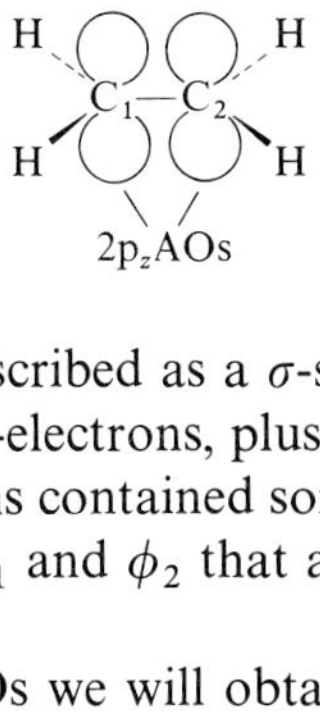

FIGURE II.1    2p$_z$ AOs

The ethylenic system can be described as a $\sigma$-skeleton (lying in the xy-plane) containing a "core" of $\sigma$-electrons, plus inner-shell electrons and nuclei, with the two $\pi$-electrons contained somehow in a $\pi$-system made up of two basis functions $\phi_1$ and $\phi_2$ that are 2p$_z$ AOs centered on carbons 1 and 2.

Since we start with two 2p$_z$ AOs we will obtain two linearly independent LCAO–MOs

$$\psi_j = C_{j1} \phi_1 + C_{j2} \phi_2 \quad (j = 1, 2)$$

The $j$ index can be omitted for now since the treatment is the same for any $j$, and can be reinserted later. The simplified MO function is then

$$\psi = C_1 \phi_1 + C_2 \phi_2$$

and

$$\varepsilon = \frac{\int (C_1\phi_1 + C_2\phi_2)\mathcal{H}(C_1\phi_1 + C_2\phi_2)\,d\tau}{\int (C_1\phi_1 + C_2\phi_2)^2\,d\tau} = \frac{N}{D}$$

Expanding the numerator $N$ and making use of the basic postulates and properties of Hamiltonian operators as before,

$$N = C_1{}^2 \int \phi_1 \mathcal{H} \phi_1\,d\tau + C_2{}^2 \int \phi_2 \mathcal{H} \phi_2\,d\tau + 2C_1 C_2 \int \phi_1 \mathcal{H} \phi_2\,d\tau$$
$$= C_1{}^2 H_{11} + C_2{}^2 H_{22} + 2C_1 C_2 H_{12}$$

where $H_{ij} = \int \phi_i \mathcal{H} \phi_j\,d\tau$, and so forth.

Expansion of the denominator $D$ gives similarly

$$D = C_1{}^2 \int \phi_1{}^2\,d\tau + C_2{}^2 \int \phi_2{}^2\,d\tau + 2C_1 C_2 \int \phi_1 \phi_2\,d\tau$$
$$= C_1{}^2 S_{11} + C_2{}^2 S_{22} + 2C_1 C_2 S_{12}$$

where $S_{ij} = \int \phi_i \phi_j\,d\tau$, and so forth. The integrals (or so-called matrix elements) are given the symbols $H$ or $S$ for simplicity. Their meaning will be explained later. Thus

$$\varepsilon = \frac{C_1{}^2 H_{11} + C_2{}^2 H_{22} + 2C_1 C_2 H_{12}}{C_1{}^2 S_{11} + C_2{}^2 S_{22} + 2C_1 C_2 S_{12}}$$

Since the denominator can never vanish,[†] we can cross-multiply before differentiating with respect to each $C_i$. This gives

$$\varepsilon(C_1{}^2 S_{11} + 2C_1 C_2 S_{12} + C_2{}^2 S_{22}) = C_1{}^2 H_{11} + 2C_1 C_2 H_{12} + C_2{}^2 H_{22}$$

and partial differentiation with respect to $C_1$ gives

$$\varepsilon(2C_1 S_{11} + 2C_2 S_{12}) + \frac{\partial \varepsilon}{\partial C_1}(C_1{}^2 S_{11} + 2C_1 C_2 S_{12} + C_2{}^2 S_{22})$$
$$= 2C_1 H_{11} + 2C_2 H_{12}$$

Setting $\partial \varepsilon / \partial C_1 = 0$ we obtain, for a minimum value of $\varepsilon$,

$$\varepsilon(2C_1 S_{11} + 2C_2 S_{12}) = 2C_1 H_{11} + 2C_2 H_{12}$$

which on rearrangement and cancellation of the common factor yields

$$C_1(H_{11} - \varepsilon S_{11}) + C_2(H_{12} - \varepsilon S_{12}) = 0$$

---

[†] At least for any reasonable choice of basis functions $\phi_i$.

## II.2 The Variation Principle

Similarly, differentiation with respect to $C_2$, followed by setting $\partial \varepsilon / \partial C_2 = 0$, would yield

$$C_1(H_{21} - \varepsilon S_{21}) + C_2(H_{22} - \varepsilon S_{22}) = 0$$

(Note that $H_{12} = H_{21}$, etc.)

These two equations are called the *secular equations* for the ethylene problem. There are two sets of these, one for $j = 1$ and one for $j = 2$, which are, on reinserting the $j$ index,

$$C_{j1}(H_{11} - \varepsilon_j S_{11}) + C_{j2}(H_{12} - \varepsilon_j S_{12}) = 0$$
$$C_{j1}(H_{21} - \varepsilon_j S_{21}) + C_{j2}(H_{22} - \varepsilon_j S_{22}) = 0$$

It is now necessary to find solutions to these sets of equations to obtain $\varepsilon_1$ and $\varepsilon_2$ corresponding to $\psi_1$ and $\psi_2$, and also sets of coefficients for each of these functions, namely, $(C_{11}$ and $C_{12})$ and $(C_{21}$ and $C_{22})$. One solution is $C_{11} = C_{12} = 0$ (or $C_{21} = C_{22} = 0$), but these are trivial solutions. It turns out that there can only be nontrivial solutions if the secular determinant associated with these secular equations is zero. Since the associated *secular determinant* for each set is the same, this means that

$$\begin{vmatrix} H_{11} - \varepsilon_j S_{11} & H_{12} - \varepsilon_j S_{12} \\ H_{21} - \varepsilon_j S_{21} & H_{22} - \varepsilon_j S_{22} \end{vmatrix} = 0$$

for a nontrivial solution. This determinant can be expanded to give a polynomial equation of the same order as the determinant, called the *secular polynomial* (or *characteristic equation*). The roots of this polynomial correspond to the "eigenvalues" or the allowed energies of the possible MOs $\psi_1$ and $\psi_2$ for the ethylenic $\pi$-system.

These roots can then be reinserted, one at a time, into the original secular equations and these can be solved to obtain the appropriate sets of MO coefficients that define the electron distribution in each MO. (Note that it is only necessary to solve the secular problem once; the resulting $j$ roots give the energies and coefficients for the $j$ MOs.)

The above problem for ethylene is easy to solve without making further approximations, but this is the simplest $\pi$-system of interest to organic chemists. We now outline the corresponding results of applying the variation method in the more general case of an $n$-$\pi$ electron system.

For an $n$-carbon system (where each carbon is $sp^2$-hybridized and contributes one $2p_z$ orbital to the $\pi$-system) there are $n$ MOs $\psi_j$ each of the form:

$$\psi_j = C_{j1}\phi_1 + C_{j2}\phi_2 + C_{j3}\phi_3 + \cdots + C_{jn}\phi_n \qquad (j = 1, 2, 3, \ldots, n)$$

Since all have the same form, the $j$ index can again be dropped for simplicity, giving

$$\psi = C_1\phi_1 + C_2\phi_2 + C_3\phi_3 + \cdots + C_n\phi_n$$

Applying the variation method to the expression for the energy as before would require differentiation of

$$\varepsilon = \frac{\int \psi \mathcal{H} \psi \, d\tau}{\int \psi^2 \, d\tau}$$

with respect to $C_1, C_2, C_3, \ldots, C_n$, in turn. Setting $\partial \varepsilon / \partial C_i = 0$ in each case would yield $n$ secular equations each containing $n$ terms:

$$C_1(H_{11} - \varepsilon S_{11}) + C_2(H_{12} - \varepsilon S_{12})$$
$$\qquad + C_3(H_{13} - \varepsilon S_{13}) + \cdots + C_n(H_{1n} - \varepsilon S_{1n}) = 0$$
$$C_1(H_{21} - \varepsilon S_{21}) + C_2(H_{22} - \varepsilon S_{22}) + \cdots + C_n(H_{2n} - \varepsilon S_{2n}) = 0$$
$$C_1(H_{31} - \varepsilon S_{31}) + \qquad\qquad\qquad \cdots + C_n(H_{3n} - \varepsilon S_{3n}) = 0$$
$$C_1(H_{n1} - \varepsilon S_{n1}) + \qquad\qquad\qquad \cdots + C_n(H_{nn} - \varepsilon S_{nn}) = 0$$

or in abbreviated form:

$$\sum_{i=1}^{n} C_i(H_{ki} - \varepsilon S_{ki}) = 0 \qquad (k = 1, 2, \ldots, n)$$

As before, each of the $n$ MO wave functions would yield a similar array of $n$ equations. However, this set of $n$ equations, which represents the secular problem for this $n$-$\pi$ system, need only be solved once, since the associated secular determinant is again exactly the same for each $\psi_j$. The determinant, on setting it equal to zero, gives all $n$ roots that yield the $n$ eigenvalues of the $n$ possible MOs:

$$\begin{vmatrix} H_{11} - \varepsilon S_{11} & (H_{12} - \varepsilon S_{12}) & \cdots & H_{1n} - \varepsilon S_{1n} \\ H_{21} - \varepsilon S_{21} & (H_{22} - \varepsilon S_{22}) & \cdots & H_{2n} - \varepsilon S_{2n} \\ & \vdots & & \\ H_{n1} - \varepsilon S_{n1} & & \cdots & H_{nn} - \varepsilon S_{nn} \end{vmatrix} = 0$$

## II.2 The Variation Principle

As before, expansion of this determinant gives a secular polynomial (this time of order $n$) whose roots can be reinserted into the $n$ secular equations to obtain $n$ sets of coefficients, one set for each of the $n$ $\psi_j$.[†]

To solve this secular problem as it now stands, and to obtain the roots for even fairly simple molecules, would require a matrix-diagonalization procedure and a computer. For more complex molecules this would be quite difficult and time consuming. Thus further approximations are needed to simplify the problem, and it is in the nature of these approximations where the HMO method differs from more advanced (or sophisticated) MO techniques.

The HMO technique, with its simplifying approximations, is suitable for calculations on a wide range of molecules of interest to organic chemists, even very large $\pi$-systems, with use of only pencil and paper, or possibly a simple desk calculator. More advanced MO techniques that attempt to solve the secular problem by making less approximations are usually limited to relatively simple systems and require large amounts of computer time.

Despite the nature of some of the approximations involved in the HMO method, which will be described in the next section, the results obtained are generally in good semiquantitative agreement with a large body of experimental data and have given extremely useful insights into the nature of organic $\pi$-systems and their reactions.

Before we describe the HMO method in some detail and illustrate its application to organic chemistry, it is perhaps useful for us to digress at this point to review briefly how the basic postulates of quantum mechanics have been used (or could have been used) to obtain results of chemical significance; and to see the extent that such results provide justification for these basic postulates.

This can be done in schematic form, as represented by Fig. II.2. The diagram illustrates two extreme approaches to the use of the basic postulates and the Schrödinger equation. By making no approximations at all, exact solutions are possible only for the simplest systems, such as the H atom. The results obtained are in excellent accord with experiment, thus providing justification of the basic postulates. Also,

---

[†] Note that this simple $n$-dimensionality is typical of all systems within the simple Hückel framework. Thus $n$ atoms in the $\pi$-system involve $n$AOs that yield $n$ LCAO–MOs. There are $n$ terms in each secular equation, and $n$ sets of $n$ equations. The secular determinant is of order $n$, as is the secular polynomial. Solution yields $n$ energies and $n$ sets of $n$ coefficients for the $n$ allowed MOs.

36  II  Hückel Molecular Orbital Theory

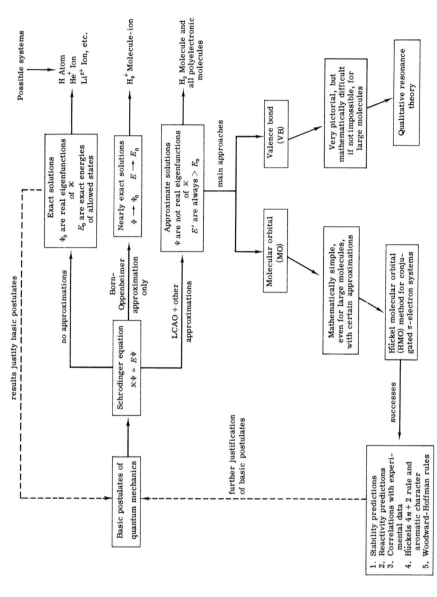

FIGURE II.2

nearly exact solutions can be obtained for a few more complex systems (such as $H_2^+$) by making only the Born–Oppenheimer approximation. Again, the agreement between calculated results and experiment is very good, providing further justification of the basic postulates.

At the other extreme, some very large and complex systems can be treated by the very approximate HMO approach after making a number of simplifying assumptions and approximations. The results of these calculations have been very successful in a number of areas, as indicated in Fig. II-2 and as will be described later. Therefore, these results, too, albeit crude and approximate, provide further justification of the basic postulates.

In addition, there are numerous other approaches that vary between these two extremes in the nature and degree of approximations involved, which also give varying degrees of success in predicting and explaining chemical properties. The general success of all these methods, ranging from exact agreement with experiment to rough semiquantitative agreement, lends considerable confidence to the use of the Schrödinger equation and the postulates on which it is based. It is difficult to see how these postulates could be fundamentally incorrect and at the same time be so successful in leading to generally satisfactory explanations of the behavior of a very wide range of chemical systems. The only really surprising aspect to the use of these basic quantum chemical ideas is that they work so well in cases where some quite drastic approximations are made in order to perform calculations on quite complex systems.

## II.3 THE BASIC HÜCKEL METHOD

After setting up the secular problem by appropriately constructing LCAO–MO wave functions from $2p_z$ AOs and subjecting these to the variation method as described in the previous section, the secular determinant is solved after making the following approximations and using certain definitions.

(i) The matrix elements $H_{ii}$ that appear on the diagonal of the determinant are called *Coulomb integrals*, and for all carbon atoms that are part of the $\pi$-framework of the molecule these integrals are set equal to some particular numerical value $\alpha$. Thus for the $i$th carbon atom

$$H_{ii} = \int \phi_i \mathcal{H} \phi_i \, d\tau = \alpha$$

These terms are taken to represent the interaction energy of an electron in an isolated $2p_z$ orbital with its own nucleus. Since all nuclei considered so far are carbon,[†] and all atoms in the $\pi$-framework are similarly hybridized, it is not unreasonable to assume that the interactions represented by $H_{ii}$ terms will be approximately equivalent for all $\pi$-centers. It turns out that the actual numerical value of $\alpha$, which is a negative energy term, is not of critical importance, but is useful as a reference point in evaluating bond energy terms.

(ii) The matrix elements $H_{ij}$ that appear in the off-diagonal terms of the determinant are called bond (or resonance) integrals. These are set equal to some particular numerical value $\beta$ if atoms $i$ and $j$ are directly $\sigma$-bonded to each other in the classical valence bond structure of the molecule, and are set equal to zero if the two atoms are not directly attached in the $\sigma$-framework. Thus

$$H_{ij} = \int \phi_i \mathcal{H} \phi_j \, d\tau = \beta \quad \text{or} \quad 0$$

These terms represent the energy of interaction of an electron with two nuclei simultaneously, arising from the overlap of $2p_z$ AOs on the atoms in question. The importance of this interaction will clearly depend on the distance of separation of the two atomic centers (or on the overlap of the two AOs involved). Hence, the assumption is not unreasonable that by far the largest energy contribution comes when centers $i$ and $j$ are directly attached in the $\sigma$-framework. In this case, the interaction should be very similar for all $ij$ pairs since all $\sigma$-bond distances are approximately the same and all $2p_z$ AOs share a common nodal plane. Therefore, it is not unreasonable to set these $H_{ij}$ terms equal to some common value $\beta$. If centers $i$ and $j$ are not directly attached, the internuclear distance will be much larger and the interaction energy represented by $H_{ij}$ will be much smaller. It is approximated that these terms are zero in order to simplify the calculations.

Methods of evaluation of $\beta$ are discussed later as well as ways of taking into account variations of $\beta$ as a function of the nuclei involved and variations in bond distances and structures. (Note that relative to

---

[†] The question of heteromolecules and the effect of changing the nuclei on the value of $\alpha$ will be considered later.

an electron at an infinite distance from the nucleus $\beta$ as well as $\alpha$ is a negative energy term.) The term $\beta$ is of more importance than $\alpha$ in assessing bonding interactions and stabilities, since it is these terms that indicate electronic stability relative to an electron in a free 2p orbital on a hypothetical isolated sp² carbon.

Neither of the above approximations is very drastic, nor are they unreasonable chemically. However, in order to simplify the calculations considerably, we now introduce a third, and much more drastic approximation.

(iii) The terms $S_{ij}$ are called *overlap integrals* and are set equal to unity if $i = j$, otherwise they are set equal to zero. Thus

$$S_{ij} = \int \phi_i \phi_j \, d\tau = 1 \quad \text{if} \quad i = j$$
$$= 0 \quad \text{if} \quad i \neq j$$

This is quite a drastic general approximation, for although $S_{ij}$ will be equal to unity where $i = j$ if we use already normalized $2p_z$ AOs, it is also known that $S_{ij}$ for adjacent centers is of the order of 0.25 for $2p_z$ overlap, which is far from zero.[2] The approximation is not bad for nonadjacent $i$ and $j$ since $S_{ij}$ falls off rapidly with distance. For example, in a linear chain of sp² hybridized carbons, the calculated values are

$$S_{11} = 1.00, \quad S_{14} \simeq 0.01$$
$$S_{12} \simeq 0.25, \quad S_{15} \simeq 0$$
$$S_{13} \simeq 0.07,$$

The above approximation is equivalent to considering that each $2p_z$ AO in the basis set is orthogonal to all others except itself, and is known as the *neglect of differential overlap* (NDO) or *zero differential overlap* (ZDO) approximation. The justification for this rather drastic approximation is that it greatly simplifies the mathematical calculations involved in solving the secular problem, and it also gives results that are in good relative (if not absolute) agreement with experimentally known facts. It will be considered later on what happens if this approximation is not made for nearest neighbors (where it should be most serious). Surprisingly, the final results are not affected very much in a relative sense and no conclusions reached on the basis of the NDO approximation are significantly altered.

It is noteworthy that the NDO approximation is also made in a number of other more sophisticated quantum chemical methods, such as the Pople–Pariser–Parr (PPP)[3] and CNDO-type methods,[4] again because it greatly simplifies the calculations. It is truly remarkable that calculations incorporating the NDO approximation are so successful, especially when it is clear that this approximation is fundamentally at variance with the principle of maximum overlap in chemical bonding.

With the inclusion of the three approximations described above, the secular equations and determinant are considerably simplified and quite easy to solve, even for complex systems. Returning to the general case of an $n$-center system, the secular equations are now reduced to the following form (omitting the $j$ index):

$$C_1(\alpha - \varepsilon) + C_2\beta_{12} + C_3\beta_{13} + \cdots + C_n\beta_{1n} = 0$$
$$C_1\beta_{21} + C_2(\alpha - \varepsilon) + C_3\beta_{23} + \cdots + C_n\beta_{2n} = 0$$
$$C_1\beta_{31} + C_2\beta_{32} + C_3(\alpha - \varepsilon) + \cdots + C_n\beta_{3n} = 0$$
$$\vdots \qquad \qquad \qquad \qquad \qquad \vdots$$
$$C_1\beta_{n1} \qquad \qquad + \cdots + C_n(\alpha - \varepsilon) = 0$$

In this array of equations all diagonal terms are of the form $C_i(\alpha - \varepsilon)$, and all off-diagonal terms are of the form $C_j\beta_{ij}$. Note that all $\beta_{ij}$ subscripts are retained for the present. Although some of these terms will be zero [by approximation (ii)], it is not yet known in this general case which atoms are bonded to which in the $\sigma$-framework. For example, we cannot set $\beta_{12} = \beta$ and $\beta_{13} = \beta_{14} = 0$, and so forth since we may be dealing with a cyclic or branched chain system.

From this simplified set of $n$ secular equations, we obtain a corresponding $n$th-order secular determinant that must vanish for a nontrivial solution.

*Generalized Hückel Secular Determinant*

$$\begin{vmatrix} \alpha - \varepsilon & \beta_{12} & \beta_{13} & \cdots & \beta_{1n} \\ \beta_{21} & \alpha - \varepsilon & \beta_{23} & & \beta_{2n} \\ \beta_{31} & \beta_{32} & \alpha - \varepsilon & & \vdots \\ \beta_{n1} & & \cdots & & \alpha - \varepsilon \end{vmatrix} = 0$$

Expansion of this determinant yields an $n$th-order secular polynomial $P_s$ that has $n$ real roots of the form

$$(\alpha - \varepsilon_j) = m_j\beta \qquad (j = 1, 2, 3, \ldots, n)$$

and $n$ associated energy values

$$\varepsilon_j = \alpha - m_j\beta$$

## II.3 The Basic Hückel Method

which are the allowed energy values for the MOs, or the so-called Hückel eigenvalues. This procedure thus yields an energy value for each of the $n$ MOs $\psi_j$ that is given as the algebraic sum of a Coulomb integral and some fraction or multiple of a bond integral. Hence the energies of the $n$ molecular orbitals can be represented as a series of energy levels above and below some arbitrary reference energy that is taken as $\varepsilon_0 = \alpha$. This immediately indicates whether the energy of an electron in a particular MO is greater or less than the energy the electron would have in an isolated $2p_z$ AO (with $\varepsilon_0 = \alpha$), and how much so in terms of $\beta$ units.

Since $\beta < 0$, negative values of the roots ($m_j < 0$) represent energy levels more stable than that of an isolated $2p_z$ orbital, and hence correspond to stable bonding levels or *bonding molecular orbitals* (BMOs). Similarly, positive roots ($m_j > 0$) correspond to high-energy antibonding levels or *antibonding molecular orbitals* (ABMOs). For any root $m_j = 0$, the energy of an electron in this particular MO is the same as the energy the electron would have in an isolated, noninteracting $2p_z$ orbital. Such levels are called nonbonding levels and orbitals (NBMOs).

This can be shown schematically (Fig. II.3) for a general case. The following should be noted:

(a) NBMOs may or may not be present, and in some cases there may be several of these,

(b) the number of BMOs may be equal to, greater than, or less than the number of ABMOs;

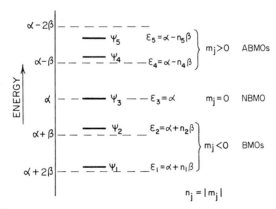

FIGURE II.3

(c) the energy levels may be symmetrically or unsymmetrically displaced about $\varepsilon = \alpha$;

(d) there may be degenerate levels of any type (BMO, ABMO, or NBMO).

At this point, all the $\pi$-electrons can be fed into these levels, starting with the lowest energy level and obeying the Pauli principle and Hund's rule. (The total number of electrons fed in may be equal to, greater than, or less than the total number of MOs obtained.) The total energy of the $\pi$-system can then be determined to see if there is a net bonding energy or not; in other words, to determine if the total array of $\pi$-electrons is more stable in the molecular orbitals of the system than it would have been in the isolated AOs that make up the basis set. It can also be determined from the electronic arrangement in these MO levels whether the ground state is predicted to have all electrons spin paired, or whether it is predicted to be a radical or diradical ground state for example.

The next step should be to reinsert these $n$ energy values (or more properly the roots), one at a time, into the original secular equations to produce sets of $n$ simultaneous equations, one set corresponding to each MO. Each set now represents a separate problem, since each root involved is different, which is to find a set of coefficients $\{C_{j1}, C_{j2}, C_{j3}, \ldots, C_{jn}\}$ for each MO $\psi_j$. This will give a description of the electron distribution in each MO.

These secular equations, with the roots inserted, can be solved to yield all values of the coefficients $C_{ji}$ in absolute numerical terms, although it may appear at first that we are using each set of original $n$ secular equations to find $(n + 1)$ unknowns for each $\psi_j$ (i.e., there are $n$ coefficients plus one energy for each $\psi_j$). However, there is an $(n + 1)$th equation available in each case. This comes from the normalization condition that

$$\int \psi_j^2 \, d\tau = 1$$

for any MO, which in effect states that an electron obeying the function $\psi_j$ must be somewhere in the region of space associated with this orbital. If we substitute the original expression for $\psi_j$,

$$\psi_j = \sum_{i=1}^{n} C_{ji} \phi_i$$

## II.3 The Basic Hückel Method

into this normalization equation, we obtain

$$\sum_{i=1}^{n}\sum_{k=1}^{n} C_{ji}C_{jk}\int \phi_i\phi_k\,d\tau = \sum_{i=1}^{n}\sum_{k=1}^{n} C_{ji}C_{jk}S_{ik} = 1$$

which involves the sum of all possible products of the basis orbitals in $\psi_j$. However, because we have taken [approximation (iii)] each $S_{ik} = 1$ if $i = k$ and $S_{ik} = 0$ if $i \neq k$, all the cross terms vanish and we are left with only

$$\sum_{i=1}^{n} C_{ji}^{2} = 1 \qquad \text{for any } j$$

In other words, for normalization, the sum of the squares of the derived coefficients for any MO must be equal to unity. This is the $(n + 1)$th equation in every case, which allows us to determine one energy and $n$ coefficients for each MO.

It is advisable at this point to summarize the steps involved in the basic Hückel procedure before going on to treat and discuss specific cases.

### Summary of Basic Procedure in HMO Calculations

(i) *Draw the basic π-electronic framework of the molecule.* It is necessary to include only those trigonally hydridized atoms that are part of the π-system, and that contribute $2p_z$ AOs to the basis set. Other atoms which do not have available $2p_z$ AOs can be omitted.

(ii) *Number the atoms of the π-skeleton arbitrarily from $1, \ldots, n$.* This is purely arbitrary for the present; later on it may be necessary to exercise some judgment in numbering more complicated systems to reduce the possibility of mechanical errors.

(iii) *Set up the Secular Determinant.* With a little practice this is easy to do by inspection. With practice it is equally simple to write down the complete set of associated secular equations. This will be illustrated for various systems.

(iv) *Expand the Secular Determinant to obtain the Secular Polynomial.* This is not difficult for π-systems with five atoms or less. For systems of six atoms or more it is advisable to make use of the symmetry properties of the system, if possible, and simplify the whole procedure by using elementary group theory. Standard procedures for expansion of determinants will be described.

(v) *Determine the roots of the Secular Polynomial ($P_s$).* Again, this is not difficult for π-systems of five atoms or less, especially if $P_s$ can be factored out. For many systems it may be necessary to use graphical or other approximate methods to determine the roots. Where possible, for larger systems, it is advisable to use elementary group theory to simplify the determinant.

(vi) *Determine the values of the MO coefficients $C_{ji}$ for each value of j.* This is done by substituting the roots $m_j$, one at a time, into the original secular equations and solving for $C_{j1}, C_{j2}, C_{j3}, \ldots, C_{jn}$ for each value of $j$. Again, this is not difficult, using standard procedures.

The first five steps give the energy levels of the Hückel MOs. The electrons are then placed in these MOs starting from the lowest energy MO, following Pauli's principle (two electrons at most in any MO) and Hund's rule (insert electrons singly into any degenerate levels before pairing any in one level). This gives a basic idea of the electronic structure of the molecule and associated quantities such as $E_\pi$ (the total π-electronic energy), $B_\pi$ (the π-bonding energy), and $DE$ (the delocalization energy) to be described later.

The last step gives a physical description of the actual spatial electron distribution in each MO and in the molecule as a whole, as well as associated quantities such as $q_r$ (the electron density), $\xi_r$ (the charge density), $\rho_{rs}$ (the bond order), and $\mathscr{F}_r$ (the free valence index), also to be described later.

## II.4 APPLICATION OF THE HMO METHOD TO SIMPLE Π-SYSTEMS

The first five of the foregoing operations will be carried out first on several simple systems, to illustrate the type of results obtained, before returning to the last step and calculation of the MO coefficients.

*Ethylene*

$$\begin{array}{c} H \\ \diagdown \\ H \diagup \end{array} C_1 - C_2 \begin{array}{c} H \\ \diagup \\ \diagdown H \end{array}$$

I

## II.4 Simple Π-Systems

The π-framework is very simple since there are only two π-centers, and can be represented as

$$\overset{\circ}{1} \underline{\hspace{2em}} \overset{\circ}{2}$$

The secular determinant is equally simple, as shown previously,

$$\begin{vmatrix} H_{11} - \varepsilon S_{11} & H_{12} - \varepsilon S_{12} \\ H_{21} - \varepsilon S_{21} & H_{22} - \varepsilon S_{22} \end{vmatrix} = 0$$

which with the Hückel approximations becomes (since $\beta_{12} = \beta_{21} = \beta$)

$$\begin{vmatrix} \alpha - \varepsilon & \beta \\ \beta & \alpha - \varepsilon \end{vmatrix} = 0$$

It is usually convenient in expanding and solving these determinants to divide each element by $\beta$ and to set $(\alpha - \varepsilon)/\beta = x$:

$$\begin{vmatrix} x & 1 \\ 1 & x \end{vmatrix} = 0$$

Thus it is easy to multiply this out and obtain the polynomial

$$P_s = x^2 - 1 = 0$$

The roots are obviously $x = +1, -1$, and since the most negative root always denotes the lowest energy these can be designated

$$x_1 = -1, \qquad x_2 = +1$$

and from $(\alpha - \varepsilon)/\beta = x$, the corresponding energies are

$$\varepsilon_1 = \alpha + \beta, \qquad \varepsilon_2 = \alpha - \beta$$

The energy level diagram for ethylene (Fig. II.4) is therefore very simple,

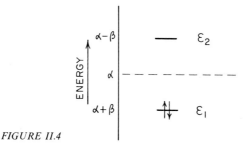

FIGURE II.4

and the two available π-electrons are both placed in the lower (BMO) orbital with paired spin. The total π-electronic energy, designated $E_\pi$, is given by

$$E_\pi = 2(\alpha + \beta) = 2\alpha + 2\beta$$

In general,

$$E_\pi = \sum_{i=1}^{n_{occ}} n_i \varepsilon_i$$

where $n_i$ is the number of electrons occupying level $\varepsilon_i$. The value of $E_\pi$ for ethylene is a very useful reference value for a completely localized (two-center) π-bond, since this is necessarily the case in ethylene.

If the energy of two electrons, one in each of two isolated non-interacting $2p_z$ AOs is $2\alpha$, then the net gain in energy on formation of the ethylene π-system is

$$B_\pi = E_\pi - E_{\substack{isol \\ atom}} = (2\alpha + 2\beta) - (2\alpha) = 2\beta$$

The quantity $2\beta$ is therefore the total π-bonding energy on forming the ethylene molecule, and can be given the symbol $B_\pi$. For carbon only π-systems, $B_\pi$ is just the $\beta$-containing term of the quantity $E_\pi$; however, this is not true in general, such as when heteroatoms are included.

*The Allyl System*

In this case, it could be the cation, radical, or anion that is being considered. However, because the electrons are not explicitly taken into account in the HMO approach until the MO levels have been determined, the calculation (at the present level of sophistication) is the same for all three systems.

II

Again, the π-framework is simple, and can be represented as shown, in linear or bent arrangement, because only the order of attachment of the π-centers is important for the calculations. The secular deter-

## II.4 Simple Π-Systems

minant is now third order, and with the Hückel simplifications [approximations (i)–(iii)] becomes

$$\begin{vmatrix} \alpha - \varepsilon & \beta_{12} & \beta_{13} \\ \beta_{21} & \alpha - \varepsilon & \beta_{23} \\ \beta_{31} & \beta_{32} & \alpha - \varepsilon \end{vmatrix} = 0$$

Since atoms 1 and 3 are not directly joined together in the σ-framework, $\beta_{13} = \beta_{31} = 0$, and by symmetry $\beta_{12} = \beta_{21} = \beta_{23} = \beta_{32} = \beta$. Thus the simplified determinant is

$$\begin{vmatrix} \alpha - \varepsilon & \beta & 0 \\ \beta & \alpha - \varepsilon & \beta \\ 0 & \beta & \alpha - \varepsilon \end{vmatrix} = 0$$

Dividing each element by $\beta$ as before and setting $(\alpha - \varepsilon)/\beta = x$ yields

$$\begin{vmatrix} x & 1 & 0 \\ 1 & x & 1 \\ 0 & 1 & x \end{vmatrix} = 0$$

Expansion is again simple, giving

$$P_s = x^3 - x - x = x^3 - 2x = 0$$

This polynomial expression obviously factors out, giving

$$P_s = x(x^2 - 2)$$
$$= x(x - \sqrt{2})(x + \sqrt{2}) = 0$$

and the roots (beginning with the most negative) are

$$x_1 = -\sqrt{2}$$
$$x_2 = 0$$
$$x_3 = +\sqrt{2}$$

Therefore, the Hückel eigenvalues for the allyl system are

$$\varepsilon_1 = \alpha + 1.414\beta$$
$$\varepsilon_2 = \alpha$$
$$\varepsilon_3 = \alpha - 1.414\beta$$

Since there are three reasonable systems to consider (with two, three, and four π-electrons, respectively) we can draw three electron occupancy

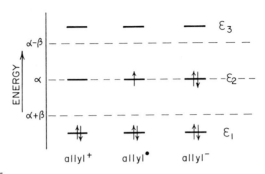

FIGURE II.5

patterns based on the same energy level diagram (Fig. II.5): The total $\pi$-energies are thus

$$E_\pi^+ = 2\alpha + 2.83\beta$$
$$E_\pi^\cdot = 3\alpha + 2.83\beta$$
$$E_\pi^- = 4\alpha + 2.83\beta$$

which differ only in the $\alpha$ term because the additional electrons in the radical and anion are placed in the nonbonding level ($\varepsilon_2 = \alpha$). Thus in each case $B_\pi$ is the same, namely, $2.83\beta$. However, the net bonding energy per electron (*BEPE*) is largest for the cation, since this system has all its $\pi$-electrons in bonding orbitals:

$$BEPE^+ = 1.415\beta$$
$$BEPE^\cdot = 0.943\beta$$
$$BEPE^- = 0.707\beta$$

The approach of comparing systems on a per electron basis is reasonable for systems with different numbers of electrons, since a system does not necessarily have a greater intrinsic electronic stability simply because it has more electrons. It can be noted that compared with ethylene, for which *BEPE* is obviously $1.0\beta$, the $\pi$-electrons in the allyl cation have a significantly lower bonding energy, whereas the values of *BEPE* for the radical and anion are less favorable than for ethylene.

The above picture is undoubtedly a gross over simplification since more electrons present in the same molecular framework should lead to significantly lower stability because of increased interelectronic repulsions. Thus the anion should be least stable. However, interelectronic repulsions are completely neglected in this naive type of MO

## II.4 Simple Π-Systems

treatment. Nevertheless, the results do suggest that no special stability is to be expected for the radical and anion since the extra electrons go into a NBMO and hence do not contribute to bonding. This is in accord with experimental evidence since the allyl cation is much better established as a reaction intermediate[5] than either the radical or anion.

The question is now, to what extent is the allyl system stabilized because of $\pi$-electron delocalization of the kind represented qualitatively by the resonance (VB) formulation III.

or

$$\{CH_2=CH-\overset{+}{C}H_2 \longleftrightarrow \overset{+}{C}H_2-CH=CH_2\}$$

$$[\overset{\delta+}{C}H_2=\!=\!=CH=\!=\!=\overset{\delta+}{C}H_2]$$

III

In the MO framework, this question is approached by comparing the energy obtained from a completely delocalized model (namely, $E_\pi$) with that which would be obtained by an MO calculation if the same system was constrained to be localized (which we can call $E_\pi^{loc}$). The value of the latter term is obtained by considering the energy that the best localized model of the system would have. In the case of the allyl cation this corresponds to IV.

$$C^1H_2=C^2H-C^3H_2{}^+$$

IV

where no interaction is permitted between $C^2$ and $C^3$. Clearly this would give the $\pi$-energy of an isolated (and localized) ethylenic fragment plus the energy of a $2p_z$ orbital on an isolated methylene fragment.[†] Thus, taking into account the number of electrons associated with each fragment,

$$E_\pi^{loc}(\text{cation}) = E_\pi(\text{ethylene}) + E(2p_z)$$
$$= 2(\alpha + \beta) + O(\alpha)$$
$$= 2\alpha + 2\beta$$

and since we have already determined that

$$E_\pi = 2\alpha + 2.83\beta$$

---

[†] This can be seen by reconstructing the secular determinant for a localized allyl system, and setting $\beta_{23} = \beta_{32} = 0$.

the net gain in energy of the delocalized structure over the localized model is defined to be

$$DE = E_\pi - E_\pi^{loc} = 0.83\beta$$

This term, the delocalization energy (or *DE* value) can be thought of as an MO measure of what we would call the resonance stabilization energy (*RSE*) in valence-bond terms. However, the two terms *DE* and *RSE* are clearly not synonymous, and are not necessarily even roughly equivalent as will be seen later on.

(Note that in calculating delocalization energies, always subtract the energy that the *best* localized valence-bond structure would be expected to have, from the HMO energy based on the completely delocalized structure; by best is generally meant that valence-bond structure with the maximum number of double bonds. Any other electrons that this localized structure possesses are given an energy corresponding to an electron in an isolated $2p_z$ orbital, namely, $\alpha$.)

Again it is clear in this particular case that *DE* for the radical or anion will be the same as that for the cation, since the additional electrons contribute nothing to stability. (This will not always be true of comparisons between different electronic occupancies of the same molecular framework.) Thus

$$DE^\cdot = E_\pi^\cdot - E_\pi^{loc}(\text{rad})$$
$$= (3\alpha + 2.83\beta) - (2\alpha + 2\beta + \alpha)$$
$$= 0.83\beta$$

where $E_\pi^{loc}$ is based on $CH_2$=CH—$CH_2$. Similarly,

$$DE^- = E_\pi^- - E_\pi^{loc}(\text{anion})$$
$$= (4\alpha + 2.83\beta) - (2\alpha + 2\beta + 2\alpha)$$

where $E_\pi^{loc}$ is based on $[CH_2$=CH—$CH_2]^-$.

However, although *DE* is the same for all three allyl systems, the important quantity for comparison is the delocalization energy per electron (*DEPE*), which again indicates that the cation is the most stabilized of the three systems:

$$DEPE^+ = 0.83\beta/2 = 0.414\beta$$
$$DEPE^\cdot = 0.83\beta/3 = 0.276\beta$$
$$DEPE^- = 0.83\beta/3 = 0.207\beta$$

## II.4 Simple Π-Systems

(It should not be assumed from this that cationic species will necessarily have the highest value of *DEPE* in general.)

*Butadiene*

V

In cases like butadiene, the *s*-cis and *s*-trans conformations are equivalent insofar as the simple HMO technique is concerned, and the π-framework can be represented as before in linear form, as shown. The HMO treatment only considers which π-centers are bonded to which, and conformational energy differences would in any event be small in comparison to $E_\pi$.

The secular determinant for butadiene is thus:

$$\begin{vmatrix} \alpha - \varepsilon & \beta_{12} & \beta_{13} & \beta_{14} \\ \beta_{21} & \alpha - \varepsilon & \beta_{23} & \beta_{24} \\ \beta_{31} & \beta_{32} & \alpha - \varepsilon & \beta_{34} \\ \beta_{41} & \beta_{42} & \beta_{43} & \alpha - \varepsilon \end{vmatrix} = 0$$

Because this is a linear (or straight-chain) conjugated system, only $\beta_{12} = \beta_{21} = \beta_{23} = \beta_{32} = \beta_{34} = \beta_{43} = \beta$. All other $\beta_{ij} = 0$. Thus in simplified form the determinant is

$$\begin{vmatrix} \alpha - \varepsilon & \beta & 0 & 0 \\ \beta & \alpha - \varepsilon & \beta & 0 \\ 0 & \beta & \alpha - \varepsilon & \beta \\ 0 & 0 & \beta & \alpha - \varepsilon \end{vmatrix} = 0$$

This form is typical of the determinants for all linear conjugated polyene systems. Note first that all Hückel determinants contain $\alpha - \varepsilon$ terms along the diagonal and that the determinant is always symmetrical about this diagonal. In addition, linear conjugated polyenes have all immediately off-diagonal terms equal to $\beta$, and all other elements equal to zero.

Making the same substitutions as before, the determinant becomes

$$\begin{vmatrix} x & 1 & 0 & 0 \\ 1 & x & 1 & 0 \\ 0 & 1 & x & 1 \\ 0 & 0 & 1 & x \end{vmatrix} = 0$$

To expand determinants of this order or greater, to obtain the appropriate secular polynomial expressions, it is necessary to use the *method of signed minors* (or *cofactors*).

*The Method of Cofactors*

For any determinant with elements designated as $A_{ij}$, such as

$$\begin{vmatrix} A_{11} & A_{12} & A_{13} & A_{14} & A_{1n} \\ A_{21} & A_{22} & A_{23} & A_{24} & A_{2n} \\ A_{31} & A_{32} & A_{33} & & \vdots \\ \vdots & & & & \\ A_{n1} & \cdots & & \cdots & A_{nn} \end{vmatrix}$$

the *minor* of any element $A_{ij}$ (the element appearing in the $i$th row and $j$th column) is equal to the determinant that is left when both row $i$ and column $j$ are struck out of the original determinant. For example the minor of $A_{23}$ in the above determinant is

$$M_{A_{23}} = \begin{vmatrix} A_{11} & A_{12} & A_{14} & \cdots & A_{1n} \\ A_{31} & A_{32} & A_{34} & \cdots & A_{3n} \\ \vdots & & & & \vdots \\ A_{n1} & A_{n2} & A_{n4} & \cdots & A_{nn} \end{vmatrix}$$

To form the *cofactor* of the element, its minor is given a sign according to the position of the element in the original determinant, which is $(-1)^{i+j}$. Thus the cofactor of element $A_{23}$ is given by

$$A_{23} = (-1)^{2+3} M_{A_{23}} = -M_{A_{23}}$$

The original determinant ($D_A$) can then be expanded and solved by taking the algebraic sum of the product of each element in any *row or column* and its own cofactor. Thus

$$D_A = \sum_{i \text{ or } j} A_{ij}(-1)^{i+j} M_{A_{ij}}$$

## II.4 Simple Π-Systems

(Although it usually does not make any difference, the first row is generally chosen to work on.) This method can be illustrated by returning to the butadiene secular determinant:

$$D_{butad} = \begin{vmatrix} A_{11} & A_{12} & A_{13} & A_{14} \\ x & 1 & 0 & 0 \\ 1 & x & 1 & 0 \\ 0 & 1 & x & 1 \\ 0 & 0 & 1 & x \end{vmatrix} = 0$$

The minor of $A_{11}$ is therefore

$$\begin{vmatrix} x & 1 & 0 \\ 1 & x & 1 \\ 0 & 1 & x \end{vmatrix}$$

and its sign is $(-1)^{1+1} = +1$. The minor of $A_{12}$ is

$$\begin{vmatrix} 1 & 1 & 0 \\ 0 & x & 1 \\ 0 & 1 & x \end{vmatrix}$$

and its sign is $(-1)^{1+2} = -1$. It is not necessary to consider the minors of $A_{13}$ and $A_{14}$ in this case, since the elements themselves are zero, hence the product (element × (cofactor)) is also zero. Thus after one expansion

$$D_{butad} = x \begin{vmatrix} x & 1 & 0 \\ 1 & x & 1 \\ 0 & 1 & x \end{vmatrix} - \begin{vmatrix} 1 & 1 & 0 \\ 0 & x & 1 \\ 0 & 1 & x \end{vmatrix}$$

This process can be repeated by appropriately taking minors of each element in these two third-order determinants to give

$$D_{butad} = x\left\{ x \begin{vmatrix} x & 1 \\ 1 & x \end{vmatrix} - \begin{vmatrix} 1 & 1 \\ 0 & x \end{vmatrix} \right\} - \begin{vmatrix} x & 1 \\ 1 & x \end{vmatrix} - \begin{vmatrix} 0 & 1 \\ 0 & x \end{vmatrix}$$

This can now be multiplied out very easily to give

$$\begin{aligned} P_s &= x[x(x^2 - 1) - (x)] - [(x^2 - 1) + 0] \\ &= x[x^3 - x - x] - [x^2 - 1] \\ &= x^4 - 2x^2 - x^2 + 1 \\ &= x^4 - 3x^2 + 1 = 0 \end{aligned}$$

(Note that if any minor determinant has all elements equal to zero in any one of its rows or columns, this minor vanishes. Thus expansion of large determinants is usually simplified by this fact, as well as by the fact that some elements in the original determinant are also zero.)

The resulting secular polynomial for butadiene is now easy to solve since it is in quadratic form with respect to $x^2$, that is,

$$P_s = x^4 - 3x^2 + 1 = (x^2)^2 - 3(x^2) + 1 = 0$$

Thus

$$x^2 = \frac{+3 \pm \sqrt{9-4}}{2}$$

and the roots $x$ can be found from the solutions for $x^2$. However, this polynomial is also factorable into

$$P_s = (x^2 - x - 1)(x^2 + x - 1) = 0$$

Thus the four roots can be obtained from these two quadratic factors, namely,

$$x = \frac{1 \pm \sqrt{1+4}}{2}, \qquad x = \frac{-1 \pm \sqrt{1+4}}{2}, \qquad \text{or} \qquad x = \frac{\pm 1 \pm \sqrt{5}}{2}$$

Therefore, the four roots and eigenvalues in order of increasing energy are:

$$x_1 = \frac{-1-\sqrt{5}}{2} = -1.618, \qquad \varepsilon_1 = \alpha + 1.618\beta$$

$$x_2 = \frac{+1-\sqrt{5}}{2} = -0.618, \qquad \varepsilon_2 = \alpha + 0.618\beta$$

$$x_3 = \frac{-1+\sqrt{5}}{2} = +0.618, \qquad \varepsilon_3 = \alpha - 0.618\beta$$

$$x_4 = \frac{+1+\sqrt{5}}{2} = +1.618, \qquad \varepsilon_4 = \alpha - 1.618\beta$$

Feeding in the four electrons of the butadiene $\pi$-system gives the energy level diagram shown in Fig. II.6. Thus for butadiene the total $\pi$-energy is

$$E_\pi = 2(\alpha + 1.618\beta) + 2(\alpha + 0.618\beta) = 4\alpha + 4.47\beta$$

and the $\pi$-bonding energy is

$$B_\pi = 4.47\beta, \qquad BEPE = 1.12\beta$$

## II.4 Simple Π-Systems

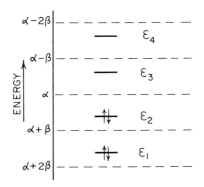

FIGURE II.6

On a per electron basis the $\pi$-bonding energy is somewhat higher than for ethylene, but distinctly lower than for the allyl cation ($BEPE = 1.414\beta$). Another way of looking at this difference is through the $DE$ values. The best localized structure for butadiene is clearly VI, which,

$$CH_2{=}CH{-}CH{=}CH_2,$$

VI

assuming no interaction between the two ethylenic moieties, would give $E_\pi^{loc} = 4\alpha + 4\beta$. Thus $DE = 0.47\beta$ and $DEPE = 0.12\beta$, whereas for the allyl cation the corresponding values are $1.414\beta$ and $0.414\beta$ per electron; therefore, even the total $DE$ for butadiene with four electrons is considerably lower than that for allyl$^+$ with only two electrons. The conclusion from these HMO results is that although butadiene is somewhat stabilized by delocalization (relative to two ethylenes), it is much less conjugated than the allyl cation. This is in accord with our qualitative expectations about the extent of resonance stabilization in these two systems, based on simple resonance arguments. The canonical structures VII for butadiene are nonequivalent, and the dipolar

$$CH_2{=}CH{-}CH{=}CH_2 \longleftrightarrow \overset{+}{C}H_2{-}CH{=}CH{-}\bar{C}H_2 \text{ etc.}$$

VII

forms are expected to contribute much less to the total resonance hybrid than in the case of the equivalent allyl structures VIII, which in addition involve no charge separation.

$$CH_2{=}CH{-}\overset{+}{C}H_2 \longleftrightarrow \overset{+}{C}H_2{-}CH{-}CH_2$$

VIII

In this particular case the simple HMO results are in accord with resonance arguments, and provide a quantitative measure of the stabilization achieved by delocalized systems. (It turns out that this kind of accord between MO results and VB arguments is not general, and it will be of particular interest to compare the predictions of the two approaches, when they disagree.)

To illustrate that the secular determinants are not always of the simple form so far encountered in linear polyene systems, a cyclic case is considered next.

*Cyclobutadiene*

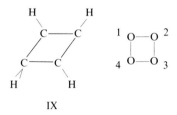

IX

The secular determinant (in simplified form) can be written down immediately by inspection as

$$\begin{vmatrix} \alpha - \varepsilon & \beta & 0 & \beta \\ \beta & \alpha - \varepsilon & \beta & 0 \\ 0 & \beta & \alpha - \varepsilon & \beta \\ \beta & 0 & \beta & \alpha - \varepsilon \end{vmatrix} = \begin{vmatrix} x & 1 & 0 & 1 \\ 1 & x & 1 & 0 \\ 0 & 1 & x & 1 \\ 1 & 0 & 1 & x \end{vmatrix} = 0$$

Multiplication by the method of cofactors gives

$$P_s = x^4 - 4x^2 = x^2(x^2 - 4) = x^2(x-2)(x+2) = 0$$

The four roots and energies are therefore

$$\begin{aligned} x_1 &= -2, & \varepsilon_1 &= \alpha + 2\beta \\ x_2 &= 0, & \varepsilon_2 &= \alpha \\ x_3 &= 0, & \varepsilon_3 &= \alpha \\ x_4 &= +2, & \varepsilon_4 &= \alpha - 2\beta \end{aligned}$$

Since in this case there are degenerate NBMOs, the energy level occupancy diagram (Fig. II.7) predicts that the cyclobutadiene ground state will have a diradical, and probably triplet structure (since electrons prefer to occupy degenerate levels with spins unpaired if possible). The total $\pi$-energy and bonding energy are given by

$$E_\pi = 2(\alpha + 2\beta) + 1(\alpha) + 1(\alpha) = 4\alpha + 4\beta$$
$$B_\pi = 4\beta$$

## II.4 Simple Π-Systems

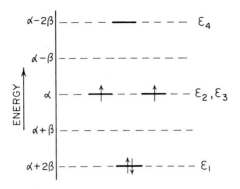

**FIGURE II.7**

Since the energy of two isolated ethylenic fragments in the localized structure is

$$E_\pi^{loc} = 2(\alpha + \beta) + 2(\alpha + \beta) = 4\alpha + 4\beta$$

the delocalization energy achieved by cyclobutadiene is zero, that is,

$$DE = E_\pi - E_\pi^{loc} = (4\alpha + 4\beta) - (4\alpha + 4\beta) = 0$$

An alternative way of expressing this lack of delocalization stabilization is to say that the net bonding energy per $\pi$-electron in cyclobutadiene is exactly the same as that in ethylene.

Thus the HMO results make two predictions that would not have been expected using VB arguments, namely:

(i) the ground state of the molecule will be a diradical, and probably a triplet state;
(ii) the molecule will have no stability due to $\pi$-delocalization.

From VB arguments, the equivalence of the two canonical structures X

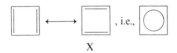

X

would have led to the prediction of a quite significant resonance stabilization energy. This would be offset by ring strain of course, but other equally strained systems are known. In fact, cyclobutadiene has never been prepared as a stable molecular species despite many attempts, although it has recently been prepared as a transient species in an inert gas matrix at very low temperatures ($\sim 4°K$).[6] Whether this marked instability is due to its lack of delocalization energy, or the very high reactivity[†] as a diradical is not clear, but nonetheless the results of HMO calculations are much more consistent with the known properties of cyclobutadiene than are VB-resonance arguments.

*Trimethylenemethane*

It is worth considering briefly one more simple $\pi$-system to illustrate an important point before going on to the calculation of the MO coefficients for these systems.

$$\begin{array}{cc} \text{CH}_2\text{—C}\begin{array}{c}\text{CH}_2\\ \\ \text{CH}_2\end{array} & \text{O—O}\begin{array}{c}\text{O}_3\\ \\ \text{O}_4\end{array}\\ 1 & 2 \end{array}$$

XI

The simplified secular determinant in terms of $x$ is

$$\begin{vmatrix} x & 1 & 0 & 0 \\ 1 & x & 1 & 1 \\ 0 & 1 & x & 0 \\ 0 & 1 & 0 & x \end{vmatrix} = 0$$

which on expansion by the method of cofactors gives

$$P_s = x^4 - 3x^2 = x^2(x^2 - 3) = 0$$

yielding the roots and energies

$$\begin{aligned} x_1 &= -\sqrt{3}, & \varepsilon_1 &= \alpha + 1.732\beta \\ x_2 &= 0, & \varepsilon_2 &= \alpha \\ x_3 &= 0, & \varepsilon_3 &= \alpha \\ x_4 &= +\sqrt{3}, & \varepsilon_4 &= \alpha - 1.732\beta \end{aligned}$$

Thus the energy level pattern is similar to that of cyclobutadiene, and the ground state has a diradical electronic structure. This is not

---

[†] There have been suggestions that a square cyclobutadiene might prefer to undergo a type of Jahn–Teller distortion to remove the degeneracy, and that it might be more stable in a rectangular structure.[7] Even so, reactive excited states would probably be easily accessible thermally. Recent spectroscopic results from matrix isolation experiments appear to support the square planar structure, however.[6]

## II.4  Simple Π-Systems

surprising, since any reasonable valence bond formula XII for trimethylenemethane is of diradical nature, for example,

$$CH_2=C\begin{matrix}\dot{C}H_2\\ \\ \dot{C}H_2\end{matrix}$$

XII

Nevertheless, it is interesting that MO theory predicts a similar electronic arrangement on completely different grounds.

The quantities $E_\pi$ and $B_\pi$ are as follows:

$$E_\pi = 2(\alpha + 1.732\beta) + 1(\alpha) + 1(\alpha) = 4\alpha + 3.46\beta$$
$$B_\pi = 3.46\beta \qquad (BEPE = 0.87\beta)$$

Note that in line with the fact that no VB formula can have more than one double bond, the MO bonding energy per electron is less than that of ethylene.

In order to calculate $DE$ for this species, it is again necessary to compare the energy of the best localized structure with $E_\pi$. Thus for trimethylenemethane.

$$E_\pi^{loc} = 2(\alpha + \beta) + 1(\alpha) + 1(\alpha) = 4\alpha + 2\beta$$

Therefore

$$DE = (4\alpha + 3.46\beta) - (4\alpha + 2\beta) = 1.46\beta \qquad (DEPE = 0.37\beta)$$

Therefore, trimethylenemethane has both a large $DE$ and $DEPE$, indicating significant stabilization due to $\pi$-electron delocalization. This result is contrary to that stated in some sources, where it has been suggested that $DE$ for this system is negative. The latter result is based on the erroneous assumption that $DE$ for an $n$-electron system is obtained by taking the difference between $E_\pi$ and $(n/2)$ times the energy of ethylene, or $(n\alpha + n\beta)$. This is clearly incorrect, since no localized structure of trimethylenemethane can have two ethylenic double bonds.

The important point is that no simple Hückel system can have a $DE$ value of less than zero, since this would imply that the best approximate energy obtained by HMO calculation is worse than another approximate energy obtained by constraining the system to form a localized set of double bonds. This is in violation of the variation principle, since both energies are based on the same set of approximations and the result obtained for $E_\pi$ should give the lowest (approximate) energy for the system. (This includes heteroatom systems as well, providing $E_\pi^{loc}$ is based on the lowest available energy for any localized structure, even though it may not contain the maximum possible number of double bonds.)

The above HMO results are in reasonable accord with what is known experimentally about the properties of the trimethylenemethane system.[8]

## II.5 CALCULATION OF THE MO COEFFICIENTS

Before going on to consider the eigenvalue distribution for more complex molecules, the problem of determining the MO coefficients and associated quantities will be discussed for the above simple systems.

For each molecule we have $n$ MO wave functions each of the form

$$\psi_j = C_{j1}\phi_1 + C_{j2}\phi_2 + C_{j3}\phi_3 + \cdots + C_{jn}\phi_n \quad (j = 1, 2, 3, \ldots, n)$$

and for each one of these $\psi_j$ (or eigenfunctions) there is associated some root $m_j$ (or $x_j$) that corresponds to the energy of an electron in that particular MO. Also for each $\psi_j$ we have a set of $n$ secular equations from which the secular determinant was obtained. The problem is now to insert each root into the appropriate set of secular equations and solve for the sets of coefficients $C_{ji}$ for each $\psi_j$. This will give us a description of the behavior (spatial distribution) of the electrons in each MO.

*Ethylene*

Taking the simple case of ethylene, we have seen that the secular equations are of the form

$$C_1(H_{11} - \varepsilon S_{11}) + C_2(H_{12} - \varepsilon S_{12}) = 0$$
$$C_1(H_{21} - \varepsilon S_{21}) + C_2(H_{22} - \varepsilon S_{22}) = 0$$

and that there are two sets of these (one for $\varepsilon_1$, one for $\varepsilon_2$). After the appropriate substitutions have been made, these equations become

$$C_1(\alpha - \varepsilon) + C_2\beta = 0$$
$$C_1\beta + C_2(\alpha - \varepsilon) = 0$$

## II.5 Calculation of the MO Coefficients

and with the substitution $x = (\alpha - \varepsilon)/\beta$

$$C_1 x + C_2 = 0$$
$$C_1 + C_2 x = 0$$

For the first MO, $\psi_1$ we have $\varepsilon_1 = \alpha + \beta$ from the root $x_1 = -1$. This root is substituted into the secular equations, yielding

$$-C_1 + C_2 = 0$$
$$C_1 - C_2 = 0 \quad \text{or} \quad C_1 = C_2$$

Applying the normalization condition that for any $\psi_j$

$$\sum_{i=1}^{n} C_i^2 = 1$$

we obtain

$$C_1^2 + C_2^2 = 1$$

which combined with the above equality gives

$$2C_1^2 = 1, \quad C_1^2 = \tfrac{1}{2}, \quad C_1 = \pm 1/\sqrt{2}$$

Arbitrarily choosing the positive root,[†] we have

$$C_1 = 1/\sqrt{2} = C_2$$

Therefore, for $\psi_1$ the coefficients are as shown,

$$\psi_1 = \frac{1}{\sqrt{2}} (\phi_1 + \phi_2) = 0.707\phi_1 + 0.707\phi_2$$

which is the bonding MO for ethylene.

Similarly for $\psi_2$ with energy $\varepsilon_2 = \alpha - \beta$, we have the root $x_2 = +1$, which on substitution into the same set of secular equations gives

$$C_1 + C_2 = 0, \quad \text{hence} \quad C_1 = -C_2$$

Again, from normalization we have

$$C_1^2 + C_2^2 = 1$$

Therefore,

$$2C_1^2 = 1 \quad \text{and} \quad C_1 = \pm 1/\sqrt{2}$$

---

[†] The choice of sign has no consequence other than to multiply the function by $-1$, which simply turns the MO upside down.

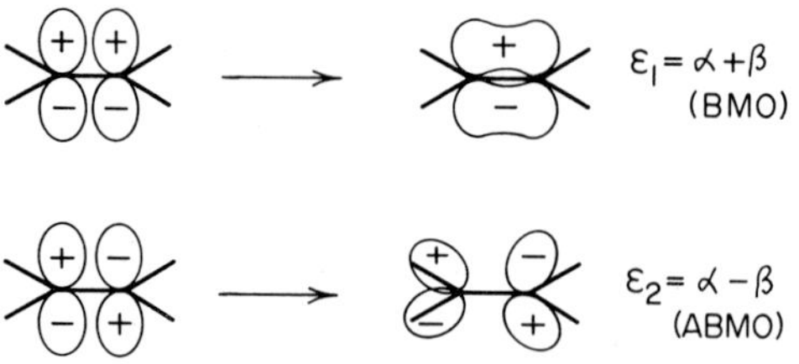

FIGURE II.8

Once more we choose the positive root arbitrarily for $C_1$ and obtain
$$C_1 = +1/\sqrt{2}, \quad C_2 = -1/\sqrt{2}$$
and
$$\psi_2 = (1/\sqrt{2})\phi_1 - (1/\sqrt{2})\phi_2 = 0.707\phi_1 - 0.707\phi_2$$
The two MOs for ethylene are now completely characterized and can be depicted schematically as shown in Fig. II.8.

The solution and description for ethylene is trivial, but it provides a useful reference point for more complex systems.

*Allyl*

For this system the secular equations are
$$\begin{aligned} C_1(\alpha - \varepsilon) + C_2\beta &= 0 \\ C_1\beta + C_2(\alpha - \varepsilon) + C_3\beta &= 0 \\ C_2\beta + C_3(\alpha - \varepsilon) &= 0 \end{aligned}$$
which on the usual substitution become
$$\begin{aligned} C_1 x + C_2 &= 0 \\ C_1 + C_2 x + C_3 &= 0 \\ C_2 + C_3 x &= 0 \end{aligned}$$
The roots already determined are $x_1 = -\sqrt{2}$, $x_2 = 0$, and $x_3 = +\sqrt{2}$. Taking $x_1$ first and substituting in the above equations gives
$$\begin{aligned} -\sqrt{2}C_1 + C_2 &= 0 \\ C_1 - \sqrt{2}C_2 + C_3 &= 0 \\ C_2 - \sqrt{2}C_3 &= 0 \end{aligned}$$
for which $C_2 = \sqrt{2}C_1 = \sqrt{2}C_3$. Hence $C_1 = C_3$. From normalization
$$C_1{}^2 + C_2{}^2 + C_3{}^2 = 1$$

## II.5 Calculation of the MO Coefficients

Thus

$$C_1^2 + 2C_1^2 + C_1^2 = 1$$
$$C_1^2 = \tfrac{1}{4} \quad \text{or} \quad C_1 = \pm\tfrac{1}{2}$$

Taking the positive root again, this gives

$$C_1 = \tfrac{1}{2} = C_3, \quad C_2 = \sqrt{2}/2 = 1/\sqrt{2}$$

Therefore, the coefficients for the first MO are as in

$$\psi_1 = 0.500\phi_1 + 0.707\phi_2 + 0.500\phi_3$$

Taking the second root $x_2 = 0$ and substituting gives

$$C_2 = 0$$
$$C_1 + C_3 = 0$$
$$C_2 = 0$$

Hence $C_1 = -C_3$ and $C_2 = 0$, which on use of the normalization gives

$$C_1 = 1/\sqrt{2}, \quad C_3 = -1/\sqrt{2}, \quad C_2 = 0$$

Therefore, for MO $\psi_2$ we have

$$\psi_2 = 0.707\phi_1 - 0.707\phi_3$$

Finally taking $x_3 = +\sqrt{2}$, a similar procedure yields $\psi_3 = 0.500\phi_1 - 0.707\phi_2 + 0.500\phi_3$. Thus the three MOs for the allyl system are completely characterized and can be represented as in Table II.1. Note that these sets of coefficients are both normalized and orthogonalized (e.g., $\int \psi_i \psi_j \, d\tau = 0$).

TABLE II.1

| $\psi_i$ | $x_i$ | $C_i$ | $C_2$ | $C_3$ | $\varepsilon_i$ |
|---|---|---|---|---|---|
| 1 | $-\sqrt{2}$ | 0.500 | 0.707 | 0.500 | $\alpha + 1.414\beta$ |
| 2 | 0 | 0.707 | 0 | $-0.707$ | $\alpha$ |
| 3 | $+\sqrt{2}$ | 0.500 | $-0.707$ | 0.500 | $\alpha - 1.414\beta$ |

### Schematic Representation of MOs of Allyl System

There are several ways of representing these MOs to give a pictorial idea of the electron distribution in allyl systems. One is simply in terms of the constituent AOs and their weights (or coefficients) in each MO,

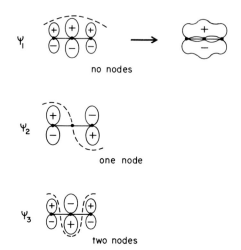

FIGURE II.9

representing the molecular framework in linear array, for simplicity. Alternatively, the resulting overlap can be depicted where appropriate, as shown for $\psi_1$ in Fig. II.9.

If the boundary of the positive lobes of the MO wave functions is traced by a dashed line, the nodal properties of the MOs can also be shown. It can be seen that apart from the nodal plane (XY) common to all simple $\pi$-systems, $\psi_1$ has no other nodal regions (i.e., where $\psi$ goes to zero). However, $\psi_2$ has one additional nodal region where the wave function changes sign (at $C_2$), and $\psi_3$ has two nodes or nodal regions in the $C_1$-$C_2$ and $C_2$-$C_3$ $\sigma$-bond axes.

*As a general statement, the number of nodes in any MO* (vertical to the nodal plane of all $2p_z$ AOs) *increases as the energy of the MO increases.*

The reason for this is clear if the MOs of the allyl system are represented schematically in terms of the radial electron density distribution (in the $+z$ direction only) as a function of distance along the (linearized) $\sigma$-bond axes (Fig. II.10). It is clear from this why $\psi_1$ is a bonding MO, since it has maximum $\pi$-electron density in the internuclear regions and no nodal regions other than the nodal plane that contains the $\sigma$-bond axes. Similarly, it is clear why $\psi_3$ is an antibonding MO, since $\pi$-electron density is at a minimum in the two internuclear regions (i.e., where the nodes occur). The MO $\psi_2$ has an electron distribution that is similar (in a bonding sense) to that of two isolated AOs on $C_1$

## II.5 Calculation of the MO Coefficients

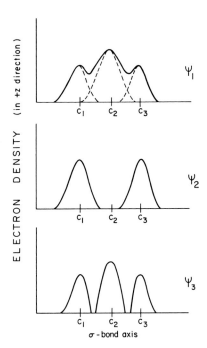

FIGURE II.10

and $C_3$ and is neither bonding nor antibonding in the internuclear regions. Thus it is categorized as a NBMO.

An estimation of the total $\pi$-electron distribution in any one of the three allyl systems considered can be arrived at by a superposition of the above distributions, taking into account the electron occupancy of each MO appropriate to the cation, radical, or anion.

One other way of assessing bonding energy from the MO coefficients in a rapid qualitative way, which is sometimes useful for comparing different MOs, is the following:

(i) If the MO function does not change sign between two adjacent nuclei, this means a net bonding interaction. Each such interaction can be assigned a rough value of $+1$.

(ii) If the MO function changes sign between two adjacent nuclei, this means a net antibonding interaction. Each of these can be assigned a value of $-1$.

(iii) If the MO function goes to zero at a particular nucleus, the interactions in the regions between that nucleus and adjacent nuclei are nonbonding and each can be assigned a value of zero.

Summation of these values for all internuclear regions in any MO gives a rapid and rough guide to the energy of that MO on a scale relative to the other MOs for the system. This method can serve as a useful check on the numerical accuracy of the calculations for both the energies and the coefficients for large and complex systems as well as simple ones like allyl.[†] This will be illustrated for various molecules. It is worth noting that although this quick evaluation of relative energies neglects the magnitudes of the coefficients and hence the importance of the various bonding and antibonding interactions, nonetheless, MOs that are degenerate will have equivalent nodal properties; on the other hand, two MOs with equivalent nodal characteristics will not necessarily be degenerate, although they should be close to each other in energy.

*Butadiene*

The coefficients for any system can be obtained by solving each set of $n$ simultaneous equations, as for allyl.

However, an easier, more generalized method that minimizes the possibility of error is available for larger systems ($n \geq 4$). This method involves the use of cofactors and automatically ensures that the MOs are normalized and orthogonal to each other. (It is important to check this for any sets of coefficients obtained.) The method is based on the fact that the value of any coefficient $C_j$ is proportional to the cofactor of the appropriate element $A_{ij}$ in any row $i$ of the secular determinant.

## GENERALIZED PROCEDURE FOR CALCULATING MO COEFFICIENTS

1. Choose any row in the secular determinant (usually the first row).
2. Calculate values of the minors of each element in this row in terms of $x$ and numbers, and give each the appropriate sign $(-1)^{i+j}$. (This includes the minors of zero as well as nonzero elements, since zero elements may have nonzero minors.)
3. Calculate numerical values of the cofactors in 2 using the values determined for the roots, one at a time (e.g., use the first root to calculate a set of numerical values for the cofactors of first row elements, then repeat later for the second root and so forth).

---

[†] For example, the three MOs of the allyl system would have total net bonding interactions of $+2$, $0$, and $-2$ for $\psi_1$, $\psi_2$, and $\psi_3$, respectively.

## II.5 Calculation of the MO Coefficients

4. Square each value in 3 and sum the squares (for a given root).
5. Take the square root of the sum of squares in 4 and divide each value of the set in 3 by this quantity.
6. The numbers obtained in 5 for any given root $x_i$ are the MO coefficients of $\psi_i$. (It may be necessary to multiply all coefficients for a given $\psi_i$ by $-1$ for convenience.) This procedure will be illustrated for the first root for butadiene, namely, $x_1 = -1.618$. The secular determinant, as before, is

$$\begin{vmatrix} A_{11} & A_{12} & A_{13} & A_{14} \\ x & 1 & 0 & 0 \\ 1 & x & 1 & 0 \\ 0 & 1 & x & 1 \\ 0 & 0 & 1 & x \end{vmatrix} \quad \text{choose first row elements arbitrarily}$$

The minor of $A_{11}$ is

$$M_{A_{11}} = \begin{vmatrix} x & 1 & 0 \\ 1 & x & 1 \\ 0 & 1 & x \end{vmatrix} = x^3 - 2x$$

and the cofactor is

$$(-1)^{1+1} M_{A_{11}} = x^3 - 2x = \mathscr{C}_1$$

For $A_{12}$ we have

$$M_{A_{12}} = \begin{vmatrix} 1 & 1 & 0 \\ 0 & x & 1 \\ 0 & 1 & x \end{vmatrix} = x^2 - 1$$

and the cofactor is

$$(-1)^{1+2} M_{A_{12}} = -x^2 + 1 = \mathscr{C}_2$$

Similarly for $A_{13}$ and $A_{14}$

$$M_{A_{13}} = \begin{vmatrix} 1 & x & 0 \\ 0 & 1 & 1 \\ 0 & 0 & x \end{vmatrix} = x; \quad \text{cofactor} = x = \mathscr{C}_3$$

$$M_{A_{14}} = \begin{vmatrix} 1 & x & 1 \\ 0 & 1 & x \\ 0 & 0 & 1 \end{vmatrix} = 1; \quad \text{cofactor} = -1 = \mathscr{C}_4$$

Therefore the numerical values of the cofactors for $x_1 = -1.618$ are

$$\mathscr{C}_1 = x^3 - 2x = -1.00$$
$$\mathscr{C}_2 = x^2 + 1 \ = -1.618$$
$$\mathscr{C}_3 = x \qquad\ = -1.618$$
$$\mathscr{C}_4 = -1.00$$

It is best at this stage to tabulate the results for $x_1 = -1.618$ (Table II.2) to avoid making errors.

TABLE II.2

| $j$ | $\mathscr{C}_j$ | $\mathscr{C}_j^2$ | $C_j$ | = | $\mathscr{C}_j/(\sum \mathscr{C}_j^2)^{1/2}$ |
|---|---|---|---|---|---|
| 1 | $-1.000$ | $+1.00$ | $-1.000/2.69$ | = | $-0.372$ |
| 2 | $-1.618$ | $+2.62$ | $-1.618/2.69$ | = | $-0.601$ |
| 3 | $-1.618$ | $+2.62$ | $-1.618/2.69$ | = | $-0.601$ |
| 4 | $-1.000$ | $+1.00$ | $-1.000/2.69$ | = | $-0.372$ |

$$\sum \mathscr{C}_j^2 = 7.24 \quad \text{and} \quad (\sum C_j^2)^{1/2} = 2.69$$

In this case all values of $C_i$ can be multiplied by $-1$ for simplicity (since this changes nothing except the orientation of the total $\pi$-system in space). Therefore, for the first MO of butadiene, we have

$$\psi_1 = 0.372\phi_1 + 0.601\phi_2 + 0.601\phi_3 + 0.372\phi_4$$

The corresponding calculation of the coefficients for $\psi_2$, $\psi_3$, and $\psi_4$ is left as an exercise for the student. The final results are as shown in Table II.3 which characterizes the butadiene $\pi$-system. Note that the sets of coefficients are very similar, except for their signs and positions in the array, because of the simple symmetry of the butadiene system.

TABLE II.3

| $\psi_i$ | $x_i$ | $C_1$ | $C_2$ | $C_3$ | $C_4$ | $\varepsilon_i$ |
|---|---|---|---|---|---|---|
| 1 | $-1.618$ | 0.372 | 0.601 | 0.601 | 0.372 | $\alpha + 1.62\beta$ |
| 2 | $-0.618$ | 0.601 | 0.372 | $-0.372$ | $-0.601$ | $\alpha + 0.62\beta$ |
| 3 | $+0.618$ | 0.601 | $-0.372$ | $-0.372$ | 0.601 | $\alpha - 0.62\beta$ |
| 4 | $+1.618$ | 0.372 | $-0.601$ | 0.601 | $-0.372$ | $\alpha - 1.62\beta$ |

The MOs can be described pictorially (Fig. II.11), as for the allyl system. As before the energy increases with the number of nodal regions in each MO. Looking at the general characteristics of the in-

## II.6 Bond Orders and Electron Densities

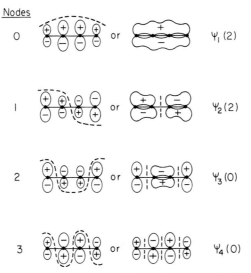

FIGURE II.11  Schematic representation of the butadiene MOs. Shaded orbitals are doubly occupied; unshaded orbitals are vacant.

dividual MOs it is clear from the pictorial representations that $\psi_1$ should be a strongly bonding MO since all interactions are favorable, and that $\psi_2$ should be a weakly bonding MO since it has two favorable and one unfavorable interaction. Similarly, $\psi_3$ should be weakly antibonding and $\psi_4$ strongly antibonding. (Using the approximate method of assessment described on p. 65, the net bonding interactions in each of the four MOs would be $+3$, $+1$, $-1$, and $-3$ for $\psi_1$ to $\psi_4$, respectively, which agrees roughly with the calculated order of bonding energies of these MOs, that is, the $\beta$-components.)

### II.6  BOND ORDERS AND ELECTRON DENSITIES

Although the foregoing schematic representations of $\psi_j$ may give a pictorial idea of the $\pi$-electron distributions that are consistent with the order of allowed energy levels, it is necessary for some purposes to have more quantitative estimates of electron distributions in molecules. These can be obtained from the calculated coefficients by introducing the important concepts of bond order and electron density in the following ways.

*Bond Order*

The quantity $\rho_{rs}{}^j$ is defined[9] to be a measure of bond order between two atoms $r$ and $s$, where

$$\rho_{rs}{}^j = C_{jr}C_{js}$$

This quantity is taken to be the partial bond order of the $r$-$s$ bond arising from an electron occupying the $j$th MO. The total bond order (in the $\pi$-system) between atoms $r$ and $s$ is then given by $\rho_{rs}$, where

$$\rho_{rs} = \sum_{j\text{occ}} \rho_{rs}{}^j = \sum_{j\text{occ}} n_j C_{jr} C_{js}$$

The summation is taken over the $j$ occupied levels and $n_j$ is the number of electrons occupying the $j$th MO.

This quantity $\rho_{rs}$ or the *bond order* is a convenient, but arbitrarily defined measure of the $\pi$-bond strength between two adjacent atoms. It has the physical significance of a bond order since the product of coefficients of two adjacent AOs $\phi_r$ and $\phi_s$, in an MO $\psi_j$, is a reasonable measure of the probability of finding an electron occupying $\psi_j$ in the $r$-$s$ region. If both $C_{jr}$ and $C_{js}$ are large and of the same sign, this means there is a substantial probability of finding an electron in this region of the MO, and therefore a large contribution to the bond order. If the coefficients are of opposite sign, this means there is a very low probability of finding electrons in this region of the MO, hence a negative contribution to bond order. If one of the coefficients is zero, the electron distribution in the region of one of the atoms will be similar to that in an isolated AO, and in the other region zero. This, in other words, is a nonbonding situation or zero contribution to bond order.

It is instructive to calculate the bond order for some of the simple systems considered so far.

*Ethylene*

$$\underset{1\phantom{--}2}{\text{O---O}}$$

**MO Coefficients**

$\psi_1: C_{11} = 1/\sqrt{2}, \quad C_{12} = 1/\sqrt{2} \quad$ (doubly occupied)
$\psi_2: C_{21} = 1/\sqrt{2}, \quad C_{22} = -1/\sqrt{2} \quad$ (vacant)

## II.6 Bond Orders and Electron Densities

Thus

$$\rho_{rs} = \sum_{j\,occ} n_j C_{jr} C_{js} = (2)(1/\sqrt{2})(1/\sqrt{2}) = 1.00 = \rho_{12}$$

Therefore, according to the above definition of bond order, there is a single $\pi$-bond in ethylene, as expected.

*Allyl cation*

$$\underset{1}{\bigcirc}\!\!-\!\!\underset{2}{\bigcirc}\!\!-\!\!\underset{3}{\bigcirc}$$

Here only $\psi_1$ is occupied (doubly) and thus it is necessary to consider only the coefficients

$$C_{11} = \tfrac{1}{2}, \qquad C_{12} = 1/\sqrt{2}, \qquad C_{13} = \tfrac{1}{2}$$

Thus

$$\rho_{12} = (2)(\tfrac{1}{2})(1/\sqrt{2}) = 0.707 = \rho_{23} \qquad \text{(by symmetry)}$$

Note that the $\pi$-bond order in each internuclear region is not 0.5 as might have been expected from the resonance formulation

$$\overset{\oplus}{C}H_2\!\!-\!\!CH\!\!=\!\!CH_2 \;\longleftrightarrow\; CH_2\!\!=\!\!CH\!\!-\!\!\overset{\oplus}{C}H_2$$

but is 0.707. Therefore, in the HMO framework the total $\pi$-bond order is 1.414 and not 1.000 as it would have been for any localized valence-bond structure. This extra bonding ability arises naturally in the MO theory from the simultaneous interaction of all three orbitals of the basis set. The stabilization by electron delocalization is associated with *DE* in the energy terms, and with extra bond order in the coefficient terms, and in corresponding ways. For polyene systems (i.e., carbon atoms only) there is a simple relationship between bond energy and bond order:

$$B_\pi = 2 \sum_{\substack{\text{all} \\ \text{bonds}}} \rho_{rs}$$

or

$$E_\pi = n\alpha + 2\beta \sum_{\substack{\text{all} \\ \text{bonds}}} \rho_{rs}$$

From the Hückel definition of delocalization energy it is obvious that twice the bond order term which is in excess of that of the best VB structure is numerically equal to *DE*, in $\beta$ units.

This simple correspondence between extra bonding ability and extra stabilization energy, both due to delocalization, is more satisfying and understandable than the VB (resonance) formulation, where it is difficult to see in bonding terms where the extra resonance stabilization energy arises.

*Butadiene*

$$\underset{1}{\circ}-\underset{2}{\circ}-\underset{3}{\circ}-\underset{4}{\circ}$$

Here the first two MOs are doubly occupied, and there are two distinct types of internuclear region (1,2) and (2,3). Thus from the coefficients for $\psi_1$ and $\psi_2$:

$C_{11} = 0.372,$ $\quad C_{12} = 0.601,$ $\quad C_{13} = 0.601,$ $\quad C_{14} = 0.372$
$C_{21} = 0.601,$ $\quad C_{22} = 0.372,$ $\quad C_{23} = -0.372,$ $\quad C_{24} = -0.601$

we have

$$\rho_{12} = n_1 C_{11} C_{12} + n_2 C_{21} C_{22}$$
$$= (2)(0.372)(0.601) + (2)(0.601)(0.372)$$
$$= 0.894 = \rho_{34} \quad \text{(by symmetry)}$$
$$\rho_{23} = n_1 C_{12} C_{13} + n_2 C_{22} C_{23}$$
$$= 2(0.601)(0.601) + 2(0.372)(-0.372)$$
$$= 0.447$$

The bond order distribution in butadiene can be represented simply as

$$\overset{0.894}{\circ}-\overset{0.447}{\circ}-\overset{0.894}{\circ}-\circ$$

Note that again the sum of the $\pi$-bond orders in the molecule is in excess of the value of 2 for the fully localized structure,

$$\underset{CH_2=CH}{\overset{1.00}{}}\overset{\displaystyle\nearrow}{}\overset{1.00}{\underset{CH=CH_2}{}}$$

$$\sum(\pi\text{-bond orders}) = 2.00$$

whereas for the delocalized MO picture

$$\sum_{\substack{\text{all} \\ \text{bonds}}} \rho_{rs} = 0.894 + 0.447 + 0.894$$

$$= 2.24 \quad \text{(excess bonding} = 2.24 - 2.00$$
$$= 0.24 \text{ due to delocalization)}$$

## II.6 Bond Orders and Electron Densities

[Recall that $B_\pi$ for butadiene is $4.48\beta$ or $2(2.24)$ in $\beta$ units. Also $DE$ is $0.48\beta$ or $2(0.24)$ in $\beta$ units.]

### Cyclobutadiene

It is instructive to consider this case, since it has been determined already that the bonding energy is no greater than that for two localized ethylenic systems and that there is no delocalization energy to be expected. The coefficients for the first three MOs of cyclobutadiene are

$C_{11} = 0.500,$  $C_{12} = 0.500,$  $C_{13} = 0.500,$  $C_{14} = 0.500$
$C_{21} = 0.500,$  $C_{22} = 0.500,$  $C_{23} = -0.500,$  $C_{24} = -0.500$
$C_{31} = 0.500,$  $C_{32} = -0.500,$  $C_{33} = -0.500,$  $C_{34} = 0.500$

(Note $\psi_1$ is doubly occupied, and $\psi_2$ and $\psi_3$ are singly occupied.[†]) Thus

$$\rho_{12} = n_1 C_{11} C_{12} + n_2 C_{21} C_{22} + n_3 C_{31} C_{32}$$
$$= 2(0.5)(0.5) + 1(0.5)(0.5) + 1(0.5)(-0.5)$$
$$= 0.500 = \rho_{23} = \rho_{34} = \rho_{41} \quad \text{(by symmetry)}$$

Thus the total bond order in the delocalized structure is 2.00, which is the same total as that in the localized VB formula XIII.

```
            0
     ┌──────────┐
 1.0 │          │ 1.0      Σ(π-bond orders) = 2.00
     │          │          excess bonding = 0
     └──────────┘
            0
          XIII
```

Therefore, despite the fact that cyclobutadiene must have a fully delocalized π-system due to its cyclic system of four interacting orbitals, it gains nothing in energy terms from this delocalization.

### Free Valence Index

One other useful quantity that can be derived from bond orders is the free valence index. This is one measure of potential reactivity at various positions in the π-system. It is perhaps not as useful a concept as bond order, but can often give valuable insights into chemical behavior.

One way of approaching reactivity (there are others as we will see later) is to consider to what extent atoms in the molecule are already

---

[†] This is only one possible representation of the degenerate MOs for cyclobutadiene.

bonded to other atoms, relative to the theoretical maximum π-bonding capability of that type of atom. Thus if any atom in the molecule is not strongly bonded to its adjacent atoms, relative to the maximum π-bond capability of that type of atom, this should represent a position of high reactivity. On the other hand if the bonding capabilities of the atoms are almost saturated, this should represent a position of low reactivity.

The free valence index $\mathscr{F}$ at any position $r$ is thus defined[10] as follows:

$$\mathscr{F}_r = \begin{pmatrix} \text{maximum possible bonding} \\ \text{power at } r\text{th atom} \end{pmatrix} - \sum_s \rho_{rs}$$

where the summation is taken over all atoms directly attached to atom $r$. (The σ-bond orders are excluded from consideration, since it is assumed that any carbon in the π-system is sp² hybridized and fully σ-bonded to three other atoms. In terms of reactivity this is reasonable since π-containing molecules usually react first through the most readily available or highest energy electrons, namely, the π-electrons.)

The above definition can be restated as

$$\mathscr{F}_r = N_{\max} - N_r$$

where $N_{\max}$ is the theoretical maximum number of π-bonds and $N_r$ is the actual total bond order to $r$. The question now is, what value to take for $N_{\max}$. Various suggestions have been made but the following approach[11] seems most reasonable and widely accepted. If we consider the trimethylenemethane system, in Fig. II.12 the central carbon atom is π-bonded to each of three other π-carbons, which in turn are not π-bonded to anything else. Therefore, the three outer carbons are capable of donating all their π-bonding ability to the central carbon. The central carbon $C_2$ is bonded to the maximum number of centers for an sp² carbon, and in each case to the maximum extent. Thus $C_2$ is taken to have zero free valence, and in fact this is the last place one would expect the system to be attacked by external reagents.

Therefore, from the definition of free valence this carbon has

$$\mathscr{F}_2 = N_{\max} - N_2 = 0$$

FIGURE II.12

## II.6 Bond Orders and Electron Densities

The three bond orders to $C_2$ in trimethylenemethane each have a value of $1/\sqrt{3}$, thus $N_2 = 3/\sqrt{3} = \sqrt{3}$. Therefore, $N_{max}$ for all carbon $\pi$-systems is defined to be $\sqrt{3}$.

Illustrating this approach for butadiene, we have

$$C^1H_2 \xrightarrow{0.894} C^2H \xrightarrow{0.447} C^3H \xrightarrow{0.894} C^4H_2$$

$\mathscr{F}_1 = 1.732 - 0.894 = 0.838 = \mathscr{F}_4$ (by symmetry)

$\mathscr{F}_2 = 1.732 - (0.894 + 0.447) = 0.391 = \mathscr{F}_3$ (by symmetry)

Thus, according to these $\mathscr{F}$ values, the terminal positions of butadiene should be by far the most highly reactive, which is in accord with experimental observation. Free valence will be compared with other reactivity indices in later sections.

One final point is that both bond orders and free valences for a given system are usually represented in shorthand notation, as illustrated for the basic butadiene skeleton XIV:

XIV

### Electron Density

Since the total probability that an electron in a given MO $\psi_j$ will be found in any general region of space $d\tau$ is given by $\psi_j^2 \, d\tau$, where the function $\psi_j$ is normalized, it should be possible to obtain a measure of the probability of finding the electron in a region of space associated with a particular nucleus from the magnitude of the coefficient of its AO in the MO. Therefore, a measure of the total probability of finding electrons (or the total electron density) in the region of any nucleus should be obtainable from the magnitudes of the appropriate coefficients in the occupied MOs. The quantity electron density, $q_r$ at the $r$th position, is defined by

$$q_r = \sum_{j_{occ}} n_j C_{jr}^2$$

where $n_j$ is the number of electrons occupying the $j$th MO. To illustrate this definition, we consider various simple systems.

For ethylene, only $\psi_1$ is occupied, and by two electrons, and its coefficients are

$$C_{11} = \frac{1}{\sqrt{2}}, \quad C_{12} = \frac{1}{\sqrt{2}}$$

Thus

$$q_1 = 2(1/\sqrt{2})^2 = 1.00 = q_2 \quad \text{(by symmetry)}$$

and according to the above definition, the electron density in the region of either nucleus is 1.00, which is expected for this trivial case. For the allyl cation, only $\psi_1$ is (doubly) occupied and its coefficients are

$$C_{11} = \tfrac{1}{2}, \quad C_{12} = 1/\sqrt{2}, \quad C_{13} = \tfrac{1}{2}$$

Thus

$$\begin{aligned} q_1 &= 2(\tfrac{1}{2})^2 = 0.500 = q_3, \\ q_2 &= 2(1/\sqrt{2})^2 = 1.000, \end{aligned} \quad \underset{CH_2-CH-CH_2}{\overset{0.5 \quad 1.0 \quad 0.5}{}}$$

giving a simple picture of the electron distribution in the allyl cation, whereby the electron density is twice as high at the central carbon as at the terminal carbons. (Note that the total of the $q_r$ values for any system must equal the total number of electrons in the occupied $\pi$-orbitals, unlike bond orders that have no fixed total, except that the sum of bond orders to any carbon must be $\leq \sqrt{3}$.) Considering one more system, butadiene, we have for the coefficients of the occupied MOs

$$\begin{aligned} \psi_1(2); \quad & C_{11} = 0.372, \quad C_{12} = 0.601, \quad \text{etc.} \\ \psi_2(2); \quad & C_{21} = 0.601, \quad C_{22} = 0.372, \quad \text{etc.} \end{aligned}$$

Thus

$$\begin{aligned} q_1 &= 2(0.372)^2 + 2(0.601)^2 = 1.00 = q_4 \\ q_2 &= 2(0.601)^2 + 2(0.372)^2 = 1.00 = q_3 \end{aligned}$$

so that the electron density is equal at each position, and the probability of finding one of the four $\pi$-electrons associated with any one nucleus at a given time is unity.

Neutral systems such as butadiene (or even charged systems), where the electron density is equally distributed to each $\pi$-center, are called

## II.6 Bond Orders and Electron Densities

*self-consistent*.[†] However, not all neutral $\pi$-systems have this property, as will be seen later.

Another quantity that is simply associated with electron density is the *charge density* $\xi$. This is defined by

$$\xi_r = n_r - q_r$$

where $n_r$ is the number of electrons that would be required for electrical neutrality by an isolated $sp^2$ hybridized atom of the type $r$. For all carbon atoms $n_r$ is clearly unity, thus

$$\xi_C = 1.00 - q_C$$

For ethylene

$$\xi_1 = \xi_2 = 0$$

and for the allyl cation

$$\xi_1 = +0.5 = \xi_3$$
$$\xi_2 = 0$$

Thus the charge distribution in this cation is naively pictured as

$$+\tfrac{1}{2} \quad 0 \quad +\tfrac{1}{2}$$
$$CH_2\text{===}CH\text{===}CH_2$$

(This is undoubtedly a gross oversimplification and methods of refining this simple picture will be described in a later section.)

For butadiene

$$\xi_1 = \xi_2 = \xi_3 = \xi_4 = 0$$

and this type of distribution is also referred to as a self-consistent charge distribution, for reasons to be discussed later. (Note that as for the sum of $q_r$ values, there is also a definite value for the sum of $\xi_r$ values, which is equal to the net electrical charge of the system. However, although for all neutral molecules $\sum \xi_r$ must equal zero, the individual $\xi_r$ values need not themselves be equal to zero.)

Final shorthand representations of the systems considered so far, in terms of the quantities described, are shown in Fig. II.13. (Electron density values are usually chosen over charge densities, and are placed over the atomic positions except where it is obvious by symmetry what their values are; bond orders are written over the bonds, and free valences are depicted by arrows.)

---

[†] This is not to be confused with self-consistent field (SCF) calculations.

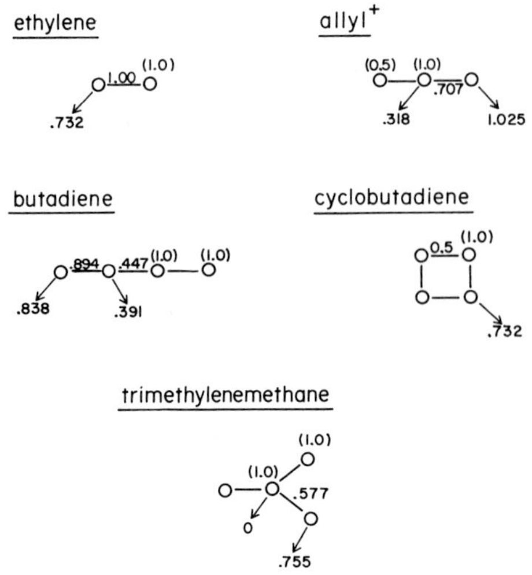

### ethylene

O—1.00—O
(1.0)
.732

### allyl⁺

(0.5) (1.0)
O—.707—O
.318      1.025

### butadiene

O—.894—O—.447—O—1.0—O
(1.0)        (1.0)
.838   .391

### cyclobutadiene

O—0.5—O
(1.0)
|         |
O——————O
                .732

### trimethylenemethane

(1.0)
O
(1.0) /
O——O—.577
     |
     O
.755

FIGURE II.13

## II.7 ALTERNANT AND NONALTERNANT HYDROCARBONS

The form taken by the roots to the secular polynomials is particularly simple for some types of hydrocarbon system, especially for linear polyenes and some monocyclic polyenes. This is because of their high degree of symmetry. In many of these cases the secular polynomials factor out, and even where they do not, they are usually easy to solve. There are some useful generalizations which can be made at this stage that help simplify the problem of determining and checking the energy levels for many systems, and that can also give useful information without doing any calculations at all.

One helpful generalization is that for all hydrocarbon systems, the algebraic sum of the roots to the secular polynomial must vanish:

$$\sum_{j=1}^{n} x_j = 0$$

This is clearly true if the Hückel energy levels are symmetrically disposed about $E = \alpha$, but it is also true for nonsymmetrical eigenvalue distributions.

Another useful generalization is that for all hydrocarbons, no root can be greater than three in absolute magnitude. Thus if $|x_j| \leq 3$,

## II.7 Alternant and Nonalternant Hydrocarbons

all Hückel eigenvalues must lie in the range

$$(\alpha - 3\beta) \geq \varepsilon_j \geq (\alpha + 3\beta)$$

The most useful generalizations that can be made about hydrocarbon systems are based on their classification into two basic types, called *alternant and nonalternant hydrocarbons*. The form of the eigenvalue distribution for certain of these systems takes on a very restricted pattern, and classification into the above types can give useful information about their properties (such as energy level distributions and electron distributions) without the necessity of doing any calculations.

Alternant (or AH) hydrocarbons are defined as those planar, conjugated hydrocarbons in which the carbon atoms of the $\pi$-systems can be divided arbitrarily into two sets $s$ (starred) and $u$ (unstarred) in such a way that each $s$-carbon has only $u$-neighbors and vice versa. (This generally means that if the $\pi$-system contains cyclic components, there can be no odd-membered rings present.) This type (AH) can be further subdivided into two types, *even-AH* and *odd-AH*.

In an even-AH system, the sum of the number of starred positions $n_s$ and of unstarred positions $n_u$ is even (and frequently $n_s = n_u$ for this type). For odd-AH systems $(n_s + n_u)$ is odd, and clearly $n_s \neq n_u$ for this type.

Considering the even-AH category first, the simple $\pi$-systems XV–XIX are all of the AH type since each $C^*$ position has only $C$-neighbors

|  XV  |  XVI  |  XVII  |  XVIII  |  XIX  |

and vice versa. Further $(n_s + n_u)$ is even in every case, thus they are all even-AH systems. In addition $n_s = n_u$ in each case (although this need not be true for an even-AH system), which means that the roots for such systems take on the restricted form,

$$x_i = \pm x_1, \pm x_2, \pm x_3, \ldots$$

that is, $2n_s$ (or $2n_u$) roots occurring in pairs of opposite sign. Furthermore, the roots for this type of system are generally nonzero, that is, even-AH systems usually do not have NBMOs. (One exception to this is cyclobutadiene, which is even-AH and yet has two zero roots or two NBMOs).

In some other even-AH systems $n_s \neq n_u$, and thus we can arbitrarily make the starred set larger than the unstarred set. Usually for these systems $(n_s - n_u) = 2$, but it can be equal to 4 or 6 in some cases. [Note: $(n_s - n_u)$ can never be equal to 1 or 3 for even-AH systems, by definition.] Where $n_s \neq n_u$, the roots take on the simple form

$$x_i = \pm x_1, \pm x_2, \pm x_3, \ldots$$

as before, but $(n_s - n_u)$ of these are equal to zero. This is the case for the systems XX–XXII.

| XX | XXI | XXII |
|---|---|---|
| $n_s = 3$ | $n_s = 4$ | $n_s = 5$ |
| $n_u = 1$ | $n_u = 2$ | $n_u = 3$ |

Therefore, both types of even-AH system have the property of having their energy levels symmetrically distributed about zero (or $E = \alpha$), and for the latter type there are necessarily $(n_s - n_u)$ degenerate NBMOs, as shown in Fig. II.14.

This means that even-AH systems for which $n_s \neq n_u$ can be described as polyradicals in cases where the species are electrically neutral. For

XXIII   $n_s = 4$   $n_4 = 2$

example, the system XXIII will have an energy level distribution of the type shown; thus for the neutral system with six electrons, the prediction of simple HMO theory is that it will have a diradical ground state (Fig. II.15).

It is very interesting to note that for systems of this type (where $n_s \neq n_u$) it is impossible to write any VB (Kekulé) structure having more than $n_u$ double bonds. Thus we can write structures XXIV–XXVII. Therefore, for the six electron system, simple VB arguments

XXIV    XXV    XXVI    XXVII    etc.

## II.7 Alternant and Nonalternant Hydrocarbons

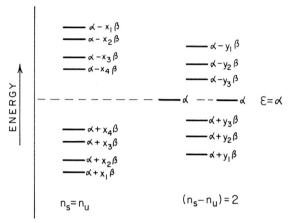

FIGURE II.14 Typical energy level patterns for even-AH systems.

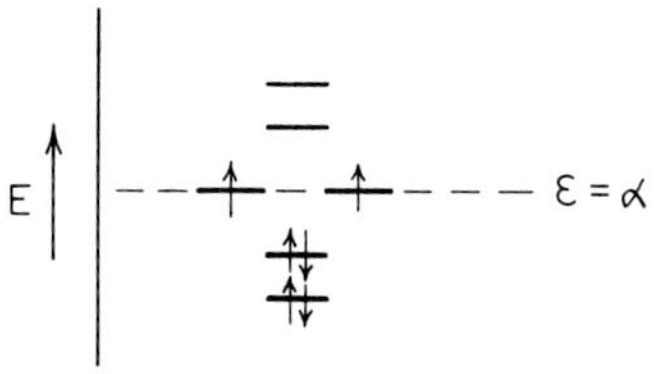

FIGURE II.15

also lead to the conclusion that the ground state will be a diradical. This is another case where the MO results correspond exactly with simple VB ideas, yet they are obtained in a completely different and independent way. The above correspondence is typical of all even-AH systems, whether $n_s = n_u$ or not; for example, if we compare the three quinodimethane systems, o-, m-, and p-xylylene, XXVIII–XXX, we

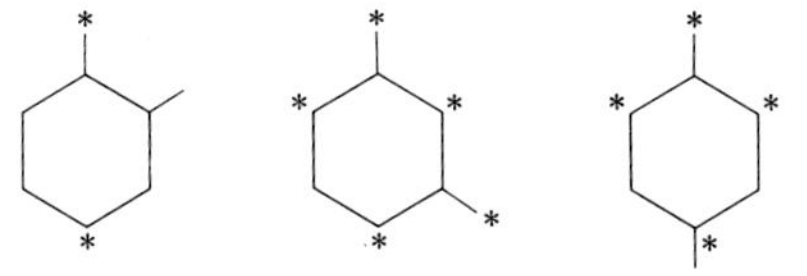

XXVIII $\quad$ XXIX $\quad$ XXX
$n_s = n_u = 4 \quad n_s = 5, n_u = 3 \quad n_s = n_u = 4$

can write VB structures for all three that have a maximum of $n_u$ double bonds, although it is only the m-isomer that is predicted to be a diradical from simple MO results (and from the VB formulas XXXI–XXXIII).

XXXI    XXXII    etc.    XXXIII

Turning to odd-AH systems, $n_s > n_u$ in every case and usually $(n_s - n_u) = 1$, but this difference can equal 3 or 5 in rare cases. For all these systems, the roots also occur in pairs except that there is always one (or three, or five) zero roots:

$$X_i = 0, \ldots, \pm X_1, \pm X_2, \pm X_3, \ldots$$

The systems XXXIV–XXXVI are all odd-AH and have the energy

(a)   XXXIV   $n_s = 3 n_u = 2$   (one zero root)

(b)   XXXV   $n_s = 6 n_u = 3$   (three zero roots)

(c)   XXXVI   $n_s = 3 n_u = 2$   (one zero root)

level patterns shown in Fig. II.16. (Note that for these systems also, no Kekulé structure can be drawn which has more than $n_u$ double bonds.)

Therefore, in general, all alternant hydrocarbons whether even-AH or odd-AH have their energy levels symmetrically disposed about zero,

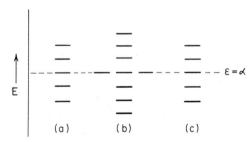

FIGURE II.16

## II.7 Alternant and Nonalternant Hydrocarbons

and where $n_s \neq n_u$ there will always be $(n_s - n_u)$ degenerate nonbonding levels. (However, nonbonding levels *may* occur even when $n_s = n_u$.)

Another generalization that can be made about all AH systems is that for the electrically neutral species, the electron density at any position will equal unity. Thus for both even-AH molecules and odd-AH radicals

$$q_r = 1.00, \qquad \xi_r = 0 \qquad (r = 1, 2, \ldots, n)$$

Since these are all uncharged systems, it might have been expected that each position would necessarily have zero charge. Therefore, all AH-systems can be said to have uniform or self-consistent charge distributions.[†] However, this is not true for many other neutral systems, for example, nonalternant systems.

Nonalternant (non-AH) systems are defined as those planar conjugated π-systems for which the positions cannot be designated *s* and *u* in any way so that each *s*-carbon has only *u*-neighbors and vice versa. This usually means that an odd-membered ring is present. Note that all odd-membered ring containing hydrocarbons are automatically non-AH, whether they also contain even-member rings or not. The non-AH systems XXXVII–XL illustrate this:

XXXVII    XXXVIII    XXXIX    XL

Such systems do not have their energy levels symmetrically distributed about zero, nor can we state whether they will have NBMOs or not. (Although in most cases there are no $x_i = 0$ for non-AH systems, one exception to this is pentalene.) Also, all neutral non-AH systems, either molecules or radicals, do not generally have all $q_r = 1.00$ (or all $\xi_r = 0$). In fact, non-AH systems generally do not have uniform charge distributions.

One useful and practical generalization that can be drawn from all this is that both even- and odd-AH systems can be expected to have zero or very small dipole moments, because of their internally uniform charge distributions; whereas non-AH systems may be expected to have quite significant dipole moments in some cases. This generalization is in good accord with experimental evidence, as illustrated by a

---

[†] The question of the self-consistency or non-self-consistency of HMO charge distributions will be considered later.

comparison of the three isomeric systems naphthalene XLI, azulene XLII, and [6.2.0]-bicyclodecapentaene XLIII. Of the three, only

azulene is non-AH; it also is the only one with a significant dipole moment ($\mu = 1.0$ D.).[12] It is worth pointing out that, in general, hydrocarbon molecules have very small or zero values of $\mu$, thus a value as high as 1.00 D. is extremely large for a hydrocarbon.

TABLE II.4

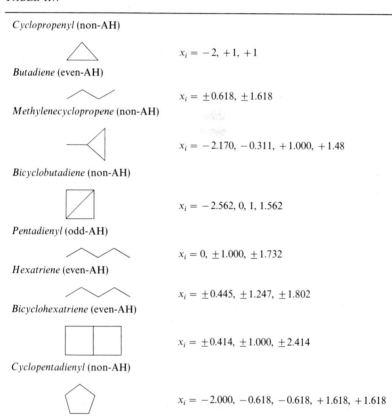

Cyclopropenyl (non-AH)

$x_i = -2, +1, +1$

Butadiene (even-AH)

$x_i = \pm 0.618, \pm 1.618$

Methylenecyclopropene (non-AH)

$x_i = -2.170, -0.311, +1.000, +1.48$

Bicyclobutadiene (non-AH)

$x_i = -2.562, 0, 1, 1.562$

Pentadienyl (odd-AH)

$x_i = 0, \pm 1.000, \pm 1.732$

Hexatriene (even-AH)

$x_i = \pm 0.445, \pm 1.247, \pm 1.802$

Bicyclohexatriene (even-AH)

$x_i = \pm 0.414, \pm 1.000, \pm 2.414$

Cyclopentadienyl (non-AH)

$x_i = -2.000, -0.618, -0.618, +1.618, +1.618$

# Problems

Finally, the generalizations concerning the energy level distributions in AH and non-AH systems can be illustrated by considering the calculated roots in a number of simple cases (Table II.4). Other generalizations concerning the eigenvalue distributions in some of the above systems will be considered later.

## PROBLEMS

1. Determine the $\pi$-energy levels for each of the following systems (in terms of $\alpha$ and $\beta$) and show the electron occupancies of the neutral species of I and II and those for the cation, radical, and anion of III, and those for the neutral species and dication of IV.

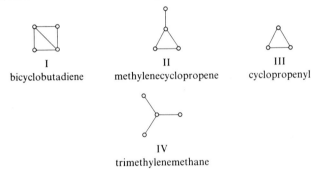

I
bicyclobutadiene

II
methylenecyclopropene

III
cyclopropenyl

IV
trimethylenemethane

Which of these species would be predicted to have a diradical ground state? Are these predictions in accord with qualitative VB (resonance) arguments?

2. For each of the above systems (I, II, III$^+$, III$^{\cdot}$, III$^-$, IV, and IV$^{++}$) calculate the total $\pi$-energy ($E_\pi$) and the delocalization energy (DE). Which species would be predicted to be particularly stable on the basis of their DE(electron) values? Are these predictions in accord with resonance theory? (Decide this by writing reasonable VB contributors for each species.)

3. Calculate the coefficients for the four MOs of trimethylenemethane (IV). Use these to calculate the $\pi$-bond order for this system.

4. The MO coefficients for the bicyclohexatriene system are as shown in the accompanying tabulation. Calculate the electron densities and free valence indices at atoms A and B and the bond orders of the A—B, A—F, and B—E bonds. Draw schematic representations of all the molecular orbitals and indicate the number of nodes (or nodal regions) in each.

| $\psi_i$ | A | B | C | D | E | F | $x_i$ |
|---|---|---|---|---|---|---|---|
| $\psi_1$ | 0.354 | 0.500 | 0.354 | 0.354 | 0.500 | 0.354 | −2.416 |
| $\psi_2$ | 0.500 | 0 | −0.500 | −0.500 | 0 | 0.500 | −1.000 |
| $\psi_3$ | 0.354 | 0.500 | 0.354 | −0.354 | −0.500 | −0.354 | −0.414 |
| $\psi_4$ | 0.354 | −0.500 | 0.354 | 0.354 | −0.500 | 0.354 | +0.416 |
| $\psi_5$ | 0.500 | 0 | −0.500 | 0.500 | 0 | −0.500 | +1.000 |
| $\psi_6$ | 0.354 | −0.500 | 0.354 | −0.354 | 0.500 | −0.354 | +2.414 |

5. Designate which of the following conjugated π-systems are even-AH, odd-AH, or non-AH. How many of these molecules (assuming all of them to be planar and electrically neutral)
   (a) would be expected to have π-energy levels symmetrically disposed about zero?
   (b) How many would be expected to have nonbonding MOs.
   (c) How many would be predicted to have triplet (diradical) ground states?
   (d) How many would be predicted to have significantly large dipole moments, and in which directions?

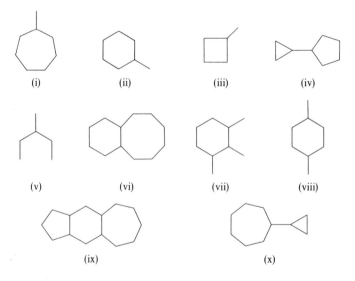

## REFERENCES

1. H. Eyring, J. Walter, and G. E. Kimball, "Quantum Chemistry," p. 99. Wiley, New York, 1944.
2. K. Ruedenberg, C. C. Roothaan, and W. Jaunzemis, *J. Chem. Phys.* **24**, 201 (1956); E. A. Boudreaux, L. P. Cusacks, and L. D. Dureau, "Numerical Tables of Two Center Overlap Integrals," pp. 1, 314. Benjamin, New York, 1970.
3. R. Pariser and R. G. Parr, *J. Chem. Phys.* **21**, 466, 767 (1953); J. A. Pople, *Trans. Faraday Soc.* **49**, 1375 (1953).
4. M. J. S. Dewar, "The Molecular Orbital Theory of Organic Chemistry," p. 444. McGraw-Hill, New York, 1969. J. A. Pople and D. L. Beveridge, "Approximate Molecular Orbital Theory," p. 57. McGraw-Hill, New York, 1970.
5. R. H. de Wolfe and W. G. Young, *Chem. Rev.* **56**, 784 (1956).
6. O. L. Chapman, C. L. McIntosh, and J. Pacansky, *J. Am. Chem. Soc.* **95**, 614 (1973).
7. D. P. Craig, *Proc. R. Soc., Ser. A* **202**, 498 (1950).
8. P. Dowd, *Acc. Chem. Res.* **5**, 242 (1972).
9. C. A. Coulson, *Proc. R. Soc., Ser. A* **169**, 413 (1939).
10. C. A. Coulson, *Disc. Faraday Soc.* **2**, 9 (1947).
11. J. D. Roberts, A. Streitwieser, Jr., and C. M. Regan, *J. Am. Chem. Soc.* **74**, 4579 (1952).
12. G. W. Wheland and D. E. Mann, *J. Chem. Phys.* **17**, 264 (1949).

## SUPPLEMENTARY READING

Coulson, C. A., and Streitwieser, A., Jr., "Dictionary of $\pi$-Electron Calculations." Freeman, San Francisco, California, 1965.

Flurry, R. L., "Molecular Orbital Theories of Bonding in Organic Molecules." Dekker, New York, 1968.

Heilbronner, E., and Bock, H., "Das HMO-Modell and seine Anwendung." Verlag Chemie, Weinheim, 1968.

Higasi, K., Baba, H., and Rembaum, A., "Quantum Organic Chemistry." Wiley (Interscience), New York, 1965.

Roberts, J. D., "Molecular Orbital Calculations." Benjamin, New York, 1962.

* Streitwieser, A., Jr., "Molecular Orbital Theory for Organic Chemists." Wiley, New York, 1961.

# III | THE USE OF SYMMETRY PROPERTIES IN SIMPLIFYING HMO CALCULATIONS

We have seen how the eigenvalues, coefficients, and related quantities such as bond order and electron density can be obtained by solving the secular problem directly for simple $\pi$-systems containing up to five $\pi$-centers. It is not difficult to find the roots of the simple determinants involved in such systems, and to find the coefficients by the method of cofactors.

If we now consider larger, and more complex, systems such as benzene or naphthalene, these would give larger secular determinants (sixth and tenth order, respectively) and larger numbers of secular equations to solve. Although there is nothing to prevent us from proceeding in exactly the same way as for simple systems to obtain solutions for these systems, the manipulations involved would be very laborious and the chances of mechanical error greatly increased. For these reasons, it is advisable to make use of the symmetry properties of the system, wherever possible, to simplify the secular problem. This involves the use of elementary group theory and results in greatly simplified secular determinants and equations.

## III.1 APPLICATION OF ELEMENTARY GROUP THEORY

This section will attempt to describe the bare essentials of the symmetry arguments used in simplifying secular problems. The reader is referred to standard works[1a,b] for a more complete discussion of the group theoretical arguments involved.

## III.1 Application of Elementary Group Theory

### III.1.1 SYMMETRY OPERATIONS

Any operation performed on an array of points that leaves the total array unchanged is called a *symmetry operation*. (The operation may change the positions and labeling of equivalent individual points, but not the total array). Similarly, any operation that can be performed on a molecular system which leaves the positions of the various *types* of atoms or groups, and the position of the molecule as a whole unchanged, can be called a symmetry operation.[2]

The only types of symmetry operation we will need to consider are the following:

(i) The *identity operation*, which is given the symbol $E$. This corresponds to no operation at all, which obviously leaves the molecule unchanged. This operation is needed for the completeness of the groups of operations, since the product of any two elements (or operations) in a group must also be an element of the group. Inclusion of the $E$ operation ensures this.

(ii) *Rotation* about a $p$-fold axis of symmetry by $360/p$ degrees, which is given the symbol $C_p$. Since in the present treatment we will only need to consider twofold axes of symmetry, rotations will be performed through $180°$. Thus only $C_2$-type operations will be involved in simplifying HMO calculations. Note that although only twofold symmetry axes are used, the molecule may actually possess higher symmetry axes, such as $C_3$ or $C_6$. However, no error will be introduced by taking the molecule to have a lower degree of symmetry than it actually possesses. In fact, if we were to try to take full advantage of the actual symmetry possessed by many molecules, the proper choice of symmetry axes would often involve more labor than it is worth. Using only twofold axes, this choice is usually obvious. If the molecule considered is placed in reference to a Cartesian coordinate system, three types of $C_2$ operation may be possible, namely, $C_2^x$ (i.e., $180°$ rotation about the $x$-axis), $C_2^y$, and $C_2^z$.

(iii) *Reflection* of all points through a mirror plane of symmetry to an equal distance on the opposite side of the plane, which is given the symbol $\sigma$. Where the mirror plane is vertical to, and contains, the principal axis of symmetry of the molecule, the reflection operation is given the symbol $\sigma_v$. Only this type of reflection will be considered. Note that some molecules can have two such planes (perpendicular to each other) so that there can be both $\sigma_v^{xz}$ and $\sigma_v^{yz}$ operations, where the $z$-axis is taken to be the principal axis of symmetry. (Again the molecule may actually possess more than two $\sigma_v$ planes of symmetry, but no error will be introduced by considering only $\sigma_v{}^{xz}$ and $\sigma_v{}^{yz}$ provided that it possesses these as well.) It should be noted at this point that the plane of symmetry $(xy)$ which includes all atoms having $p_z$ orbitals will not generally be considered as an element of symmetry. The reason for this is that all $\pi$-systems possess the same symmetry with respect to this plane (i.e., they are all antisymmetric in this sense), therefore no simplification is gained by considering $\sigma^{xy}$ as an element of symmetry.

The above symmetry operations $E$, $C_2$, and $\sigma_v$ will be illustrated by considering a few simple molecules.

*Butadiene*

By placing all atoms of butadiene (in the *s*-trans conformation) in a common plane, which is taken to be the $xy$-plane, it is clear that the molecule possesses a twofold axis of symmetry, which is the $z$-axis (Fig. III.1). Rotation about this axis by 180° transforms any atom or point into a point equivalent to itself, and leaves the molecule as a whole unchanged (E.g., $H_1$ is transformed into $H_6$, $C_1$ into $C_4$, and so on.) Therefore, $C_2$ is one operation that can be performed on butadiene; the only other useful symmetry operation that leaves the molecule unchanged is the identity operation $E$. (Note that reflection in the $xy$-plane of symmetry is ignored, since this is common to all planar $\pi$-systems, and adds nothing useful.) There is one group of operations that contains only $C_2$ and $E$, which is called the $C_2$ point group; therefore, we can say that butadiene belongs to the $C_2$ symmetry group, or more simply that butadiene has $C_2$ symmetry.

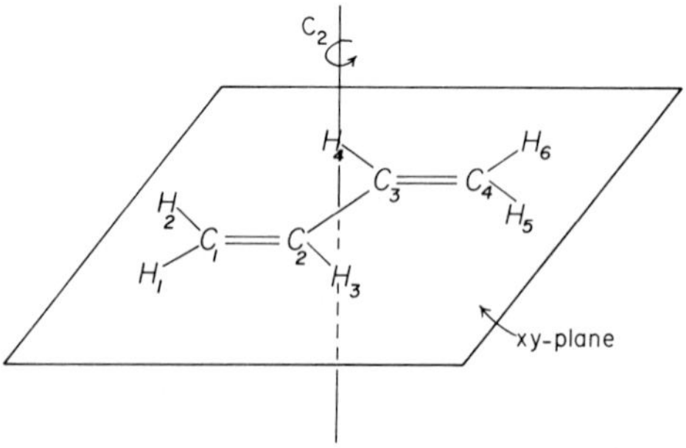

FIGURE III.1

### III.1 Application of Elementary Group Theory

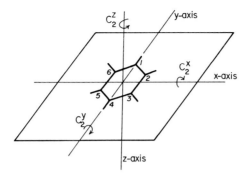

FIGURE III.2

#### Benzene

Again all atoms of the benzene molecule are placed in the $xy$-plane, as shown by the skeletal representation in Fig. III.2. (It is only really necessary to consider atoms of the $\pi$-system in determining and using symmetry operations to simplify HMO calculations.) Although benzene actually has a sixfold axis of symmetry, we will consider only the twofold axes shown. Thus $C_2^z$ transforms $C_1$ into $C_4$, $C_2$ into $C_5$, $C_3$ into $C_6$, and so on, which are equivalent positions or environments. Similarly $C_2^x$ transforms $C_1$ into $C_4$, $C_2$ into $C_3$, $C_5$ into $C_6$, and so forth. Also $C_2^y$ transforms $C_1$ into $C_1$, $C_2$ into $C_6$, $C_3$ into $C_5$, and so on. Therefore, $C_2^z$, $C_2^x$, and $C_2^y$ are appropriate symmetry operations on benzene. One other operation that can be performed is, as before, the $E$ operation.

The group of operations $E$, $C_2^x$, $C_2^y$, and $C_2^z$ is called the $D_2$ point group, therefore we can say that benzene has $D_2$ symmetry. (In actual fact, benzene belongs to a much higher symmetry group, $D_{6h}$, but the $D_2$ point group is a proper subgroup of this, and hence can be used for the present purposes without introducing any error.[†]) An alternative way of considering the symmetry properties of benzene is to use $\sigma_v$ operations. Suppose that instead of considering all three $C_2$ axes, benzene had been represented as having one $C_2$ axis, the $z$-axis (Fig. III.3). Then there are two planes $xz$ and $yz$ each of which contain, and are vertical to, the principal axis of symmetry of the molecule. Then leaving $E$ and $C_2^z$ as before, reflection of any point to an equal distance

---

[†] Note that great care must be taken not to assign a molecule a *higher* degree of symmetry than it actually possesses.

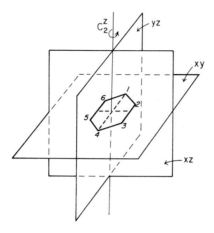

FIGURE III.3

through the $xz$-plane would transform $C_1$ into $C_4$, $C_2$ into $C_3$, $C_5$ into $C_6$, and so on. Thus $\sigma_v^{xz}$ is an appropriate symmetry operation for benzene. Similarly, reflection through the $yz$-plane transforms $C_1$ into $C_1$, $C_2$ into $C_6$, $C_3$ into $C_5$, and so on. Hence $\sigma_v^{yz}$ is also a symmetry operation. (Note that $\sigma_v^{yz}$ achieves the same result as for $C_2^y$ previously considered, in terms of the atoms involved, as does $\sigma_v^{xz}$ for $C_2^x$.)

Therefore, an equivalent set of operations (at least for many molecules like benzene) would be $E$, $C_2^z$, $\sigma_v^{xz}$, and $\sigma_v^{yz}$. This constitutes the $C_{2v}$ point group and thus we could equally well have said that benzene possesses $C_{2v}$ symmetry.

For most simple molecules, and particularly for the planar $\pi$-systems of principal interest, the $C_{2v}$ and $D_2$ groups are equivalent[†]; therefore, for simplicity it is advisable to stick to one of these in carrying out all HMO calculations. The choice is purely arbitrary, but there is an advantage in using $C_{2v}$ operations in that it is possible to use planar projections of the molecules and to perform all symmetry operations in the plane of the paper. Thus $C_2^v$ operations may be easier to visualize in two dimensions.

This will be illustrated on naphthalene. In planar projection onto the $xy$-plane, the $z$-axis becomes a point, which is placed at the center

---

[†] To illustrate that these two groups are not equivalent for all molecules, consider *cis*-1,3-dimethylcyclobutane and *trans,trans,trans*-1,2,3,4-tetramethylcyclobutane. Only one of these actually has $D_2$ symmetry (verify this for yourself), whereas both could be said to have $C_{2v}$ symmetry. The $D_2$ is a higher symmetry group than $C_{2v}$.

### III.1 Application of Elementary Group Theory

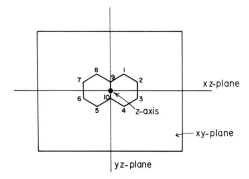

**FIGURE III.4**

of the 9,10-bond, and is the principal axis of symmetry of the molecule (Fig. III.4). In projection, the $yz$- $xz$-planes become lines, as shown, that intersect at the position of the $z$-axis. Therefore, naphthalene can be subjected to the $E$, $C_2^z$, $\sigma_v^{xz}$, and $\sigma_v^{yz}$ operations, all in the plane of the paper, since $C_2^z$ becomes rotation about a point, and the $\sigma_v$ operations become reflections through lines. Naphthalene thus clearly belongs to the $C_{2v}$ point group since each of the four types of operation transforms each point into a completely equivalent point.

In carrying out symmetry operations it is usually better to label each atomic position by a letter $A, B, C, \ldots$ rather than by a number, for reasons that will become clear later, and when dealing with $\sigma_v$ operations it is simpler to designate one as $\sigma_v$ and the other as $\sigma_v'$. The choice is arbitrary but it is simpler to designate as $\sigma_v$ that operation which leaves the greater number of points completely unaffected (in label as well as position). Note that $\sigma_v^{yz}$ leaves the bridgehead positions unchanged in label, whereas $\sigma_v^{xz}$ changes the labeling of all positions.

The results of performing symmetry operations can be illustrated in simple form for naphthalene (I) to determine what each *labeled* position is transformed into, under each symmetry operation of the $C_{2v}$ group. First label each position of the planar projection arbitrarily

TABLE III.1

| Position | Operation | | | |
|---|---|---|---|---|
| | $E$ | $C_2$ | $\sigma_v$ | $\sigma_v'$ |
| A | A | F | I | D |
| B | B | G | H | C |
| C | C | H | G | B |
| D | D | I | F | A |
| E | E | J | E | J |
| F | F | A | D | I |
| G | G | B | C | H |
| H | H | C | B | G |
| I | I | D | A | F |
| J | J | E | J | E |
| | $10^a$ | $0^a$ | $2^a$ | $0^a$ |

[a] Number of positions unaffected both in label and in type.

by a letter, and designate the appropriate symmetry elements. Now perform each of the operations on the array of labeled positions, and indicate in tabular form (see Table III.1) the result of each operation (i.e., although each position will be transformed into an *equivalent* position, the initially designated label may have changed).

Table III.1 is called the *transformation table* for naphthalene under the $C_{2v}$ group operations. These transformation tables have important uses in simplifying the calculations based on the secular equations. Note that as indicated at the foot of the table, different operations leave different numbers of positions completely unaffected. These numbers are also important in that they determine how the secular equations and determinant will eventually simplify after making appropriate use of the symmetry of the system. For example, in the case of naphthalene, these numbers can be used to show that the secular problem will be reduced to solving two third-order and two second-order determinants, instead of the much more laborious and difficult problem of solving a full tenth-order determinant.

The only other symmetry group that we need to consider is the simpler $C_2$ group, which contains only the operations $E$ and $C_2$ shown in Fig. III.5. Use of the $C_2$ operations to obtain the transformation table (Table III.2) for butadiene can again be illustrated in planar projection (Fig. III.6).

## III.1 Application of Elementary Group Theory

|   | E |   |   | E |
|---|---|---|---|---|
| $C_2$ | $C_2$ | | $C_{2v}$ | $C_2$ |
| | | | | $\sigma_v$ |
| | | | | $\sigma_v'$ |

FIGURE III.5

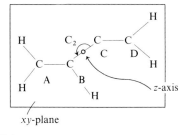

FIGURE III.6

TABLE III.2

|   | E | $C_2$ |
|---|---|---|
| A | A | D |
| B | B | C |
| C | C | B |
| D | D | A |
|   | 4 | 0 |

Before going on to make use of symmetry properties in HMO calculations, it is worthwhile considering several simple molecules to illustrate several useful points.

### Butadiene

This molecule (IIa) has $C_2$ symmetry only and must be placed in the $C_2$ group. In dealing with linear polyene systems like butadiene, it makes no difference if the molecular skeleton is represented in linear form for simplicity, provided only $C_2$ group operations are used, e.g., (IIb).

$$\begin{array}{cc} C_2 \diagup^{C-C} & C-C \overset{\frown}{\bullet} C-C \\ C-C & C_2 \\ \text{IIa} & \text{IIb} \end{array}$$

All linear polyenes belong to the $C_2$ symmetry group.

### Cyclobutadiene

This molecule (III) actually has $C_{2v}$ symmetry, but it is sufficient to place it in the $C_2$ group, since this will simplify the calculations satisfactorily.

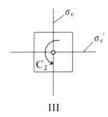

III

## Bicyclohexatriene

This molecule (IV) possesses $C_{2v}$ symmetry and should be placed in that group.

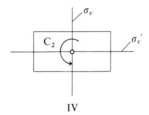

IV

## Cyclooctatetraene

Although this molecule (V) has higher symmetry than $C_{2v}$, placing it in this group will result in adequate simplification and introduce no error. Note however that there is a choice between placing the orthogonal $\sigma_v$ planes through the centers of bonds, or through nuclear positions, as shown (VI).

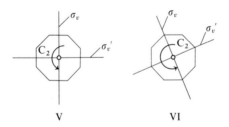

V           VI

## p-Xylylene-type systems

Although p-xylylene VII itself has $C_{2v}$ symmetry and should be placed in that group, the pyridazine analog VIII has only $C_2$ symmetry.

## III.2 Simplifying Secular Determinants

In general, introduction of heteroatoms will result in a lower degree of symmetry being imposed on the system.

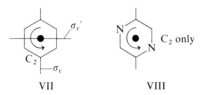

VII  VIII

In connection with the foregoing, the following general points can be made:

(i) If the molecule is simple (i.e., up to five π-centers) adequate simplification will be achieved by using $C_2$ symmetry even though the molecule may also have $C_{2v}$ symmetry.

(ii) If the molecule has six or more π-centers, it should be placed in the $C_{2v}$ group, if possible.

(iii) If there is a choice, try to avoid placing $\sigma_v$ planes through atomic positions wherever possible; greater simplification will be achieved in this way.

(iv) Try to avoid using the $xy$-plane as a symmetry element, if at all possible. If this plane is used for $\sigma_v$-type operations in a $C_{2v}$ system, equal simplification, with less labor, will usually be achieved by ignoring this element and placing the molecule in the simpler $C_2$ group.

### III.2 USE OF SYMMETRY TO SIMPLIFY SECULAR DETERMINANTS

Instead of using $n$ atomic orbitals of the $2p_z$-type as a basis set to build up LCAO–MO functions giving an $n$th-order secular determinant, we could use certain simple linear combinations of AOs as a starting basis set and then use these combinations to construct LCAO–MO functions. If these simple linear combinations had been suitably chosen with respect to the symmetry properties of the system, a much simpler secular determinant would be obtained. This can be illustrated by means of a very simple case.

Suppose the basis set of simple AO functions were $\phi_1$, $\phi_2$, $\phi_3$ and $\phi_4$. Using this set without modification to form LCAO–MO such as

$$\psi_j = C_{j1}\phi_1 + C_{j2}\phi_2 + C_{j3}\phi_3 + C_{j4}\phi_4 \quad (j = 1, 2, 3, 4)$$

would eventually give a fourth-order secular determinant to solve, such as generally represented by

$$\begin{vmatrix} A_{11} & A_{12} & A_{13} & A_{14} \\ A_{21} & A_{22} & A_{23} & A_{24} \\ A_{31} & A_{32} & A_{33} & A_{34} \\ A_{41} & A_{42} & A_{43} & A_{44} \end{vmatrix} = 0$$

Although some of the elements in this determinant might turn out to be zero, there is no guarantee this would result in any significant simplification.

Suppose instead the starting AOs had first been taken in simple linear combinations such as

$(C_1\phi_1 + C_2\phi_2)$, $(C_1'\phi_1 + C_2'\phi_2)$, $(C_3\phi_3 + C_4\phi_4)$, and $(C_3'\phi_3 + C_4'\phi_4)$

where the $C_i$ and $C_i'$ are simple normalization factors such as $\pm(1/\sqrt{2})$ (since two AOs are used in each combination). These could then used in the same way as before to construct LCAOs of the form

$$\psi_j = C_{ji}(C_1\phi_1 + C_2\phi_2) + C_{j2}(C_1'\phi_1 + C_2'\phi_2)$$
$$+ C_{j3}(C_3\phi_3 + C_4\phi_4) + C_{j4}(C_3'\phi_3 + C_4'\phi_4) \qquad (j = 1, 2, 3, 4)$$

Following the standard procedure to obtain the secular determinant would result in a much simpler determinant,

$$\begin{vmatrix} A_{11} & A_{12} & 0 & 0 \\ A_{21} & A_{22} & 0 & 0 \\ 0 & 0 & A_{33} & A_{34} \\ 0 & 0 & A_{43} & A_{44} \end{vmatrix} = 0$$

if the symmetry properties of the molecule had been properly used in choosing the starting combinations of AOs. These blocked out determinants are clearly much easier to solve, since the above could be written as

$$\begin{vmatrix} A_{11} & A_{12} \\ A_{21} & A_{22} \end{vmatrix} \begin{vmatrix} A_{33} & A_{34} \\ A_{43} & A_{44} \end{vmatrix} = 0$$

Therefore, each smaller determinant can be set equal to zero in turn and solved separately to obtain the roots. A similar simplification results in solving the sets of secular equations to obtain the coefficients.

In the foregoing simple case, instead of having to solve a fourth-order determinant and sets of four secular equations, each containing four

## III.2 Simplifying Secular Determinants

terms, the problem would simplify to solving only second-order determinants and sets of two secular equations each with only two terms.

These simple linear combinations of starting orbitals are called *symmetry orbitals*, and once chosen, the procedure involved in setting up the secular problem is *exactly the same as before*, except that the matrix elements $H_{ii}$ and $H_{ij}$ will not in general be simply $\alpha$ or $\beta$ or zero, but will have different values that can easily be worked out.

### III.2.1 PROCEDURE INVOLVED IN USING SYMMETRY ORBITALS

The steps involved in this procedure will be stated first, then illustrated by choosing simple examples and solving the secular problem.

(i) Choose the symmetry orbitals so that they are eigenfunctions of the appropriate symmetry operators for the molecule. (These operators and their eigenvalues will be determined by the appropriate point group.) Each symmetry operation can be thought of as involving an eigenvalue equation of form

$$\hat{O} \cdot f = \lambda \cdot f$$

where $\hat{O}$ the symmetry operator operates on $f$, the symmetry orbital, and produces the function $f$ times an eigenvalue $\lambda$. It turns out that all eigenvalues for operators of the $C_2$ and $C_{2v}$ groups are either $+1$ or $-1$. (This is because we are using only simple twofold symmetry; any symmetry operation either reproduces the original function unchanged, or its negative.) Thus for a set of symmetry orbitals $S_1, S_2, S_3, \ldots, S_n$, we have for the $C_2$ and $C_{2v}$-type operations

$$\hat{E} S_i = \lambda_i S_i$$
$$\hat{C}_2 S_i = \lambda_i S_i$$
$$\hat{\sigma}_v S_i = \lambda_i S_i$$
$$\hat{\sigma}_v{}^1 S_i = \lambda_i S_i$$

where $\lambda_i$ always equals $+1$ for the $E$ operation, and can only equal $+1$ or $-1$ for the other three operations.

(ii) Use these symmetry orbitals as a new basis set for a Hückel MO calculation using exactly the same procedure as before, where the $\psi_j$ are now

$$\psi_j = \sum_n C_{jn} S_n$$

(instead of $\psi_j = \sum_n C_{jn}\phi_n$ as before). Note that the coefficients $C_{jn}$ will not generally have the same values as previously found.

(iii) Write down the secular determinant in terms of these new molecular orbitals. The general form of the determinant is *exactly* the same as before [i.e., all diagonal elements will be of the form $(H_{ii} - \varepsilon)$ and all off-diagonal elements will be $H_{ij}$].

$$\begin{vmatrix} H_{11} - \varepsilon & H_{12} & \cdots & H_{1n} \\ H_{21} & H_{22} - \varepsilon & & \\ \vdots & & \ddots & \\ H_{n1} & \cdots & & H_{nn} - \varepsilon \end{vmatrix} = 0$$

However the matrix elements $H_{ii}$ and $H_{ij}$ will no longer have their original values

$$H_{ii} = \int \phi_i H \phi_i \, d\tau = \alpha$$

$$H_{ij} = \int \phi_i H \phi_j \, d\tau = \beta \quad \text{or} \quad 0$$

but will now be

$$H_{ii} = \int S_i H S_i \, d\tau$$

$$H_{ij} = \int S_i H S_i \, d\tau$$

whose values are to be determined later by expansion in terms of the previous integrals.[†]

(iv) Make use of a quantum mechanical theorem which states that *all* matrix elements of the form $H_{ij} = \int S_i H S_j \, d\tau$ must vanish if $S_i$ and $S_j$ have different eigenvalues for *any* symmetry operator in the group. (A proof of this theorem will not be given here, but this result is easy to verify for any element of the above type.) It is this step that greatly simplifies the solution of both the secular determinant and secular equations.

(v) The simplified secular determinant is solved in the same way as before to obtain the eigenvalues, and the roots are substituted one at a time into the simplified equations to obtain the coefficients, again exactly as before.

The above procedure will be illustrated by going through all necessary steps for a simple $C_2$ molecule, butadiene. Actually, since

---

[†] Note these terms $S_i$ for symmetry orbitals should not be confused with the previous $S_{ij}$ for overlap integrals.

## III.2 Simplifying Secular Determinants

butadiene is only a four-orbital system, the use of symmetry is not really worthwhile, but the steps are easier to illustrate in this simple case. Following this the method will be illustrated for a $C_{2v}$ system (benzene), where it really is worthwhile to avoid having to solve a sixth-order secular problem. (Note that expansion of a sixth-order determinant by the method of cofactors would give 6!/2 second-order determinants.)

### Butadiene

*Choice of Symmetry Orbitals* Write down the basic $\pi$-structure in planar projection (IX) and label each carbon by a letter. Thus the simple AO basis set can be designated as $\{\phi_A \phi_B \phi_C \phi_D\}$.

IX

Determine which point group the system belongs to (in the case of butadiene it is $C_2$) and obtain the appropriate *character table* (Table III.3) for this group.[2] In a sense this character table is a table of eigenvalues for the appropriate symmetry operations of the group. The character table for the $C_2$ group is quite simple and is given in Table III.3.

TABLE III.3

|   | E | $C_2$ |
|---|---|-------|
| A | 1 | 1     |
| B | 1 | -1    |

The complete set of original AOs $\phi_A, \phi_B, \phi_C, \phi_D$ can be said to form a *reducible representation* of the total array of points (the butadiene $\pi$-skeleton). The problem is to find simpler subsets of these orbitals that form a series of *irreducible representations* (which belong to the A and B types of irreducible representation), which under the appropriate symmetry operations transform with the eigenvalues listed in the character table. These irreducible representations are the simplest subsets that possess the overall symmetry characteristics of the total array (or basis set). For the $C_2$ group there is only one A type (symmetrical) and one B type (unsymmetrical) although each may contain more than one representation or subset. However, each subset of a

TABLE III.4

|   | E       | $C_2$   |
|---|---------|---------|
|   | $\phi_A$ | $\phi_A$ | $\phi_D$ |
|   | $\phi_B$ | $\phi_B$ | $\phi_C$ |
|   | $\phi_C$ | $\phi_C$ | $\phi_B$ |
|   | $\phi_D$ | $\phi_D$ | $\phi_A$ |
|   | 4       | 0       |

given type must possess the same eigenvalue under any symmetry operation as any other subset of that type.

To find these subsets, we first construct a *transformation table* (Table III.4) by writing down a list of the starting AOs, apply each symmetry operation to each AO in turn, and indicate the result of this transformation. As shown, this gives $m$ columns each containing $n$ orbitals, where $m$ is the number of symmetry operations in the group. The first column (under the $E$ operation) is always simply a list of the starting AOs. At the foot of each column is written the total number of AOs that is unchanged in position and label by each operation. For the $E$ operation this number is obviously just the total number of AOs in the set. Under the $C_2$ operation each AO is transformed to an equivalent but nonidentical orbital, thus the number left totally unchanged is zero.

These numbers from the transformation table can now be used, in conjunction with the character table, to determine how many independent subsets (or symmetry orbitals) there will be in each type of irreducible representation. (At the same time, this will give the size or order of the determinants that will finally result from use of these symmetry orbitals in an HMO treatment.)

Each of the two types of irreducible representation ($A$ and $B$) has associated with it a certain number of independent subsets or symmetry orbitals. These two groups of symmetry orbitals can be called $\Gamma_A$ and $\Gamma_B$. In $\Gamma_A$ there will be $N_A$ independent symmetry orbitals, given by

$$N_A = \frac{(4 \cdot 1 + 0 \cdot 1)}{2} = 2$$

In this expression the dot product is taken of the number of simple AOs unchanged in the transformation table under a given operation and the value for that operation in the character table. This product is then divided by the total number of operations in the group.

## III.2 Simplifying Secular Determinants

Thus there are two independent $S_i$ belonging to $\Gamma_A$ and each of these transforms with the eigenvalues 1,1 under the $C_2$ symmetry operations. These two $S_i$ are obtained as follows. Write down the dot product of the appropriate row of the character table (for $\Gamma_A$ this is row $A$) with *each* row of the transformation table. Thus for $\Gamma_A$ each row of the transformation table is multiplied by 1,1.

$$1 \cdot \phi_A + 1 \cdot \phi_D$$
$$1 \cdot \phi_B + 1 \cdot \phi_C$$
$$1 \cdot \phi_C + 1 \cdot \phi_B$$
$$1 \cdot \phi_D + 1 \cdot \phi_A$$

Of these, it is clear that there are only two linearly independent combinations of the starting AOs, namely,

$$(\phi_A + \phi_D) \quad \text{and} \quad (\phi_B + \phi_C)$$

After normalization, the two symmetry orbitals of the $A$ type of irreducible representation $\Gamma_A$ are

$$S_1 = \frac{1}{\sqrt{2}}(\phi_A + \phi_D)$$

$$S_2 = \frac{1}{\sqrt{2}}(\phi_B + \phi_C)$$

The normalization factor for any $S_i$ is always $1/\sqrt{N}$, where $N$ is the number of simple AOs in the symmetry orbital (since the original $\phi_i$ are already normalized). It is best to normalize at this stage so that

$$\int S_i^2 \, d\tau = 1$$

for any $S_i$; then every subsequent step in the calculation will be the same as before.

Turning to the $\Gamma_B$ type, the number of $S_i$ in this set is obtained in the same way. Thus

$$N_B = \frac{(4 \cdot 1 + 0 \cdot -1)}{2} = 2$$

Therefore there are also two independent $S_i$ in $\Gamma_B$. (However, in general there will not be the same number of $S_i$ in each $\Gamma$.) To find these $S_i$ the previous procedure is repeated using the $B$ row of the character table:

$$1 \cdot \phi_A - 1 \cdot \phi_D$$
$$1 \cdot \phi_B - 1 \cdot \phi_C$$
$$1 \cdot \phi_C - 1 \cdot \phi_B$$
$$1 \cdot \phi_D - 1 \cdot \phi_A$$

Again, only two of these combinations are linearly independent, since for example $(\phi_A - \phi_D)$ is *physically* equivalent to $(\phi_D - \phi_A)$. Arbitrarily choosing the first two and normalizing gives

$$S_3 = \frac{1}{\sqrt{2}}(\phi_A - \phi_D)$$

$$S_4 = \frac{1}{\sqrt{2}}(\phi_B - \phi_C)$$

TABLE III.5

| | $C_2$ | Operation | | Character |
|---|---|---|---|---|
| | | E | $C_2$ | |
| $S_1$ | | $S_1$ | $S_1$ | 1, 1 |
| $S_2$ | | $S_2$ | $S_2$ | 1, 1 |
| $S_3$ | | $S_3$ | $-S_3$ | 1, $-1$ |
| $S_4$ | | $S_4$ | $-S_4$ | 1, $-1$ |

## III.2 Simplifying Secular Determinants

Although it is obvious in this simple case, it is as well to check that each $S_i$ generated does in fact transform under the group symmetry operations with the appropriate character. This can be done by representing the $S_i$ pictorially, showing the $C_2$ axis as in Table III.5.

Note that since $\Gamma_A$ contains two $S_i$ and $\Gamma_B$ also contains two $S_i$, then instead of having to deal with a fourth-order secular determinant, a product of two independent second-order determinants will result. The reason for this is shown by the following step.

### III.2.2  USE OF SYMMETRY ORBITALS AS BASIS SET

The foregoing $S_i$ are now used to form LCAOs of the type

$$\psi_j = C_{j1}S_1 + C_{j2}S_2 + C_{j3}S_3 + C_{j4}S_4 \qquad (j = 1, 2, 3, 4)$$

This will give exactly the same form of Hückel secular determinant as before, except that the matrix elements $H_{ij}$ (or $H_{ii}$) will now be of the form

$$H_{ij} = \int S_i \mathcal{H} S_j \, d\tau$$

Thus for butadiene we have (omitting the $j$ index for simplicity as before)

$$\begin{vmatrix} H_{11} - \varepsilon & H_{12} & H_{13} & H_{14} \\ H_{21} & H_{22} - \varepsilon & H_{23} & H_{24} \\ H_{31} & H_{32} & H_{33} - \varepsilon & H_{34} \\ H_{41} & H_{42} & H_{43} & H_{44} - \varepsilon \end{vmatrix} = 0$$

Now, because $S_1$ and $S_2$ belong to $\Gamma_A$ with character (1,1) and $S_3$ and $S_4$ belong to $\Gamma_B$ with character $(1,-1)$, all matrix elements between these two sets of $S_i$ vanish.[†] Therefore, the secular determinant can immediately be rewritten in simplified block form as

$$\begin{vmatrix} H_{11} - \varepsilon & H_{12} & 0 & 0 \\ H_{21} & H_{22} - \varepsilon & 0 & 0 \\ 0 & 0 & H_{33} - \varepsilon & H_{34} \\ 0 & 0 & H_{43} & H_{44} - \varepsilon \end{vmatrix} = 0$$

or alternatively as

$$\begin{vmatrix} H_{11} - \varepsilon & H_{12} \\ H_{21} & H_{22} - \varepsilon \end{vmatrix} \begin{vmatrix} H_{33} - \varepsilon & H_{34} \\ H_{43} & H_{44} - \varepsilon \end{vmatrix} = 0$$

[†] This can easily be verified by expanding the elements in question.

or in abbreviated form as
$$D_A \cdot D_B = 0$$

The roots (and eigenvalues) for this secular problem can now be obtained by setting each smaller determinant equal to zero in turn. Therefore, the problem of obtaining the energies and the coefficients for butadiene requires the solution of only second-order determinants and equations.

Considering $\Gamma_A$ first, we have

$$D_A = \begin{vmatrix} H_{11} - \varepsilon & H_{12} \\ H_{21} & H_{22} - \varepsilon \end{vmatrix} = 0$$

It is now necessary to evaluate each matrix element $H_{ij}$ in terms of the original values of $\alpha$ and $\beta$. This is done quite simply by expanding the new integrals (in terms of $S_i$) as follows:

$$H_{11} = \int S_1 \mathcal{H} S_1 \, d\tau$$

$$= \int \frac{1}{\sqrt{2}} (\phi_A + \phi_D) \mathcal{H} \frac{1}{\sqrt{2}} (\phi_A + \phi_D) \, d\tau$$

$$= \tfrac{1}{2} \left[ \int \phi_A \mathcal{H} \phi_A \, d\tau + \int \phi_D \mathcal{H} \phi_D \, d\tau + 2 \int \phi_A \mathcal{H} \phi_D \, d\tau \right]$$

$$= \tfrac{1}{2} [H_{AA} + H_{DD} + 2H_{AD}]$$

Since we have not changed the basic HMO method or assumptions, but only the way of approaching the secular problem, these terms $H_{AA}$, $H_{DD}$, and $H_{AD}$ will have the same values as before. Thus $H_{AA} = H_{DD} = \alpha$ and $H_{AD} = 0$ since atoms $A$ and $D$ are not directly joined in the $\sigma$-framework. Thus

$$H_{11} = \tfrac{1}{2}[\alpha + \alpha + 0] = \alpha$$

Similarly,

$$H_{22} = \int S_2 \mathcal{H} S_2 \, d\tau$$

$$= \int \frac{1}{\sqrt{2}} (\phi_B + \phi_C) \mathcal{H} \frac{1}{\sqrt{2}} (\phi_B + \phi_C) \, d\tau$$

$$= \tfrac{1}{2}[H_{BB} + H_{CC} + 2H_{BC}]$$
$$= \tfrac{1}{2}[\alpha + \alpha + 2\beta]$$
$$= \alpha + \beta$$

## III.2 Simplifying Secular Determinants

and
$$H_{12} = H_{21} = \tfrac{1}{2}[H_{AB} + H_{AC} + H_{DB} + H_{DC}]$$
$$= \tfrac{1}{2}[\beta + 0 + 0 + \beta] = \beta$$

Therefore,
$$D_A = \begin{vmatrix} \alpha - \varepsilon & \beta \\ \beta & \alpha + \beta - \varepsilon \end{vmatrix} = 0$$

In terms of the previous substitution $(x = (\alpha - \varepsilon)/\beta)$ this becomes
$$D_A = \begin{vmatrix} x & 1 \\ 1 & x+1 \end{vmatrix} = 0$$

which expands to give
$$P_A = x^2 + x - 1 = 0$$

Thus two of the roots are
$$x = \frac{-1 \pm \sqrt{1+4}}{2} = \frac{-1 \pm \sqrt{5}}{2}$$

or $x = -1.618, +0.618$. These are identical with roots $x_1$ and $x_3$ previously determined without use of symmetry. (Note that use of $S_1$ and $S_2$ does not give $x_1$ and $x_2$.)

Now considering $\Gamma_B$ briefly, we have
$$D_B = \begin{vmatrix} H_{33} - \varepsilon & H_{34} \\ H_{43} & H_{44} - \varepsilon \end{vmatrix} = 0$$

and evaluating the new matrix elements in terms of the old integrals $\alpha$ and $\beta$ gives

$$H_{33} = \int \frac{1}{\sqrt{2}}(\phi_A - \phi_D)\mathcal{H}\frac{1}{\sqrt{2}}(\phi_A - \phi_D)\,d\tau$$
$$= \tfrac{1}{2}[H_{AA} + H_{DD} - 2H_{AD}] = \alpha$$
$$H_{44} = \tfrac{1}{2}[H_{BB} - H_{BC} - H_{CB} + H_{CC}] = \alpha - \beta$$
$$H_{34} = H_{43} = \tfrac{1}{2}[H_{AB} - H_{AC} - H_{DB} + H_{DC}] = \beta$$

Thus
$$D_A = \begin{vmatrix} \alpha - \varepsilon & \beta \\ \beta & \alpha - \beta - \varepsilon \end{vmatrix} = \begin{vmatrix} x & 1 \\ 1 & x-1 \end{vmatrix} = 0$$

This gives on expansion

$$P_B = x^2 - x - 1 = 0$$

which yields the roots

$$x = \frac{+1 \pm \sqrt{1+4}}{2} = \frac{1 \pm \sqrt{5}}{2}$$

or $x = -0.618, +1.618$, which correspond exactly to roots $x_2$ and $x_4$ previously determined.

Thus the four eigenvalues determined by use of symmetry are the same as those already obtained using an unmodified basis set.

Note that use of $S_1$ and $S_2$ gives energies $\varepsilon_1$ and $\varepsilon_3$, and $S_3$ and $S_4$ give energies $\varepsilon_2$ and $\varepsilon_4$. Neglecting the numerical values of the coefficients for the moment, it is clear that these combinations should give the energies found. For example, $S_1$ and $S_2$ can only be combined in two independent ways, as can $S_3$ and $S_4$:

$$(S_1 + S_2) \quad \text{or} \quad (S_1 - S_2)$$
$$(S_3 + S_4) \quad \text{or} \quad (S_3 - S_4)$$

Taking account of only the *signs* of the AOs in each combination gives

$$(S_1 + S_2) \equiv (+A + B + C + D)$$

which corresponds to the lowest energy MO $\psi_1$ with energy $\varepsilon_1$. Similarly,

$$(S_1 - S_2) \equiv (+A - B - C + D)$$

which corresponds to $\psi_3$ with energy $\varepsilon_3$ and

$$(S_3 + S_4) \equiv (+A + B - C - D)$$

which corresponds to $\psi_2$ with energy $\varepsilon_2$, and finally

$$(S_3 - S_4) \equiv (+A - B + C - D)$$

which corresponds to $\psi_4$ with energy $\varepsilon_4$.

### III.2.3 CALCULATION OF THE MO COEFFICIENTS FOR BUTADIENE

Because a number of the matrix elements are automatically zero when symmetry is used, the secular equations are also simplified in a

## III.2 Simplifying Secular Determinants

similar way to the determinant. Thus for butadiene we have

$$C_1(H_{11} - \varepsilon) + C_2H_{12} = 0$$
$$C_1H_{21} + C_2(H_{22} - \varepsilon) = 0$$
$$C_3(H_{33} - \varepsilon) + C_4H_{34} = 0$$
$$C_3H_{43} + C_4(H_{44} - \varepsilon) = 0$$

Since the first two equations involve only $C_1$ and $C_2$ and the last two involve only $C_3$ and $C_4$, these sets can be treated completely separately. To obtain the coefficients $C_1$ and $C_2$ for $\psi_1$ and $\psi_3$ the roots obtained from $D_A$ are substituted into the first two equations, one at a time. With the usual substitution these equations become

$$C_1 x + C_2 = 0$$
$$C_1 + C_2(x + 1) = 0$$

Taking $x_1 = -1.618$, substitution gives

$$-1.618 C_1 + C_2 = 0$$

or

$$C_2 = 1.618 C_1$$

Since the symmetry orbitals $S_1$ and $S_2$ have been normalized already, we also have

$$C_1^2 + C_2^2 = 1$$

Thus

$$C_1^2 + (1.618)^2 C_1^2 = 1$$

$$C_1^2 = \frac{1}{3.62} = 0.276$$

$$C_1 = 0.525$$

and

$$C_2 = 1.618(0.525) = 0.850$$

Therefore, the coefficients for $\psi_1$ (in terms of the $S_i$) are given as

$$\psi_1 = 0.525 S_1 + 0.850 S_2$$

Now recalling that

$$S_1 = \frac{1}{\sqrt{2}}(\phi_A + \phi_D), \qquad S_2 = \frac{1}{\sqrt{2}}(\phi_B + \phi_C)$$

the first MO becomes

$$\psi_1 = 0.525\left[\frac{1}{\sqrt{2}}(\phi_A + \phi_D)\right] + 0.850\left[\frac{1}{\sqrt{2}}(\phi_B + \phi_C)\right]$$

$$= 0.372\phi_A + 0.601\phi_B + 0.601\phi_C + 0.372\phi_D$$

which is exactly the same result as that obtained without using symmetry properties.

Similarly, substitution of $x_3 = +0.618$ into the same two equations gives the coefficients for $\psi_3$ in terms of $S_1$ and $S_2$, which followed by substitution for these two functions in terms of the original AOs gives

$$\psi_3 = 0.601\phi_A - 0.372\phi_B - 0.372\phi_C + 0.601\phi_D$$

again exactly as before.

The roots $x_2$ and $x_4$ obtained from solution of $D_B$ are then substituted into the second set of secular equations,

$$C_3 x + C_4 = 0$$
$$C_3 + C_4(x - 1) = 0$$

to obtain the coefficients for $\psi_2$ and $\psi_4$ in a similar way.

It is clear that if the system has a reasonable degree of symmetry, the simplification obtained will result in having to solve sets of only two or three simultaneous equations at a time. The coefficients are in general very easy to obtain, and the method of cofactors is not usually needed when using symmetry orbitals.

The solution of the butadiene secular problem is trivial, using $C_2$ symmetry. The more difficult procedure for a $C_{2v}$ system will now be illustrated for benzene, where the use of symmetry really is necessary to reduce the labor involved.

### III.2.4 APPLICATION OF SYMMETRY PROPERTIES TO THE BENZENE SECULAR PROBLEM

For the present purposes, it is sufficient to place benzene (X) in the $C_{2v}$ symmetry class. Applying the group operations, the transformation table is as shown in Table III.6. The character table for the $C_{2v}$ group

## III.2 Simplifying Secular Determinants

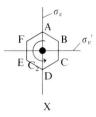

**TABLE III.6**

|   | $E$ | $C_2$ | $\sigma_v$ | $\sigma_v'$ |
|---|---|---|---|---|
| $\phi_A$ | $\phi_A$ | $\phi_D$ | $\phi_A$ | $\phi_D$ |
| $\phi_B$ | $\phi_B$ | $\phi_E$ | $\phi_F$ | $\phi_C$ |
| $\phi_C$ | $\phi_C$ | $\phi_F$ | $\phi_E$ | $\phi_B$ |
| $\phi_D$ | $\phi_D$ | $\phi_A$ | $\phi_D$ | $\phi_A$ |
| $\phi_E$ | $\phi_E$ | $\phi_B$ | $\phi_C$ | $\phi_F$ |
| $\phi_F$ | $\phi_F$ | $\phi_C$ | $\phi_B$ | $\phi_E$ |
|   | 6 | 0 | 2 | 0 |

**TABLE III.7**

|   | $E$ | $C_2$ | $\sigma_v$ | $\sigma_v'$ |
|---|---|---|---|---|
| $A_1$ | 1 | 1 | 1 | 1 |
| $A_2$ | 1 | 1 | $-1$ | $-1$ |
| $B_1$ | 1 | $-1$ | 1 | $-1$ |
| $B_2$ | 1 | $-1$ | $-1$ | 1 |

is somewhat more complicated than for $C_2$ but is still quite simple (Table III.7). There are now two types of symmetrical (or $A$-type) irreducible representation, and two types of unsymmetrical (or $B$-type) representation. However, the problem of determining the number of independent symmetry orbitals in each type is approached in the same way: by taking the dot product of the numbers in the last row of the transformation table and the numbers in the appropriate row of the character table. This time each such product is divided by four, since there are four operations in the $C_{2v}$ group. Thus in $\Gamma_{A_1}$ there are

$$N_{A_1} = \frac{(6 \cdot 1) + (0 \cdot 1) + (2 \cdot 1) + (0 \cdot 1)}{4} = 2$$

independent symmetry orbitals. Similarly, for the other three types there are

$$N_{A_2} = \frac{6 \cdot 1 + 0 \cdot 1 + 2 \cdot -1 + 0 \cdot -1}{4} = 1$$

$$N_{B_1} = \frac{6 \cdot 1 + 0 \cdot -1 + 2 \cdot 1 + 0 \cdot -1}{4} = 2$$

$$N_{B_2} = \frac{6 \cdot 1 + 0 \cdot -1 + 2 \cdot -1 + 0 \cdot 1}{4} = 1$$

Proceeding as before to find the independent linear combinations in $\Gamma_{A_1}$ we take the dot product of the $A_1$ row of the character table with each row of the transformation table, in succession:

$$1 \cdot \phi_A + 1 \cdot \phi_D + 1 \cdot \phi_A + 1 \cdot \phi_D$$
$$1 \cdot \phi_B + 1 \cdot \phi_E + 1 \cdot \phi_F + 1 \cdot \phi_C$$

and so on. It is clear that these two are the only possible linearly independent combinations; therefore, the two symmetry orbitals in $\Gamma_{A_1}$, after normalization, become

$$S_1 = \frac{1}{\sqrt{2}}(\phi_A + \phi_D)$$

$$S_2 = \tfrac{1}{2}(\phi_B + \phi_C + \phi_E + \phi_F)$$

Proceeding in the same way for $\Gamma_{A_2}$, we have

$$1 \cdot \phi_A + 1 \cdot \phi_D + (-1) \cdot \phi_A + (-1) \cdot \phi_D$$
$$1 \cdot \phi_B + 1 \cdot \phi_E + (-1) \cdot \phi_F + (-1) \cdot \phi_C$$

and so on. The first of these is clearly zero, hence the one symmetry orbital that belongs to $\Gamma_{A_2}$ is

$$S_3 = \tfrac{1}{2}(\phi_B - \phi_C + \phi_E - \phi_F)$$

For $\Gamma_{B_1}$ a similar procedure gives the two independent symmetry orbitals

$$S_4 = \frac{1}{\sqrt{2}}(\phi_A - \phi_D)$$

$$S_5 = \tfrac{1}{2}(\phi_B - \phi_C - \phi_E + \phi_F)$$

and finally for $\Gamma_{B_2}$ we obtain the one symmetry orbital

$$S_6 = \tfrac{1}{2}(\phi_B + \phi_C - \phi_E - \phi_F)$$

## III.2 Simplifying Secular Determinants

Summarizing, the six symmetry orbitals that will form the new basis set for benzene are

$$S_1 = \frac{1}{\sqrt{2}}(\phi_A + \phi_D)$$

$$S_2 = \tfrac{1}{2}(\phi_B + \phi_C + \phi_E + \phi_F)$$
$$S_3 = \tfrac{1}{2}(\phi_B - \phi_C + \phi_E - \phi_F)$$

$$S_4 = \frac{1}{\sqrt{2}}(\phi_A - \phi_D)$$

$$S_5 = \tfrac{1}{2}(\phi_B - \phi_C - \phi_E + \phi_F)$$
$$S_6 = \tfrac{1}{2}(\phi_B + \phi_C - \phi_E - \phi_F)$$

The easiest way to see what this new basis set looks like, and to verify that each $S_i$ transforms correctly under the group symmetry operations, is to draw these schematically, in planar projection, from above the plane of the benzene ring. In this way only the upper lobe of each $2p_z$ AO will be shown, and this will either be of positive sign (shaded) or of negative sign (unshaded). Now perform the operations $E$, $C_2$, $\sigma_v$, and $\sigma_v'$ on each one, indicating how each transforms. (See Table III.8.) Thus each $S_i$ has the appropriate character with respect to the classification into the four types of irreducible representation.

The MOs for benzene are now set up as before:

$$\psi_j = C_{j1}S_1 + C_{j2}S_2 + \cdots + C_{j6}S_6 \qquad (j = 1, 2, \ldots, 6)$$

and application of the standard Hückel procedure leads to the secular determinant

$$\begin{vmatrix} H_{11} - \varepsilon & H_{12} & H_{13} & H_{14} & H_{15} & H_{16} \\ H_{21} & H_{22} - \varepsilon & H_{23} & H_{24} & H_{25} & H_{26} \\ H_{31} & H_{32} & H_{33} - \varepsilon & H_{34} & H_{35} & H_{36} \\ H_{41} & H_{42} & H_{43} & H_{44} - \varepsilon & H_{45} & H_{46} \\ H_{51} & H_{52} & H_{53} & H_{54} & H_{55} - \varepsilon & H_{56} \\ H_{61} & H_{62} & H_{63} & H_{64} & H_{65} & H_{66} - \varepsilon \end{vmatrix} = 0$$

Now each element can be eliminated where $i$ and $j$ refer to $S_i$, which belong to different $\Gamma$s. For example, since only $S_1$ and $S_2$ belong to $\Gamma_{A_1}$ with character $(1,1,1,1)$, they must have at least one symmetry eigenvalue which is different from that of every other $S_i$. Therefore, all elements such as $H_{1j}$ and $H_{2j}$, where $j \neq 1$ or $2$, must vanish. This

TABLE III.8

| | | E | $C_2$ | $\sigma_v$ | $\sigma_v'$ | |
|---|---|---|---|---|---|---|
| $S_1$ | | 1 | 1 | 1 | 1 | $\Gamma_{A_1}$ |
| $S_2$ | | 1 | 1 | 1 | 1 | |
| $S_3$ | | 1 | 1 | −1 | −1 | $\Gamma_{A_2}$ |
| $S_4$ | | 1 | −1 | 1 | −1 | $\Gamma_{B_1}$ |
| $S_5$ | | 1 | −1 | 1 | −1 | |
| $S_6$ | | 1 | −1 | −1 | 1 | $\Gamma_{B_2}$ |

leads directly to the blocked out determinant:

$$D = \begin{vmatrix} H_{11}-\varepsilon & H_{12} & 0 & 0 & 0 & 0 \\ H_{21} & H_{22}-\varepsilon & 0 & 0 & 0 & 0 \\ 0 & 0 & H_{33}-\varepsilon & 0 & 0 & 0 \\ 0 & 0 & 0 & H_{44}-\varepsilon & H_{45} & 0 \\ 0 & 0 & 0 & H_{54} & H_{55}-\varepsilon & 0 \\ 0 & 0 & 0 & 0 & 0 & H_{66}-\varepsilon \end{vmatrix} = 0$$

which can be expressed more simply and usefully as

$$D = \begin{vmatrix} H_{11}-\varepsilon & H_{12} \\ H_{21} & H_{22}-\varepsilon \end{vmatrix} \begin{vmatrix} H_{33}-\varepsilon \end{vmatrix} \begin{vmatrix} H_{44}-\varepsilon & H_{45} \\ H_{54} & H_{55}-\varepsilon \end{vmatrix} \begin{vmatrix} H_{66}-\varepsilon \end{vmatrix}$$

$$= D_{A_1} \cdot D_{A_2} \cdot D_{B_1} \cdot D_{B_2}$$

Each element in these simplified determinants can now be worked out from the expression

$$H_{ij} = \int S_i \mathscr{H} S_j \, d\tau$$

### III.2 Simplifying Secular Determinants

by substituting the full expression for each $S_i$ and $S_j$ and evaluating in terms of the original integrals $\alpha$ and $\beta$, keeping in mind which $\beta_{ij} = 0$. For example, $H_{45}$ will be given by

$$H_{45} = \int S_4 \mathcal{H} S_5 \, d\tau$$

$$= \int \frac{1}{\sqrt{2}} (\phi_A - \phi_D) \mathcal{H} \tfrac{1}{2} (\phi_B - \phi_C - \phi_E + \phi_F) \, d\tau$$

$$= \frac{1}{2\sqrt{2}} [H_{AB} - H_{AC} - H_{AE} + H_{AF} - H_{DB} + H_{DC} + H_{DE} - H_{DF}]$$

$$= \frac{1}{2\sqrt{2}} [\beta - 0 - 0 + \beta - 0 + \beta + \beta - 0]$$

$$= \frac{1}{2\sqrt{2}} [4\beta] = \sqrt{2}\beta$$

and $H_{66}$ will be given by

$$H_{66} = \int S_6 \mathcal{H} S_6 \, d\tau$$

$$= \int \tfrac{1}{2}(\phi_B + \phi_C - \phi_E - \phi_F) \mathcal{H} \tfrac{1}{2}(\phi_B + \phi_C - \phi_E - \phi_F) \, d\tau$$

$$= \tfrac{1}{4}[H_{BB} + H_{BC} - H_{BE} - H_{BF} + H_{CB} + H_{CC} - H_{CE} - H_{CF}$$
$$\quad - H_{EB} - H_{EC} + H_{EE} + H_{EF} - H_{FB} - H_{FC} + H_{FE} + H_{FF}]$$

$$= \tfrac{1}{4}[4\alpha + 4\beta]$$

$$= \alpha + \beta$$

After evaluating each element[†] the finally simplified secular determinant becomes

$$D = \begin{vmatrix} \alpha - \beta & \sqrt{2}\beta \\ \sqrt{2}\beta & \alpha + \beta - \varepsilon \end{vmatrix} \begin{vmatrix} \alpha - \beta - \varepsilon \end{vmatrix} \begin{vmatrix} \alpha - \varepsilon & \sqrt{2}\beta \\ \sqrt{2}\beta & \alpha - \beta - \varepsilon \end{vmatrix} \begin{vmatrix} \alpha + \beta - \varepsilon \end{vmatrix} = 0$$

or with the usual substitution:

$$D = \begin{vmatrix} x & \sqrt{2} \\ \sqrt{2} & x+1 \end{vmatrix} \begin{vmatrix} x-1 \end{vmatrix} \begin{vmatrix} x & \sqrt{2} \\ \sqrt{2} & x-1 \end{vmatrix} \begin{vmatrix} x+1 \end{vmatrix} = 0$$

---

[†] Note that although some elements in these blocked out determinants may turn out to be zero, every element of the type $H_{ij}$, where $i$ and $j$ refer to different classes of $S_i$, must be zero. Verify that this is the case for $H_{25}$ for example.

It is now simple to extract the six roots for benzene by setting each small determinant equal to zero, in turn. Thus

$$D_{A_1} = \begin{vmatrix} x & \sqrt{2} \\ \sqrt{2} & x+1 \end{vmatrix} = 0$$

which gives on expansion

$$P_{A_1} = x^2 + x - 2$$
$$= (x+2)(x-1) = 0$$

yielding $x = -2, +1$. Similarly,

$$D_{A_2} = |x - 1| = 0$$

gives

$$P_{A_2} = x - 1 = 0$$

with root $x = +1$. Also,

$$D_{B_1} = \begin{vmatrix} x & \sqrt{2} \\ \sqrt{2} & x-1 \end{vmatrix} = 0$$

$$P_{B_1} = x^2 - x - 2$$
$$= (x-2)(x+1) = 0$$

giving $x = 2, -1$ and

$$D_{B_2} = x + 1 = 0$$
$$P_{B_2} = x + 1 = 0$$

giving

$$x = -1$$

Thus the six roots and eigenvalues for benzene are

$$\begin{aligned}
x_1 &= -2, & \varepsilon_1 &= \alpha + 2\beta \\
x_2 &= -1, & \varepsilon_2 &= \alpha + \beta \\
x_3 &= -1, & \varepsilon_3 &= \alpha + \beta \\
x_4 &= +1, & \varepsilon_4 &= \alpha - \beta \\
x_5 &= +1, & \varepsilon_5 &= \alpha - \beta \\
x_6 &= +2, & \varepsilon_6 &= \alpha - 2\beta
\end{aligned}$$

## III.2 Simplifying Secular Determinants

Note again that there is no correspondence between the numbering of the $S_i$ which are combined and the resulting roots. For example,

$S_1$ and $S_2$ combine to give $x_1$ and $x_4$ (or $x_5$)
$S_3$ gives $x_5$ (or $x_4$)
$S_4$ and $S_5$ combine to give $x_2$ (or $x_3$) and $x_6$
$S_6$ gives $x_3$ (or $x_2$)

The overall MO energy level diagram for benzene, after feeding in the six $\pi$-electrons, is shown in Fig. III.7. Note that there are pairs of degenerate BMOs and ABMOs, and that the former are completely filled with all electron spins being paired. (This type of electronic arrangement is referred to as a closed-shell structure since each level, or set of degenerate levels, is completely filled; this is unlike the situation already encountered for cyclobutadiene or trimethylenemethane, which are open-shell arrangements.)

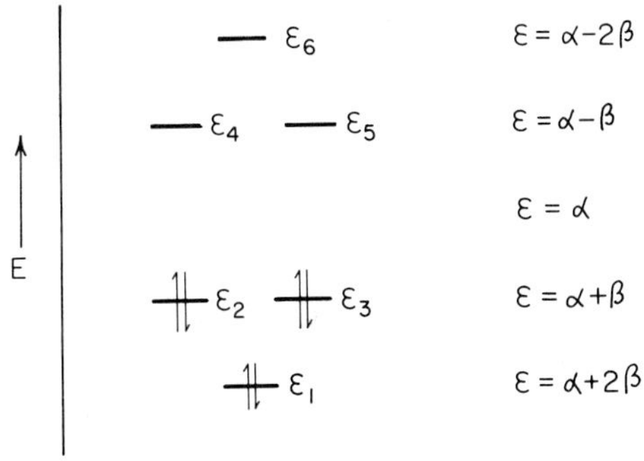

FIGURE III.7

$$\varepsilon_6 \quad \varepsilon = \alpha - 2\beta$$
$$\varepsilon_4 \quad \varepsilon_5 \quad \varepsilon = \alpha - \beta$$
$$\varepsilon = \alpha$$
$$\varepsilon_2 \quad \varepsilon_3 \quad \varepsilon = \alpha + \beta$$
$$\varepsilon_1 \quad \varepsilon = \alpha + 2\beta$$

We are now in a position to calculate the various $\pi$-electronic energy terms for the benzene system:

$$E_\pi = 2(\alpha + 2\beta) + 2(\alpha + \beta) + 2(\alpha + \beta) = 6\alpha + 8\beta$$
$$B_\pi = 8\beta$$

Note that the $\pi$-bonding energy per electron (BEPE) for benzene is quite high at $1.33\beta$ compared with values of 1.00 and 1.12 for the neutral systems ethylene and butadiene, respectively. To obtain DE for benzene, the localized cyclohexatriene structure is taken as a model

$E_\pi^{loc}$. Thus

$$DE = E_\pi - E_\pi^{loc}$$
$$= (6\alpha + 8\beta) - (6\alpha + 6\beta)$$
$$= 2\beta$$

and thus the delocalization energy per electron is

$$DEPE = 0.33\beta$$

It is worthwhile digressing at this point to discuss the significance of this value of $2\beta$ for the delocalization energy in benzene. Since benzene is the prototypical aromatic system with a well-known resonance stabilization energy of 36 kcal per mole,[3] it is tempting to equate the $DE$ value of $2\beta$ with this thermochemical value of the resonance energy and thus to conclude that each $\beta$ unit for a $\pi$-system is approximately equal in value to 18 kcal. However, this gives a distorted idea of the value of $\beta$, since the ordinary resonance stabilization energy of benzene is the difference between the experimental heat of formation ($\Delta H_f$) of a real molecule, benzene, and a calculated value of $\Delta H_f$ for a hypothetical molecule, cyclohexatriene. This hypothetical cyclohexatriene is taken to have normal olefinic single and double bonds for which thermochemical data are available, thus it has a different geometry from benzene itself. In other words, the value of 36 kcal for the resonance energy is a nonvertical one, and the true or vertical resonance stabilization energy of benzene must be significantly larger than 36 kcal.

This is shown in Fig. III.8 in which the energy of the system is plotted as a function of varying the molecular geometry of the six-membered ring system. The extremities of the plot correspond to two alternative cycylohexatriene geometries with alternating bonds of length 1.32 and 1.54 Å, respectively. The center of the diagram corresponds to the perfectly hexagonal benzene geometry with all bonds at 1.39 Å. It is clear that in order to evaluate the resonance stabilization of benzene, as represented by the formulation (XI), a comparison should be made with cyclohexatrienes possessing the benzene geometry. However, no thermochemical data appropriate to such species are available, hence

XI

### III.2 Simplifying Secular Determinants

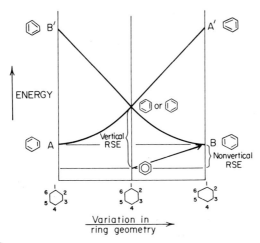

FIGURE III.8

the vertical *RSE* cannot be evaluated directly. Nevertheless, it is clear that as cyclohexatriene *A* with normal olefin geometry is distorted toward *A'* (with double bonds of 1.54 Å and single bonds of 1.32 Å), considerable energy would have to be expended; similarly for a distortion of *B* toward *B'*. Thus at the point of intersection of the two curves representing the increases in energy with these distortions, the energy of a cyclohexatriene with benzenelike geometry must be considerably higher than the value estimated thermochemically. Thus the vertical *RSE*, as shown, must be considerably higher than the nonvertical *RSE* of 36 kcal. Hence the value of $\beta$ is probably considerably in excess of 18 kcal. (Other methods of directly estimating the value of the $\beta$ unit will be described later.)

Despite the above arguments, the value of $2\beta$ for *DE* of benzene is frequently roughly equated to 36 kcal, when discussing resonance or delocalization energies, and comparing different systems. This is all right providing it is borne in mind that it is only a rough guide (in terms of nonvertical *RSEs*) and that the distortion energies of the approximate valence-bond structures involved, which correspond to the correct resonance contributors for any system, may not parallel those for benzene in any quantitative way.

Returning to the HMO results for benzene, the value of $0.33\beta$ for *DEPE* is quite substantial in comparison with most of the values already obtained for simple neutral systems, as shown in Table III.9.

TABLE III.9

| System | DE ($\beta$ units) | DEPE ($\beta$ units) |
|---|---|---|
| ethylene | 0 | 0 |
| allyl radical | 0.83 | 0.28 |
| butadiene | 0.48 | 0.12 |
| cyclobutadiene | 0 | 0 |
| pentadienyl radical | 1.464 | 0.29 |
| hexatriene | 0.99 | 0.17 |
| benzene | 2.00 | 0.33 |

This partly explains why benzene is such a stable $\pi$-system, in addition to the fact that it possesses the closed-shell structure already mentioned. The factors involved in conferring special stability on polyene systems will be discussed more fully later under the topic of Aromaticity and Hückel's rule.

### III.2.5 CALCULATION OF THE MO COEFFICIENTS FOR BENZENE

The problem of calculating the sets of coefficients for the six MOs of benzene is now greatly simplified, since there is a simpler set of secular equations associated with each of the small determinants. This will be illustrated for the $\Gamma_{A_1}$ type.

Since $D_{\Gamma_{A_1}}$ is given by

$$\begin{vmatrix} x & \sqrt{2} \\ \sqrt{2} & x+1 \end{vmatrix} = 0$$

the appropriate secular equations are

$$C_1 x + \sqrt{2} C_2 = 0, \qquad \sqrt{2} C_1 + C_2(x+1) = 0$$

The two associated roots are $x_1 = -2$, $x_4 = +1$. Substitution of $x_1$ into these equations yields

$$-2C_1 + \sqrt{2} C_2 = 0 \quad \text{or} \quad C_2 = \sqrt{2} C_1$$

Again applying the normalization condition, we have

$$C_1^2 + C_2^2 = 1$$

thus

$$C_1^2 + 2C_1^2 = 1 \quad \text{or} \quad C_1^2 = \tfrac{1}{3}$$

## III.2 Simplifying Secular Determinants

Hence

$$C_1 = 1/\sqrt{3} \quad \text{and} \quad C_2 = \sqrt{\tfrac{2}{3}}$$

Therefore, the first MO becomes

$$\psi_1 = C_1 S_1 + C_2 S_2$$

$$= \frac{1}{\sqrt{3}}\left[\frac{1}{\sqrt{2}}(\phi_A + \phi_D)\right] + \sqrt{\tfrac{2}{3}}\left[\tfrac{1}{2}(\phi_B + \phi_C + \phi_E + \phi_F)\right]$$

$$= \frac{1}{\sqrt{6}}(\phi_A + \phi_B + \phi_C + \phi_D + \phi_E + \phi_F)$$

Substitution of $x_4$ into the same equations gives

$$C_1 + \sqrt{2}C_2 = 0 \quad \text{or} \quad C_2 = -\frac{1}{\sqrt{2}}C_1$$

From normalization we have

$$C_1^2 + \tfrac{1}{2}C_1^2 = 1 \quad \text{or} \quad C_1^2 = \tfrac{2}{3}$$

Hence

$$C_1 = \sqrt{\tfrac{2}{3}}, \quad C_2 = -1/\sqrt{3}$$

Thus the fourth MO becomes

$$\psi_4 = C_1 S_1 + C_2 S_2$$

$$= \sqrt{\tfrac{2}{3}}\left[\frac{1}{\sqrt{2}}(\phi_A + \phi_D)\right] - \frac{1}{\sqrt{3}}\left[\tfrac{1}{2}(\phi_B + \phi_C + \phi_E + \phi_F)\right]$$

$$= \frac{1}{\sqrt{3}}\phi_A - \frac{1}{\sqrt{12}}\phi_B - \frac{1}{\sqrt{12}}\phi_C + \frac{1}{\sqrt{3}}\phi_D - \frac{1}{\sqrt{12}}\phi_E - \frac{1}{\sqrt{12}}\phi_F$$

The coefficients for the other four MOs can easily be calculated in a similar way using the appropriate simplified secular equations. The complete set of MO coefficients[†] for benzene is given in Table III.10.

The general appearance and nodal properties of these MOs can best be illustrated in schematic form (Fig. III.9), as for butadiene, except that it is now better to take planar projections from above the π-system. Thus only the upper lobes of the MOs are represented with

---

[†] Verify that these sets of coefficients are both normalized and mutually orthogonal.

**TABLE III.10**

| $\psi_i$ | $x_i$ | $C_A$ | $C_B$ | $C_C$ | $C_D$ | $C_E$ | $C_F$ | $\varepsilon_i$ |
|---|---|---|---|---|---|---|---|---|
| $\psi_1$ | $-2$ | 0.408 | 0.408 | 0.408 | 0.408 | 0.408 | 0.408 | $\alpha + 2\beta$ |
| $\psi_2$ | $-1$ | 0.577 | 0.289 | $-0.289$ | $-0.577$ | $-0.289$ | 0.289 | $\alpha + \beta$ |
| $\psi_3$ | $-1$ | 0 | 0.500 | 0.500 | 0 | $-0.500$ | $-0.500$ | $\alpha + \beta$ |
| $\psi_4$ | $+1$ | 0.577 | $-0.289$ | $-0.289$ | 0.577 | $-0.289$ | $-0.289$ | $\alpha - \beta$ |
| $\psi_5$ | $+1$ | 0 | 0.500 | $-0.500$ | 0 | 0.500 | $-0.500$ | $\alpha - \beta$ |
| $\psi_6$ | $+2$ | 0.408 | $-0.408$ | 0.408 | $-0.408$ | 0.408 | $-0.408$ | $\alpha - 2\beta$ |

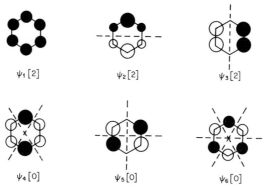

FIGURE III.9  $\psi_1$, no nodes (except nodal plane $xy$ common to all planar $\pi$-systems); $\psi_2$, one nodal plane or two nodal regions; $\psi_3$, one nodal plane or two nodal regions; $\psi_4$, two nodal planes or four nodal regions; $\psi_5$, two nodal planes or four nodal regions; $\psi_6$, three nodal planes or six nodal regions.

the appropriate sign of the wave function in each region (shaded-wave function positive, unshaded negative). The electron occupancy is given in brackets. Note again that the energies of the MOs increase with the number of nodes. (It is simpler in general to refer to nodal regions in the $\pi$-framework, since the number of nodal planes is not always simply related to the number of these nodes, which are more important.) Since each node can be thought of as a region where there is destructive interference between the components of the wave function on adjacent centers, the nodes refer to antibonding interactions. Considering $\psi_1$, there are no such interactions, thus $\psi_1$ is a strongly bonding orbital. Using the approximate method of evaluating the bonding contributions from electrons occupying this orbital, there is a set of six bonding interactions.

## III.2 Simplifying Secular Determinants

In the case of $\psi_2$ and $\psi_3$ there are two nodes in each. This is as expected since these two MOs are degenerate. For $\psi_2$ the net bonding contribution is two bonding interactions (four bonding minus two antibonding). For $\psi_3$ the net contribution is also two (two bonding and four nonbonding regions). Thus each of these MOs is a bonding orbital, but much less strongly so than $\psi_1$.

In the case of $\psi_4$ and $\psi_5$, each has four nodal regions and a set of minus-two bonding contributions. Therefore, these two are weakly antibonding orbitals. Finally, $\psi_6$ has six nodal regions and a net of minus-six bonding contributions, which makes it a strongly antibonding orbital.

Turning to the quantities associated with the MO coefficients, the electron densities for benzene can be calculated as before, from

$$q_r = \sum_{j=1}^{3} n_j C_{jr}^2$$

Thus

$$q_1 = (2)(0.408)^2 + 2(0.577)^2 + 2(0)^2 = 1.00$$

The same result would be obtained for $q_2 - q_6$ by symmetry. Therefore, all $q_r = 1.00$ and all charge densities $\xi_r = 0$, which is as expected for an even-AH moleculelike benzene.

The bond orders are given by

$$p_{rs} = \sum_{j=1}^{3} n_j C_{jr} C_{js}$$

Thus

$$p_{12} = (2)(0.408)(0.408) + 2(0.577)(0.289) + 2(0)(0.500)$$
$$= 0.667$$

The same result would be given for each of the $\pi$-bond orders in benzene, by symmetry. Therefore, again the large value of $DE$ is reflected in the extra bond order over a cyclohexatriene arrangement, since for benzene

$$\sum p_{rs} = 6(0.667) = 4.000$$

whereas for the cyclohexatriene-like resonance contributors (XII) either

XII

equivalent structure has a total of only 3.00; or an average $\pi$-bond order in any region of only 0.500.

The free valence indices are also instructive;

$$F_r = F_{max} - \sum \rho_{rs}$$
$$= 1.732 - [0.667 + 0.667]$$
$$= 0.399$$

for any position. Therefore, all positions in benzene have low values of the free valence (compare the $F_r$ value of the highest position in butadiene, for example, or other simple polyenes). Thus benzene is predicted to be relatively unreactive at any position, compared with the most reactive centers in simple olefins. This is in good accord with experimental evidence, since benzene is much less reactive towards most reagents than its acyclic analog hexatriene, for example.

Therefore, both the calculated high stabilization energy and low reactivity are in good accord with the well-known properties of benzene.

Finally, the above information on benzene can be summarized in the usual shorthand notation by XIII.

XIII

Using the methods illustrated above for butadiene and benzene, similar calculations can easily be carried out on even fairly complex systems, providing a reasonable degree of symmetry is available.

## PROBLEMS

1. Using the symmetry properties of the following systems, obtain the $\pi$-energy levels and use them to calculate $E_\pi$ and $DE$. (Compare these with $DE$ for benzene.) Show schematically the electron

I  II

Problems

distribution in the π-levels and decide the following:
(a) whether or not the neutral molecules would be expected to be radicals,
(b) if they would be easier to excite spectroscopically ($\pi \to \pi^*$) than benzene,
(c) which, if any, species of these two systems would be expected to be particularly stable.

2. (a) Calculate the π-energy levels, total π-energy, and delocalization energy ($DE$) for fulvene(III) and its monoanion. Compare the stabilities of these two.



III

(b) Calculate sets of coefficients for the first *four* MOs of fulvene and show how these can be used to depict the electron distribution in fulvene schematically (indicate nodes). Use these coefficients to calculate the electron densities at all positions and the bond orders of the 1-6 and 2-3 bonds.

3. Use symmetry to obtain the energy levels $E_\pi$ and $DE$ for the pentalene system (IV). Calculate the coefficients of the occupied MOs in pentalene, and the bond orders and charge densities.

IV

Depict the occupied MOs schematically.

4. Calculate the energy levels $E_\pi$, and π-bonding energy and $DE$ for napthalene. Compare its stability with that of benzene.

5. The 2,2′-bisallyl radical has been implicated as an intermediate in the reactions of allene dimers. Calculate the energy levels to determine whether this system is stabilized relative to two isolated

bisallyl structure

V

allyl radicals. Also calculate the (1,2) and (2,2′) bond orders and compare the former with the (1,2) bond order in the allyl radical. What can you say about ease of rotation about the bonds in this intermediate compared with rotations about ordinary single bonds and allylic bonds.

## REFERENCES

1a. H. H. Jaffe and M. Orchin, "Symmetry in Chemistry." New York, 1965.
1b. F. A. Cotton, "Chemical Applications of Group Theory," 2nd Ed. Wiley, New York, 1963.
2. D. S. Urch, "Orbitals and Symmetry," p. 88. Penguin Books, London, 1970.
3. G. W. Wheland, "Resonance in Organic Chemistry," p. 86 ff. Wiley, New York, 1955.

## SUPPLEMENTARY READING

*Coulson, C. A., and Streitwieser, A., Jr., "Dictionary of $\pi$-Electron Calculations." Freeman, San Francisco, California, 1965.
Cotton, F. A., "Chemical Applications of Group Theory." Wiley (Interscience), New York, 1963.
Daudel, R., Lefebvre, R., and Moser, C., "Quantum Chemistry." Wiley (Interscience), New York, 1959.
Eyring, H., Walter, J., and Kimball, G. E., "Quantum Chemistry." Wiley, New York, 1944.
Heilbronner, E., *Helv. Chim. Acta* **37**, 913 (1954).
Roberts, J. D., "Molecular Orbital Calculations." Benjamin, New York, 1962.
*Streitwieser, A., Jr., "Molecular Orbital Theory for Organic Chemists." Wiley, New York, 1961.

# IV | POLYENE STABILITIES, HÜCKEL'S RULE, AND AROMATIC CHARACTER

The question of the relative stabilities of different acyclic polyene systems, as well as more the interesting topic of aromatic character, can be approached within the simple Hückel framework in many cases without doing any formal HMO calculations. This is because the energy levels (or roots) for many of these systems are expressible by simple formulas. The following simple rules are applicable to both AH and non-AH systems.

(i)  *Linear Polyenes*  For straight chain $\pi$-systems of $n$ atoms the roots (and hence energies) are given by

$$x_j = -2\cos\frac{j\pi}{(n+1)} \quad (j = 1, 2, 3, \ldots, n)$$

where $\pi$ is in radians.

(ii)  *Cyclic Polyenes*  For monocyclic polyene systems containing $n$ $\pi$-centers the roots (and energies) are given by

$$x_k = -2\cos\frac{2k\pi}{n}, \quad \begin{cases} k = 0, \pm 1, \pm 2, \ldots, +(n-1)/2 \\ \quad \text{if } n \text{ is odd} \\ k = \pm 1, \pm 2, \ldots, \pm n/2 \\ \quad \text{if } n \text{ is even} \end{cases}$$

(iii) *The Polygon Rule* The above formula for cyclic polyenes is expressible in a very simple graphical rule, which can be used to obtain very rapidly the energy level pattern for many monocyclic π-systems of interest. This can be stated as follows:

To obtain the Hückel MO energy levels for any *planar monocyclic system containing n π-centers in the ring and no side-chain π-carbons*, inscribe a regular $n$-fold polygon inside a circle of radius $2\beta$, with its center at $\alpha$, so that one apex of the polygon touches the lowest point of the circle (i.e., at $\alpha + 2\beta$). Each point where an apex of the polygon touches the circle corresponds to one energy level.

This rule can be illustrated on two systems (Fig. IV.1) for which the energy levels have already been calculated. For cyclobutadiene and benzene the regular $n$-fold polygons are a square and hexagon, respectively. Thus drawing circles centered at $\alpha$ and inscribing these polygons gives the energy level patterns shown in the figure.

These simple rules can now be used to compare the electronic stabilities of

(i) acyclic and cyclic polyenes,
(ii) even- and odd-numbered straight chain polyenes,
(iii) linear and branched chain conjugated polyenes, and
(iv) cyclic systems containing $(4n + 2)$ π-electrons and other cyclic systems.

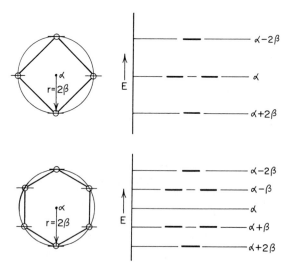

FIGURE IV.1

## IV.1 ACYCLIC AND CYCLIC POLYENES

It is of interest to determine whether HMO results make any general predictions about the relative stabilities of isoelectronic[†] cyclic and acyclic systems. As a typical example, compare the pentadienyl and cyclopentadienyl anions.

For the pentadienyl system, the roots are given by

$$x_j = -2\cos\frac{j\pi}{n+1}, \quad \text{where} \quad n = 5 \quad \text{and} \quad j = 1, 2, 3, 4, 5$$

Thus

$$x_1 = -2\cos\left(\frac{\pi}{6}\right) = -2\cos 30° = -2\left(\frac{\sqrt{3}}{2}\right) = -\sqrt{3}$$

$$x_2 = -2\cos\left(\frac{2\pi}{6}\right) = -1$$

$$x_3 = -2\cos\left(\frac{3\pi}{6}\right) = 0$$

$$x_4 = -2\cos\left(\frac{4\pi}{6}\right) = +1$$

$$x_5 = -2\cos\left(\frac{5\pi}{6}\right) = +\sqrt{3}$$

For the cyclopentadienyl system the roots are given by $x_k = -2\cos(2k\pi/n)$, where $n = 5$ and $k = 0, \pm 1, \pm 2$. Thus

$$x_1 = -2\cos\left(\frac{0\pi}{5}\right) = -2.00$$

$$x_2 = -2\cos\left(\frac{2\pi}{5}\right) = -0.618$$

$$x_3 = -2\cos\left(\frac{-2\pi}{5}\right) = -0.618$$

$$x_4 = -2\cos\left(\frac{4\pi}{5}\right) = 1.618$$

$$x_5 = -2\cos\left(\frac{-4\pi}{5}\right) = 1.618$$

[†] In the sense of the $\pi$-electrons only.

Therefore, the energy level diagrams for the two anionic systems are as shown in Fig. IV.2.

Thus HMO theory predicts that both systems are reasonably stable, as judged by their delocalization energies, but that the cyclic anion should be considerably more stable than the acyclic. This is in good agreement with experimental evidence, since the pentadienyl anion has at best transient existence as a reaction intermediate, whereas the well-known cyclopentadienyl anion[1] is directly observable spectroscopically and is stable indefinitely in suitable solvents.

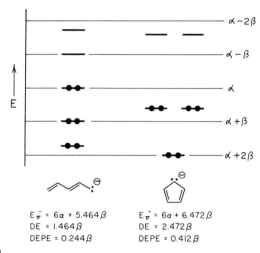

$E_\pi = 6\alpha + 5.464\beta$
$DE = 1.464\beta$
$DEPE = 0.244\beta$

$E_\pi = 6\alpha + 6.472\beta$
$DE = 2.472\beta$
$DEPE = 0.412\beta$

FIGURE IV.2

If the same type of comparison is made between other cyclic and acyclic polyenes, HMO results in some cases predict that the cyclic species is the more stable, but in others the acyclic species is predicted to have greater $\pi$-electronic stability. Although no general prediction is obtained from HMO calculations, the predictions of relative stabilities, as illustrated by the examples in Table IV.1, are generally in excellent agreement with available evidence.

It is interesting to note that these predictions would not have been arrived at by VB (resonance) arguments, since the numbers of important equivalent contributing structures that could be written for each pair would have led to the prediction that the cyclic partner would be the more stable in every case.

## IV.2 Even- and Odd-Numbered Linear Polyenes

**TABLE IV.1**

| Systems | DE ($\beta$ units)[a] | HMO Prediction |
|---|---|---|
| allyl cation (acyclic) | 0.83 | both are stabilized but cyclic system considerably more so |
| cyclopropenyl cation | 2.00 | |
| butadiene | 0.48 | acyclic system is clearly more stabilized |
| cyclobutadiene | 0 | |
| pentadienyl cation | 1.464 | acyclic system is somewhat more stabilized |
| cyclopentadienyl cation | 1.236 | |
| pentadienyl anion | 1.464 | cyclic system is much more stabilized |
| cyclopentadienyl anion | 2.472 | |
| hexatriene | 0.99 | cyclic system is much more stabilized |
| benzene | 2.00 | |

[a] It is not necessary to use $DEPE$ values since pairs of isoelectronic systems are being compared.

### IV.2 EVEN- AND ODD-NUMBERED LINEAR POLYENES

The HMO results based on the above simple rules always predict greater stability due to $\pi$-electron delocalization for odd-membered linear polyenes, compared with similar even-membered systems. This is illustrated in Table IV.2. It can be noted that for the odd systems, whether cation, radical, or anion, the delocalization energy per electron is always greater than for the nearest even systems. This is true no matter how many $\pi$-centers are present. It is also interesting that all neutral linear polyenes, whether odd or even, have lower stabilization energy per electron than benzene.

TABLE IV.2

| System | n | DE (β units) | | DEPE (β units) |
|---|---|---|---|---|
| ethylene | 2 | 0 | | 0 |
| allyl | 3 | 0.83 | (cation) | 0.414 |
| | | 0.83 | (radical) | 0.276 |
| | | 0.83 | (anion) | 0.207 |
| butadiene | 4 | 0.47 | | 0.118 |
| pentadienyl | 5 | 1.46 | (cation) | 0.366 |
| | | 1.46 | (radical) | 0.293 |
| | | 1.46 | (anion) | 0.244 |
| hexatriene | 6 | 0.99 | | 0.165 |
| benzene | 6 | 2.00 | | 0.33 |

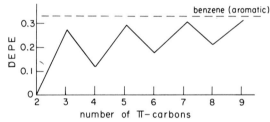

FIGURE IV.3

This alternation in stability and its relationship to the case of benzene is be shown schematically (for electrically neutral systems) in Fig. IV.3. It can be seen that each series (odd or even) will approach the intrinsic stability of benzene in the limit as the polyene chain length is increased, but that the odd-membered series approaches this limit more rapidly.[†]

[†] The foregoing is not meant to imply that the odd-series polyenes have greater absolute stabilities, since many of these are ions whose absolute stability cannot be determined, or else free radicals that are too highly reactive to study thermochemically. What is referred to solely is the *extra* electronic stabilization arising from conjugation or delocalization, over what would be expected if these systems were simply as shown by a Kekulé formulation. For example, the allyl cation, although electrostatically unstable, is nonetheless a well-recognized reaction intermediate and is considerably more stable than the n-propyl cation that is unable to achieve similar charge delocalization.

## IV.3 Linear and Branched Chain Polyenes

These results and trends are precisely what would be expected from qualitative resonance arguments. First, the odd members can always achieve resonance stabilization or charge delocalization without taking account of contributions from high-energy dipolar forms, which are necessary to achieve resonance for the even members. Thus it would be expected from resonance arguments that the odd members would be more stabilized, for example,

$$CH_2=CH-CH=CH-\overset{+}{C}H_2 \longleftrightarrow CH_2=CH-\overset{+}{C}H-CH=CH_2$$
$$\longleftrightarrow \overset{+}{C}H_2-CH=CH-CH=CH_2$$

—three roughly equivalent contributors, no high-energy dipolar structures; significant resonance stabilization

$$CH_2=CH-CH=CH_2 \longleftrightarrow \overset{+}{C}H_2-CH=CH-\overset{-}{C}H_2 \longleftrightarrow \text{etc.}$$

—only significant contributor is nonpolar VB structure, all other structures are either dipolar or contain lower numbers of valence bonds; very small resonance stabilization.

Second, as the chain length increases for both the odd- and even-series, a greater number of canonical forms (whether dipolar or not) can be written. Therefore, from resonance arguments, delocalization is expected to increase in each series with chain length, albeit more effectively for the odd members.

Finally, in terms of the complete equivalence of the two important canonical forms of benzene, it might be expected from simple resonance arguments that this system would achieve greatest stabilization.

It is interesting that in this regard the HMO results again give a quantitative expression of qualitative resonance predictions, in terms of MO energy level spacings and electron occupancies. This is frequently found to be the case, but where the MO results and VB arguments are not in accord (which also occurs in important cases), it is always the MO results that are in better agreement with the available experimental evidence.

## IV.3 LINEAR AND BRANCHED CHAIN POLYENES

If straight-chain and branched polyene systems are compared, the results of HMO calculations ($E_\pi$ values) always predict that the linear conjugated systems are somewhat more stable than the branched systems containing the same numbers of carbon atoms. The examples given in Fig. IV.4 illustrate this.

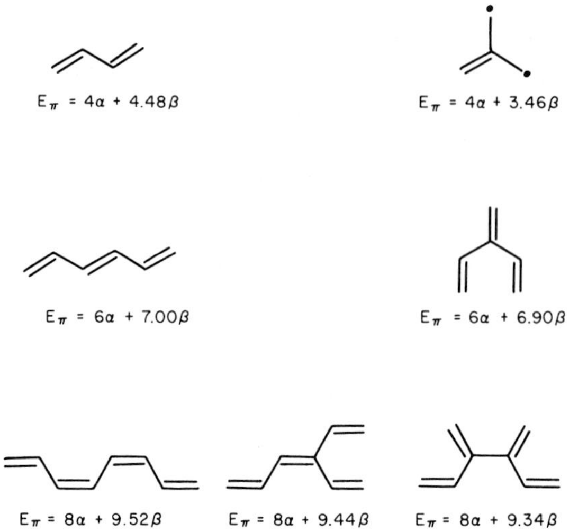

$E_\pi = 4\alpha + 4.48\beta$

$E_\pi = 4\alpha + 3.46\beta$

$E_\pi = 6\alpha + 7.00\beta$

$E_\pi = 6\alpha + 6.90\beta$

$E_\pi = 8\alpha + 9.52\beta$  $E_\pi = 8\alpha + 9.44\beta$  $E_\pi = 8\alpha + 9.34\beta$

FIGURE IV.4

The large difference between butadiene and trimethylenemethane is as expected, but it appears that, in general, increased branching leads to lower $\pi$-energies. This is in agreement with the experimental observation that terminal double bonds are somewhat less stable than internal double bonds,[2] and also with the generally held opinion of organic chemists that cross-conjugated systems are less stable than linear conjugated systems. However, it is difficult to find any convincing experimental evidence on this latter point.

### IV.4 CYCLIC SYSTEMS CONTAINING $(4n + 2)$ $\pi$-ELECTRONS

Hückel stated the following rule[3] to explain the great stability of certain cyclic conjugated systems:

Those *monocyclic planar*[†] conjugated systems of trigonally hybridized atoms that contain exactly $(4n + 2)$ $\pi$-electrons (where $n = 0, 1, 2, \ldots$) will possess special or extra stability relative to similar systems that do not have exactly this number of electrons.

[†] The condition of planarity (or near planarity) is a rigorous requirement for special stabilization, but the monocyclic requirement need not be met in every case, as will be seen later.

## IV.4 Cyclic (4n + 2) Systems

This special electronic stabilization and associated properties has come to be known as aromatic character or aromaticity, which will be treated as a special topic.

### Hückel's Rule and Aromatic Character

It is not easy to define in any absolute terms what is meant by "aromatic" as applied to $\pi$-systems. Originally it was taken to mean benzenelike properties arising from the possession of a so-called aromatic sextet of electrons as in benzene itself. This equation of aromaticity with benzenelike properties is still largely true, but it is possible to define several useful independent criteria for aromatic character:

(i) *Unusual thermodynamic stability*, especially compared with ordinary olefins and polyenes. This is shown by the experimental $\Delta H_f$ values,[4] which are considerably greater than would be expected from calculated $\Delta H_f$'s based on a summation of thermochemical data for the standard bond types possessed by the simple Kekulé structures.

(ii) *Unusual reactivity*, compared with simple olefins and even with conjugated olefins. Aromatic systems are unusual in that they generally react with electrophiles (and some nucleophiles) by substitution rather than by addition. Also, their reactions frequently require catalysis, whereas ordinary olefins react spontaneously. In general, aromatic systems are less reactive, and for example are less easily oxidized or polymerized than most olefins. This indicates a low reactivity towards free radical species.

(iii) *Unusual spectroscopic properties*, especially the ultraviolet (uv) and nuclear magnetic resonance (NMR) spectra. This will be discussed later.

It is not usually possible to obtain information in all three of the above areas in classifying systems as aromatic or nonaromatic. In general, though, a system must clearly possess at least one and preferably two of the unusual types of property listed above to be considered as aromatic. With many species that are difficult to isolate in significant quantity, the third criterion of unusual uv or NMR spectra has come to be depended on very heavily in assigning aromatic character to a given system.

Returning to Hückel's rule, it seems clear that whatever theoretical basis this rule has must lie in the pattern on energy level spacings in typical cyclic π-systems. Therefore, the polygon rule will be used (where necessary) to obtain these energy levels and compare the stabilities of various simple systems, both with and without $(4n + 2)$ electrons, in the light of available experimental evidence.

First, systems with $k = 3$ to 8 carbon atoms and appropriate numbers of electrons $(n_k)$ will be considered. Following this, larger systems with $k$ in the range 10–30 will be considered.

***Cyclopropenyl*** $(k = 3, n_k = 2, 3, \text{ or } 4)$  Use of the polygon rule for this system gives the energy level pattern shown in Fig. IV. 5.

If each level or *set* of degenerate levels is considered as a subshell, as in atoms, then, of the three systems with $n_k = 2, 3,$ and 4, only the cation with $n_k = 2$ has a closed-shell configuration (Fig. IV. 6). Further, the cation has all its π-electrons in BMOs and has by far the largest value of *DEPE*. Thus the cyclopropenyl cation should be the only very strongly stabilized one of the three systems. In addition, it is the only one that obeys Hückel's rule (with $n = 0$).

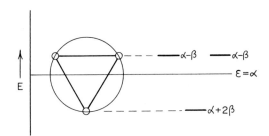

FIGURE IV.5

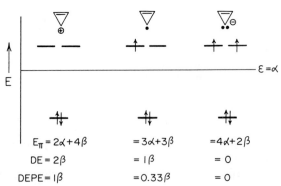

FIGURE IV.6

## IV.4 Cyclic (4n + 2) Systems

The experimental evidence fully supports this, since not only has the triphenylcyclopropenium cation[5] been known for a long time, but more recently Breslow[6] has prepared the parent ion I by the route shown.

$$\text{H}_2\text{C}(\text{Cl})\text{–CH=CH} + \text{SbF}_5 \longrightarrow \text{cyclopropenyl cation} + \text{SbF}_5\text{Cl}^{\ominus}$$

I

The NMR evidence that the resulting cation is in fact cyclopropenium is unequivocal. This is truly remarkable in view of the fact that it is (a) a cation (and a secondary carbonium ion!), and (b) subject to considerable ring strain (the interior bond angles must of necessity be 60° whereas the normal $sp^2$ trigonal bond angle is 120°!). The cyclopropenyl radical and anion are as yet unknown. Thus the available evidence suggests Hückel's rule is valid for $n = 0$. (Note that simple resonance arguments would predict equivalent stabilization for all three cyclopropenyl systems.)

**Cyclobutadienyl** ($k = 4$, $n_k = 2$ or 4)   The energy levels for this system have already been determined (Fig. IV. 7). Of these two systems, the neutral molecule has already been discussed. It has zero delocalization energy, an open-shell configuration, and two electrons in NBMOs. It is therefore not expected to be stable and does not obey Hückel's rule. All the experimental evidence is in accord with the expectation that

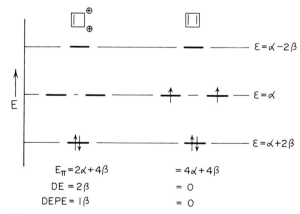

FIGURE IV.7

cyclobutadiene is a very unstable species.† Like cyclopropenyl cation it is subject to considerable ring-strain, but has no offsetting electronic stabilization factor.

The dication on the other hand does have a closed-shell structure, $(4n + 2)$ electrons, all of its electrons in BMOs, and possesses a very large *DEPE*. However, it is a dicationic species (and organic dications are rare) and is subject to considerable ring-strain. Nonetheless, there is some evidence for metal-cyclobutadiene complexes involving a dicationic moiety,[8] and also evidence that a stable dication has been observed spectroscopically.[9]

Again the available experimental evidence appears to support the predictions of Hückel's rule. (As pointed out previously, resonance arguments would predict that cyclobutadiene should achieve considerable stabilization.)

***Cyclopentadienyl*** ($k = 5$, $n_k = 4$, 5, or 6)   In this case, it is the anion that possesses a closed-shell structure and the highest value of *DEPE*.

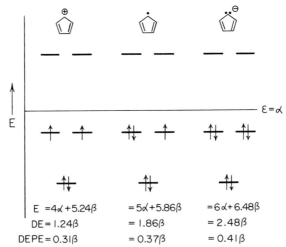

FIGURE IV.8

† Species such as II and III do exist, and are stable molecules. However, they are not truly cyclobutadiene derivatives, since the evidence is that the six-membered aromatic systems are not significantly perturbed by the fused four-membered rings.[7]

## IV.4 Cyclic (4n + 2) Systems

Therefore, the HMO results predict that this is the most stable of the three, which fits Hückel's rule with $n = 1$ (Fig. IV.8).

Experimentally, the anion is a well-known species, and although highly reactive, is stable indefinitely under suitable conditions. The available evidence is that the radical and cation are probably much less stable species. Thus Hückel's rule is again verified experimentally.

**Benzene** ($k = 6$, $n_k = 6$)  This system is so well-established as an aromatic system that it merits only brief discussion. The energy level pattern (Fig. IV. 9) shows that benzene has a closed-shell structure and a reasonably large *DEPE*; furthermore, it is subject to no ring strain and is a neutral species. It obeys Hückel's rule (also with $n = 1$) and thus the rule is again validated.

**Cycloheptatrienyl** ($k = 7$, $n_k = 6$, 7, or 8)  The polygon rule can be used to determine the energy level pattern for this system (Fig. IV.10).

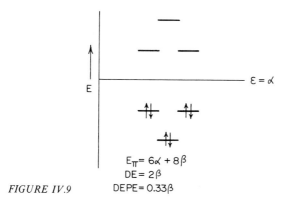

FIGURE IV.9

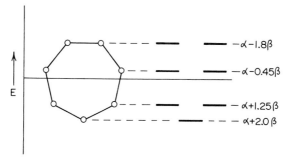

FIGURE IV.10

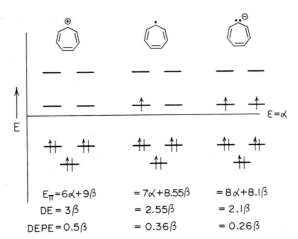

FIGURE IV.11

Therefore, for the three cycloheptatrienyl systems, the electron occupancies and energies are as shown in Fig. IV.11. Thus the cation is predicted to be the most stable of the three systems, since it has a closed-shell arrangement, all electrons in BMOs, and the highest value of $DEPE$. It also obeys Hückel's rule with $n = 1$. This rule is therefore in good accord with experiment, since the tropylium ion is a well-established entity[10a,b,c] and there is no evidence for either a stable radical or anion. Despite the very high $DEPE$, which is greater than the value for benzene, it is a cation, however, and also has ring strain that is not present in benzene. Nonetheless, 7-bromo-1,3,5-cycloheptatriene (IV) exists as a salt, tropylium bromide (V), with a planar conjugated

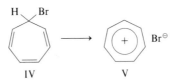

cation.[11] Also tropones (VI) have structures that more closely resemble the dipolar form VII.[12]

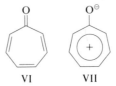

## IV.4 Cyclic $(4n + 2)$ Systems

**Cyclooctatetraenyl** ($k = 8$, $n_k = 6$, 8, or 10)  The energy level pattern for this system is shown in Fig. IV.12. Three possible and reasonable systems are the dication, the neutral molecule, and the dianion (Fig. IV.13). Of these, only the dication has a closed-shell structure and all its electrons in BMOs. It also has the highest value of *DEPE*. The dianion also has a closed-shell structure, but with four of its electrons in NBMOs has a lower *DEPE*. Both of these species obey Hückel's rule (with $n = 1$ and 2, respectively) and should both be stabilized, with the cation being more stable. The neutral molecule has an open-shell

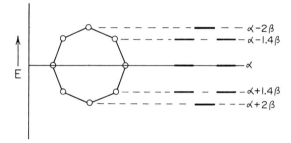

FIGURE IV.12

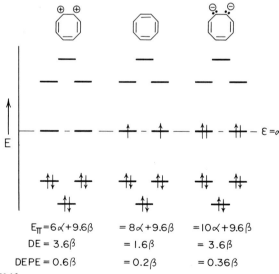

FIGURE IV.13

arrangement, low *DEPE*, and does not fit Hückel's rule; thus it should be least stable and also exist as a diradical ground state.

Considering the experimental evidence, the cyclooctatetraene molecule is well known and quite stable as a singlet ground state, in apparent disagreement with the Hückel predictions. However, the rule and the calculations are only applicable to planar conjugated systems. It is well known that cycloctatetraene is nonplanar and nonconjugated.[13] Evidently the very low *DE* achieved by a planar cyclooctatetraene molecule is not sufficient to offset the ring strain in a fully coplanar arrangement. Thus the molecule is puckered and cannot achieve the continuous $\pi$-overlap necessary for the above energy level picture to be applicable.

Of the other two species, both should be aromatic and reasonably stabilized. Surprisingly, while there is considerable evidence for a planar conjugated dianion in complexes and salts,[14] there is as yet no firm evidence for the dication, which should be even more stabilized. It will be of interest to see if such evidence is obtained in future investigations.

Nonetheless, Hückel's rule is in good accord with available evidence for $k = 8$, and it is worth noting that the cyclooclatetraenyl dianion is the first case where $n = 2$ in the $(4n + 2)$ series.

In summary, Hückel's rule appears to work remarkably well for these simple cyclic polyene systems with $k = 3-8$ carbon atoms and $n = 0, 1,$ or $2$, as shown in Table IV.3.

Finally, in assessing whether a given cyclic system is likely to be planar and possess special stability or aromatic properties, it is worth listing the following criteria for consideration:

(i) The system should obey the $(4n + 2)$ rule.

(ii) It should also have a closed-shell electron arrangement. (The first two criteria are frequently, but not necessarily, the same.)

(iii) There should be a large value of *DE*, and more particularly *DEPE*.

(iv) A fully coplanar arrangement should preferably be free of considerable bond angle deformation from the normal $sp^2$ (trigonal) value of 120°.

(v) Electrically neutral organic systems are inherently likely to be more stable and less reactive than charged species, particularly dications and dianions.

(vi) The planar system should be free of serious nonbonded interactions.

IV.5  Hückel's Rule and the Annulenes

TABLE IV.3

Summary of Results of HMO Calculations and the Applicability of Hückel's Rule for Cyclic Polyenes with 3–8 Ring Carbons[a]

| System | Species | DE ($\beta$ units) | DE per electron | Obeys ($4n+2$) rule | Angular deformation from $sp^2$ trigonal | Experimental evidence for aromatic properties |
|---|---|---|---|---|---|---|
| triangle | cation | 2.00 | 1.0 | YES | 60° | YES |
| | radical | 1.00 | 0.33 | NO | | NO |
| | anion | 0 | 0 | NO | | NO |
| square | dication | 2.00 | 1.0 | YES | 30° | ? |
| | molecule | 0 | 0 | NO | | NO |
| pentagon | cation | 1.24 | 0.31 | NO | 12° | NO |
| | radical | 1.86 | 0.37 | NO | | NO |
| | anion | 2.48 | 0.41 | YES | | YES |
| hexagon | molecule | 2.00 | 0.33 | YES | 0° | YES |
| heptagon | cation | 3.00 | 0.50 | YES | 9° | YES |
| | radical | 2.55 | 0.36 | NO | | NO |
| | anion | 2.10 | 0.26 | NO | | NO |
| octagon | dication | 3.6 | 0.6 | YES | 15° | NO |
| | molecule | 1.6 | 0.2 | NO | | NO |
| | dianion | 3.6 | 0.36 | YES | | YES |

[a] All calculations apply to fully coplanar structures.

The more of these criteria the system meets, the more likely it is to be a stable, aromatic species. (It can be seen that of all the systems considered so far, only benzene meets all the criteria fully.)

The question now is, can this rule be extended to larger cyclic systems with $n \geq 2$, where it might be expected that because of ring strain the possibility of nonplanar nonconjugated systems would be increased; also if the rule is indeed applicable to these systems, where can it be expected to break down.

## IV.5  HÜCKEL'S RULE AND THE ANNULENES

The larger monocyclic conjugated systems are called *annulenes* if they can be represented formally as even-numbered rings with alternating single and double bonds, as is the case for cyclobutadiene, benzene, and cyclooctatetraene. While there is no a priori reason to

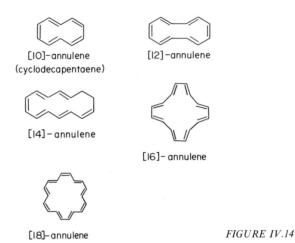

FIGURE IV.14

restrict discussion to this type rather than odd-numbered rings, it is the even annulenes that have received the greatest attention, principally through the investigations of Sondheimer and his co-workers.[15]

Neglecting the actual geometries of these species for the moment, typical annulene systems, which have been investigated for aromatic properties, are shown in Fig. IV.14.

It is of interest to consider these and other annulenes in turn, and to determine to what extent Hückel's rule is applicable in the light of available evidence.

**[10]-Annulene** This system fits Hückel's rule with $n = 2$, and the molecule is known to exist. However, the evidence is that it is nonplanar and nonaromatic; hence for the first time Hückel's rule seems to fail completely. However, if the geometry of [10]-annulene VIII is considered, it is easy to see that it would be virtually impossible for such a system to achieve coplanarity because of the severe repulsion this would cause between the hydrogens at the "9,10" positions. This can be seen by comparison with naphthalene IX. In planar naphthalene the 9,10 carbon-carbon distance is only 1.4 Å, and in order for [10]-annulene

## IV.5 Hückel's Rule and the Annulenes

to achieve a similar planar geometry without considerable distortion of the ring angles, the two C—H bonds at the "9,10" positions would have to be seriously deformed to avoid severe H—H repulsions. Therefore, steric effects would preclude coplanarity, no matter how much delocalization energy were involved. The molecule probably adopts a geometry as shown (X), which effectively rules out continuous $\pi$-orbital overlap; hence Hückel's rule cannot be expected to apply.

X

Two derivatives of [10]-annulene are known, however,[16] where these nonbonded repulsions have been removed by insertion of a bridgehead atom or group in place of the two C—H bonds. These are the methylene- and ether-bridged species (XI) and (XII). Both are

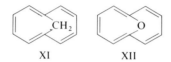

XI        XII

stable and possess aromatic properties, as judged by the close similarity of their uv and NMR spectra to those of naphthalene. There is also one example of a [10]-annulene system without bridging atoms, which could also be planar, stable, and show aromatic properties Fig. IV.15. This is [10]-bisdehydroannulene,[17] where the unfavorable "9,10" repulsions have been removed by introduction of acetylenic-type linkages.† This introduces new ring strain because of the presence of the two isolated $\pi$-bonds (in the xy-plane), Nevertheless, the orthogonal

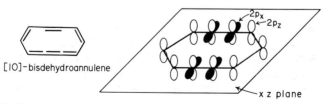

FIGURE IV.15

† Note that it is impossible to write any Kekulé structure for this system that does not have a cumulene arrangement as shown.

(and presumably noninteracting) $\pi$-system made up of the $2p_z$-type interactions contains $(4n + 2)$ electrons and still obeys Hückel's rule.

**[12]-Annulene** This system is not expected to show any aromatic properties since it is a $4n$ system. The molecule (XIII) is known and is in fact nonaromatic.[18] However, there would also be quite severe nonbonded repulsions in this system as well as in [10]-annulene if coplanarity were to be achieved, even though the larger ring is somewhat more flexible. Therefore, this is not really an adequate test of Hückel's rule. A [12]-bisdehydroannulene XIV is known, however,[19] where these nonbonded repulsions are removed in a fully coplanar system. The molecule shows no evidence of aromatic properties, unlike the [10]-bisdehydro derivative. Thus Hückel's rule appears valid.

XIII      XIV

**[14]-Annulene** This system obeys Hückel's rule (with $n = 3$) and is known.[20] From spectroscopic evidence it seems to be aromatic despite apparently fairly severe nonbonded interactions. (For example, its uv spectrum is very similar to that of anthracene.) The larger ring size presumably allows the system XV to minimize these interactions more effectively than the [10]- or [12]-systems.

XV

The most convincing evidence for the aromatic character of [14]-annulene comes from its NMR spectrum.[21] The ring protons fall into two distinct groups: one group (10H) with chemical shift at low field (1.2τ) in the so-called aromatic region and the other (4H) at very high field (11.9τ). These are clearly the *exo* and *endo* protons, respectively. Thus there must be strong ring current effects operating in [14]-annulene. This can only be the case if the system is coplanar, or very nearly coplanar, and there is continuous electron delocalization around the $\pi$-framework, as in benzene and other aromatic systems.

## IV.5 Hückel's Rule and the Annulenes

It is interesting that these systems possess *endo* as well as exocyclic ring protons. The *exo* protons are subject to the diamagnetic anistropy of the ring electrons in a similar way to benzene protons, and are thus strongly deshielded. However, the *endo* protons experience an opposite effect since they are apparently within or near to the shielding zone, and thus occur at even higher field than normally used NMR standard such as TMS.

This is even more striking in [14]-bisdehydroannulene[22] (XVI), where there are three different proton signals, which occur at 0.5$\tau$, 1.46$\tau$, and 14.54$\tau$! The two *endo* protons are evidently very strongly shielded by the aromatic ring current. Thus there is ample evidence that the [14]-annulene system obeys Hückel's rule.

XVI

**[16]-Annulene** This is a $4n$ system (XVII), and therefore special electronic properties are not expected. In fact, the *exo* protons do not appear in the NMR spectrum in the aromatic region, but give signals in the normal olefinic region.[23] However, the *endo* proton signals

XVII

appear at *lower* field. This is a curious and as yet not satisfactorily explained phenomenon, since if the *endo* protons were subject to a strong ring current effect, they would have been expected to be shielded and thus occur at high field. This assumes that the ring current effects are due to diamagnetic anisotropy. However, it has been suggested that these unusual *endo* signals (which occur in other 4n systems) might be due to *paramagnetic* effects.

If the simple HMO picture is correct, every 4n system would have singly occupied degenerate NBMOs and thus possess a diradical ground state. It is possible in this case that there might be paramagnetic effects operating. However, the theory of NMR paramagnetic effects is not yet clear on this point. One problem is that if the *endo* protons are deshielded by such a paramagnetic ring current contribution, one might have expected the *exo* protons to be correspondingly shielded, yet these occur in [14]-annulene in the normal olefinic region.

Nonetheless, it seems clear that [14]-annulene does not have normal aromatic properties, and thus Hückel's rule still holds.

**[18]-Annulene and higher annulenes**  [18]-annulene is known and appears to be aromatic, as judged by its NMR properties.[24] The six *endo*

TABLE IV.4

*Summary of Results for Annulene Systems*

| Annulene system | Compounds known | $n$ in $(4n + 2)$ | Evidence for aromatic properties | Hückel's rule prediction |
|---|---|---|---|---|
| 10-A | A[a] | 2 | NO | not applicable |
| 12-A | A | 4n systems | NO | not applicable |
|  | bisdehydro-A |  | NO | valid |
| 14-A | A | 3 | YES |  |
|  | bisdehydro-A | 3 | YES | valid |
| 16-A | A | −4n system | NO | valid |
| 18-A | A | 4 | YES | valid |
| 20-A | A | −4n system | NO | valid |
| 22-A | A | 5 | YES | valid |
|  | monodehydro-A | 5 | YES | valid |
| 24-A | A | −4n system | NO | valid |
| 26-A | A | 6 | YES | valid |
|  | monodehydro-A | 6 | NO | not valid |
| 28-A | A | −4n system | NO | valid |
| 30-A | A | 7 | NO | not valid |

[a] Parent annulene.

## IV.5 Hückel's Rule and the Annulenes

protons give a signal at $12\tau$ and the twelve *exo* protons at $1.1\tau$. Thus Hückel's rule holds for $n = 4$.

A number of the higher annulenes (or their dehydro derivatives) are also known[25] and the general situation as to their aromatic properties is summarized in Table IV.4. It appears that Hückel's rule begins to break down somewhere in the region of [26]- to [30]-annulene, or where $n$ in $(4n + 2)$ approaches 6 or 7. Nevertheless, the rule is remarkably successful in predicting special properties for a wide range of $(4n + 2)$ polyene systems (with $n = 0$ to 5), particularly in view of the simplicity of the model on which the rule is based.

It is hardly surprising that the rule eventually breaks down, since for the larger ring systems true coplanarity must be very difficult to achieve. Further, as the number of $\pi$-electrons is increased, the energy levels will become closely spaced and small ring distortions could easily remove the degeneracy of the highest level in the $4n$ systems, and thus remove one of the differences between the $4n$ and $(4n + 2)$ series. Also, as $n$ increases, the difference in $\pi$-electronic energy between a $4n$ and the next $(4n + 2)$ system must becomes very small. This is illustrated by Fig. IV.16, where the delocalization energy per electron for the annulene-type system is plotted as a function of ring size, yielding separate curves for the $(4n)$ and $(4n + 2)$ series. It can be seen that the

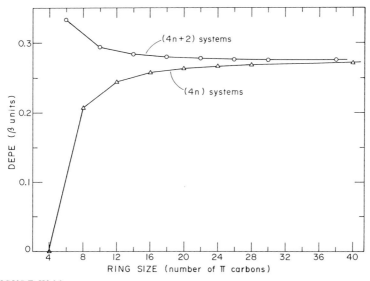

FIGURE IV.16

two curves converge rapidly as the ring size exceeds 20. Thus it is to be expected that any fundamental differences between the two series will eventually disappear, as the experimental results indicate.

## IV.6  POLYCYCLIC SYSTEMS

Although Hückel's rule is only strictly applicable to monocyclic conjugated systems, many polycyclic systems that contain $(4n + 2)$ $\pi$-electrons also show special stability and properties. For example, naphthalene with 10 $\pi$-electrons is a very stable system with a resonance stabilization energy of more than 30 kcal.[26] It also has aromatic-type spectroscopic properties and, in some ways, benzene-type reactivity. The HMO results for naphthalene show that it has a *DE* of $3.68\beta$, and hence a *DEPE* of $0.37\beta$. Thus naphthalene can be considered to be pseudoaromatic and the 9,10 bond regarded as a small perturbation of a peripheral $\pi$-system of 10 electrons.

Other polycyclic systems can be regarded as pseudoaromatic in a similar way if their periphery contains $(4n + 2)$ electrons. For example, both anthracene (with $DE = 5.314\beta$ and $DEPE = 0.38\beta$) and phenanthrene (with $DE = 5.448\beta$ and $DEPE = 0.39\beta$) have $(4n + 2)$ peripheries and are considered to be pseudoaromatic.

Even larger perturbations than that provided by the 9,10 bond in naphthalene can be accommodated within pseudoaromatic systems, provided that the periphery contains $(4n + 2)$ electrons. For example, although pyrene (XVIII) possesses a total of 16 (or $4n$) electrons, the

XVIII

periphery (analogous to the monocyclic [14]-annulene) contains only $(4n + 2)$ and the system is pseudoaromatic.[27]

However, there is an important proviso in considering whether such $(4n + 2)$-peripheral $\pi$-systems will show aromatic properties or not. That is, that the perturbing bond or bonds that produce the polycyclic framework should divide up the total π-system in such a way that *each* cyclic portion is of the (4n + 2) type. For example, if the three isomeric 10-electron systems, naphthalene XIX, azulene XX, and [6.2.0]-bicyclodecapentaene XXI are compared, all three have a

XIX      XX      XXI

(4n + 2) periphery. However, only the first two are pseudoaromatic, since the perturbing bridge bonds in each case subdivide the total system in such a way that each cycle would itself be aromatic. Thus, for

XXII

azulene, the system can be regarded as a fused cyclopentadienyl anion and tropylium cation XXII. (Azulene has a *DE* of 3.364$\beta$ or *DEPE* of 0.34$\beta$.) However, [6.2.0]-bicyclodecapentaene is a fused cyclooctatetraene and cyclobutadiene, neither of which would be aromatic. Therefore, despite the fact that the system has a (4n + 2) periphery, it is nonaromatic. (Its *DE* value is 2.6$\beta$, and hence *DEPE* = 0.26$\beta$.) The evidence for this statement is that several derivatives of the parent molecule are known and the ring protons of the [6.2.0]-bicyclic systems give signals in the normal vinyl region.[28]

Similarly, [6.6.0]-bicyclotetradecaheptaene XXIII is a perturbed [14]-annulene, but is nonaromatic since the bridging perturbation subdivides the (4n + 2) peripheral system into two fused (4n) cycles.

XXIII

Both the UV and NMR spectra of its derivatives indicate that the eight-membered rings are nonplanar and nonconjugated.[29]

The importance of the above proviso can be illustrated further by considering the perturbed [18]-annulene (XXIV). Despite the fact that it contains (4n + 2) total π-electrons, and the perturbations divide

XXIV

the system in such a way that one cyclic portion is clearly benzenoid, the evidence is that the system as a whole is nonplanar and nonaromatic. Although the 3,6-proton signal appears in the aromatic region of the NMR spectrum at $3.41\tau$, and is a singlet, the remaining protons all appear at higher field and are unsplit by the (3,6) protons.

From all the above evidence, it can be concluded that if a polyclic system possesses $(4n + 2)$ $\pi$-electrons, and the bridge bonds perturb the system in such a way that *each* cyclic fragment would obey Hückel's rule in isolation, the total system will be pseudoaromatic. Also, a system may even be pseudoaromatic where the total number of $\pi$-electrons is not $(4n + 2)$, provided the perturbing bonds effectively subdivide it into individual fragments that would each obey Hückel's rule (see pyrene, for example).

## PROBLEMS

1. Use the simple cosine expressions to calculate the energy levels for the heptatrienyl and cycloheptatrienyl systems. Calculate $DE$ and $DEPE$ for the cation, radical, and anion of each system and compare stabilities for each pair (cyclic and acyclic) of a given charge type. Which is intrinsically the most stabilized of the six systems in a $\pi$-electronic sense?
2. Use the simple polygon rule to estimate the energy level distributions for [14]- and [16]-annulene, assuming each to be fully coplanar $\pi$-systems. Compare the $DE$ and $DEPE$ values for these two molecules.
3. Predict the general features of the nmr spectrum of [22]-annulene.
4. The NMR spectrum of [24]-annulene shows a single proton resonance at 40°C, which occurs at $2.75\tau$. Cooling to $-80°$ separates the spectrum into three signals, occuring at $-2.9\tau(3H)$, $-1.2\tau(6H)$, and $5.27\tau(15H)$, respectively. Give a reasonable explanation of this change, and try to explain why all these signals occur at such low field.

5. Which of the following polycyclic systems would be predicted to show aromatic (or pseudoaromatic) properties. Which protons would be expected to give signals in the normal aromatic region, and which in the normal vinyl region of the NMR spectrum.

## REFERENCES

1. J. Thiele, *Ber. Dtsch. Chem. Ges.* **34**, 68 (1901); W. E. Doering and C. H. DePuy, *J. Am. Chem. Soc.* **75**, 5955 (1953); P. J. Garratt, "Aromaticity," p. 70. McGraw-Hill, New York, 1971.
2. T. L. Cottrell, "The Strengths of Chemical Bonds," 2nd Ed. Butterworth, London, 1958.
3. E. Hückel, *Z. Phys.* **70**, 204 (1931); **76**, 628 (1932); *Z. Elektrochem.* **43**, 752 (1937).
4. F. D. Rossini, K. S. Pitzer, R. L. Arnett, R. M. Braun, and G. C. Pimentel, "Selected Values of Physical and Thermodynamic Properties of Hydrocarbons and Related Compounds." Carnegie Press, Pittsburgh, Pennsylvania, 1953.
5. R. Breslow and C. Yuan, *J. Am. Chem. Soc.* **80**, 5991 (1958); R. Breslow, H. Höver, and H. W. Chan, *J. Am. Chem. Soc.* **84**, 3168 (1962).
6. R. Breslow, J. T. Groves, and G. Ryan, *J. Am. Chem. Soc.* **89**, 5048 (1967).
7. J. Waser and C. S. Ly, *J. Am. Chem. Soc.* **66**, 2035 (1944); J. Waser and V. Schomaker, *J. Am. Chem. Soc.* **65**, 1451 (1943); T. C. W. Mark and J. Trotter, *Proc. R. Soc., Ser. A* **261**, 163 (1961).
8. R. Criegee and G. Schröder, *Justus Liebigs Ann. Chem.* **623**, 1 (1959).
9. G. A. Olah, J. M. Bollinger, and A. M. White, *J. Am. Chem. Soc.* **91**, 3667 (1969); G. A. Olah and G. D. Mateescu, *J. Am. Chem. Soc.* **92**, 1430 (1970).
10a. G. Merling, *Ber. Dtsch. Chem. Ges.* **24**, 3108 (1891).
10b. W. E. Doering and H. L. Knox, *J. Am. Chem. Soc.* **76**, 3203 (1954).

10c. H. J. Dauben, F. A. Gadecki, K. M. Harmon, and D. L. Pearson, *J. Am. Chem. Soc.* **79**, 4557 (1957).
11. Reference 10b.
12. W. E. Doering and F. L. Detert, *J. Am. Chem. Soc.* **73**, 876 (1951); M. Kubo, T. Nozoe, and Y. Kurita, *Nature (London)* **167**, 688 (1951).
13. R. Wilstätter and E. Waser, *Ber. Dtsch. Chem. Ges.* **44**, 3423 (1911); *Ber. Dtsch. Chem. Ges.* **46**, 517 (1913); W. B. Person, G. C. Pimentel, and K. S. Pitzer, *J. Am. Chem. Soc.* **74**, 3437 (1952).
14. T. J. Katz, *J. Am. Chem. Soc.* **82**, 3784 (1960); H. P. Fritz and H. Keller, *Chem. Ber.* **95**, 158 (1962).
15. F. Sondheimer, *Acc. Chem. Res.* **5**, 81 (1972).
16. E. Vogel and H. D. Roth, *Angew. Chem.* **76**, 145 (1964); E. Vogel, M. Biskup, W. Pretzer, and W. A. Böll, *Angew. Chem.* **76**, 785 (1964); F. Sondheimer and A. Shani, *J. Am. Chem. Soc.* **86**, 3168 (1964).
17. J. T. Sworski, *J. Chem. Phys.* **16**, 550 (1948).
18. J. F. M. Oth, H. Röttele, and G. Schröder, *Tetrahedron Lett.* **1**, 61 (1970).
19. R. Wolofsky and F. Sondheimer, *J. Am. Chem. Soc.* **87**, 570 (1965).
20. F. Sondheimer, *Pure Appl. Chem.* **7**, 363 (1963).
21. F. Sondheimer, *Proc. R. Soc., Ser. A* **297**, 173 (1967); J. F. M. Oth, *Pure Appl. Chem.* **25**, 573 (1971).
22. F. Sondheimer and Y. Gaoni, *J. Am. Chem. Soc.* **82**, 5765 (1960); F. Sondheimer, Y. Gaoni, L. M. Jackman, N. A. Bailey, and R. Mason, *J. Am. Chem. Soc.* **84**, 4595 (1962).
23. F. Sondheimer and Y. Gaoni, *J. Am. Chem. Soc.* **83**, 4863 (1961); G. Schröder and J. F. M. Oth, *Tetrahedron Lett.* p. 4083 (1966).
24. F. Sondheimer, R. Wolofsky, and Y. Amiel, *J. Am. Chem. Soc.* **84**, 274 (1962); Y. Gaoni, A. Melera, F. Sondheimer, and R. Wolofsky, *Proc. Chem. Soc., London* p. 397 (1964).
25. See Ref. 15. See also R. M. McQuilkin, B. W. Metcalf, and F. Sondheimer, *Chem. Commun.* p. 338 (1971); F. Sondheimer and Y. Gaoni, *J. Am. Chem. Soc.* **84**, 3520 (1962); R. M. McQuilkin and F. Sondheimer, *J. Am. Chem. Soc.* **92**, 6341 (1970).
26. G. W. Wheland, "Resonance in Organic Chemistry," p. 98. Wiley, New York, 1955.
27. M. J. S. Dewar and R. Pettit, *J. Chem. Soc.* p. 1617 (1954); D. Peters, *J. Chem. Soc.* p. 1023 (1958).
28. G. Schröder and H. Röttele, *Angew. Chem.* **80**, 665 (1968); P. J. Garratt and R. H. Mitchell, *Chem. Commun.* p. 719 (1968).
29. P. J. Garratt, "Aromaticity," p. 157. McGraw-Hill, New York, 1971.

## SUPPLEMENTARY READING

Badger, G. M., "Aromatic Character and Aromaticity." Cambridge Univ. Press, London and New York, 1969.
Coulson, C. A., and Streitwieser, A., Jr., "Dictionary of $\pi$-Electron Calculations." Freeman, San Francisco, California, 1965.

## Supplementary Reading

* Garratt, P. J. "Aromaticity." McGraw-Hill, New York, 1971.
Heilbronner, E., and Bock, H., "Das HMO-Modell und seine Anwendung." Verlag Chemie, Veinheim, 1968.
Hückel, E., *Z. Phys.* **70**, 204 (1931); **76**, 628 (1932).
Hückel, W., "Theoretical Principles of Organic Chemistry," Vol. 1. Elsevier, Amsterdam, 1955.
*Sondheimer, F., *Acc. Chem. Res.* **5**, 81 (1972).
*Streitwieser, A., Jr., "Molecular Orbital Theory for Organic Chemists." Wiley, New York, 1961.
Wheland, G. W., "Resonance in Organic Chemistry." Wiley, New York, 1955.

# V EXTENSIONS AND IMPROVEMENTS OF THE SIMPLE HÜCKEL METHOD

The simple Hückel approach has so far been applied exclusively to polyene systems. There are a number of modifications that can be made to the simple HMO method which either extend it to a wider range of systems, or attempt to improve it by taking account of certain factors that were neglected in the simple approach. The most obvious extension of the method is to take account of the presence of atoms other than carbon in the $\pi$-systems.

## V.1 SYSTEMS INVOLVING HETEROATOMS

Many molecules of interest to organic chemists involve atoms such as nitrogen, oxygen, sulfur, and halogens, which can participate in delocalized $\pi$-systems. These atoms, which are in a similar state of hybridization to carbon, can contribute vacant, singly occupied, or doubly occupied $2p_z$ orbitals to the $\pi$-system of which they are part. This raises two questions: first, what values of the Coulomb integrals should be used in the calculations for these heteroatoms: second, how should the various bond integrals be modified to take into account the different types of bonding interaction that arise when heteroatoms are included?

## V.1 Systems Involving Heteroatoms

### V.1.1 CHOICE OF COULOMB INTEGRAL VALUES

Since only carbon atoms have been considered in the $\pi$-frameworks up to this point, the reasonable assumption was made that every atom in the framework had the same intrinsic ability to attract electrons. For example, all nuclei considered had the same nuclear charge and each was subject to effectively the same shielding due to the inner shell and $\sigma$-electrons. Thus all Coulomb integrals $H_{ii}$ were initially set equal to some numerical value $\alpha$, which we can call $\alpha_0$. However, as soon as heteroatoms such as nitrogen or oxygen are considered, then even in a simple Hückel treatment we obviously cannot use the same value for the Coulomb integrals for all these atoms as was previously used. Because of the variation in nuclear charge, and also possibly because of different inner shell electron distributions, the Coulombic attraction of these nuclei for $\pi$-electrons will vary considerably from atom to atom, independently of the particular structural environment of the atom. In other words, the different electronegativities of atoms must be taken into account in assigning Coulomb integral values.

What is normally done is to scale all Coulomb integrals in terms of the standard value $\alpha_0$, plus some fraction of a standard bond integral $\beta_0$, which serves only as a unit of energy in these Coulomb integral terms. Thus for an atom X we set the Coulomb integral $\alpha_x$ equal to

$$\alpha_x = \alpha_0 + h_x \beta_0$$

where $h_x$ is an empirical and dimensionless constant that is characteristic of X, and depends on the electronegativity difference between $X$ and carbon.[1] One method of arriving at the parameter $h_x$ is to use Pauling electronegativity values ($\chi$) and set

$$h_x = k(\chi_x - \chi_c)$$

where $k$ is frequently taken as being close to unity.

This is reasonable, since along any row of the periodic table (and particularly for the first row, which is of most interest) the effective "core" potential due to the nucleus, inner-shell, and $\sigma$-electrons will increase steadily from left to right. Therefore, $h_x$ should increase in magnitude along the series Li, Be, B, C, N, O, F, for example. Thus for the first row elements we have

$$h_{Li} < h_{Be} < h_B < h_C < h_N < h_{Ox} < h_F$$

and therefore, because all Coulomb integrals are negative energy quantities, we will have

$$\alpha_{Li} > \alpha_{Be} > \alpha_B > \alpha_C(=\alpha_0) > \alpha_N > \alpha_{O_x} > \alpha_F$$

Thus π-electrons will be more strongly attracted by, or be more stable in the region of, a fluorine nucleus than the other nuclei. Since in the simple HMO method $\alpha_C = \alpha_0$ is the reference point of energy, $h_C = 0$. Hence any atom more electronegative than carbon will have a value of $h_x > 0$ and any atom less electronegative will have $h_x < 0$. This can be shown schematically in terms of atomic energy levels (Fig. V.1).

Although the above assignment of $h_x$ values (based on $\chi_x$ values) takes into account the intrinsic electron attracting ability of a particular nucleus, there is one other very important factor that must be considered before finally assigning $h_x$ values; that is, how many electrons can or does atom X contribute to the delocalized π-system? This could be zero, one, or two, depending both on the particular atom and the way it is incorporated into the σ-framework. In other words, the question is, does a particular heteroatom *automatically* acquire any *formal charge* by participating in the π-system through use of its $2p_z$ orbitals. The answer to this can be illustrated by using pictorial representations for simple systems containing boron, nitrogen, or oxygen atoms. First, vinylborane can be represented as a localized structure in either way shown [(I) and (II)]. However, to show boron partici-

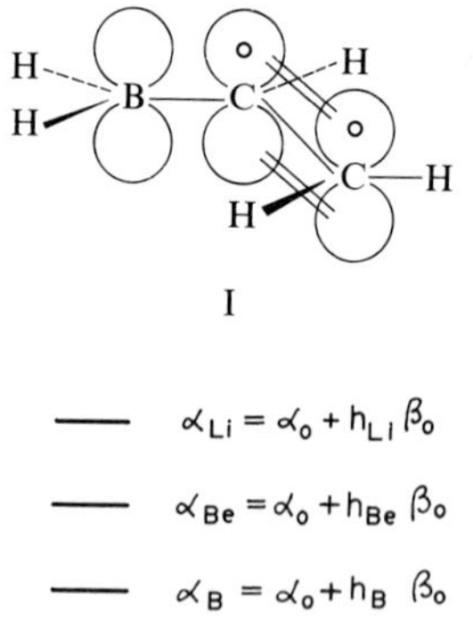

or H$_2$B—CH=CH$_2$

I   II

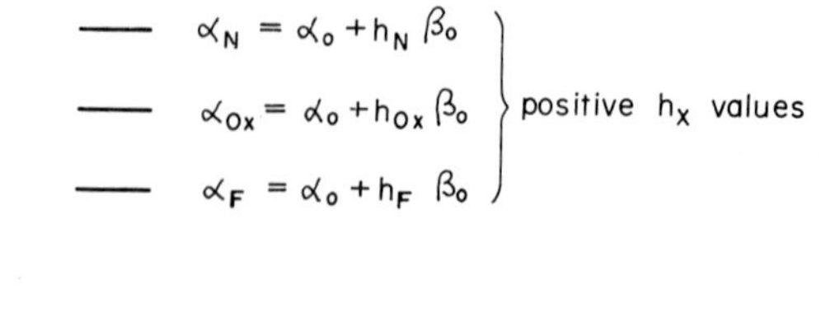

$\alpha_{Li} = \alpha_0 + h_{Li}\beta_0$
$\alpha_{Be} = \alpha_0 + h_{Be}\beta_0$  $\Big\}$ negative $h_x$ values
$\alpha_B = \alpha_0 + h_B\beta_0$

$\alpha_C = \alpha_0$

$\alpha_N = \alpha_0 + h_N\beta_0$
$\alpha_{O_x} = \alpha_0 + h_{O_x}\beta_0$  $\Big\}$ positive $h_x$ values
$\alpha_F = \alpha_0 + h_F\beta_0$

FIGURE V.1

## V.1 Systems Involving Heteroatoms

pating in a delocalized $\pi$-system through its $2p_z$ orbital, the structure must be represented as in (III). Thus any VB structure that shows boron

$$\text{or} \quad H_2\bar{B}=CH-\overset{+}{C}H_2$$

III

participation in the $\pi$-system must result in a formal negative charge on boron, or in MO terms since boron only contributes a vacant $2p_z$ orbital to the $\pi$-system, any delocalization automatically results in an increased electron density on boron. Note that although carbon atoms in a $\pi$-system *may* acquire net negative or positive charge, as in a non-self-consistent charge system, this does not occur automatically. This latter type of charge acquisition can be taken account of by a modification of the HMO technique to be described later, but in the case of atoms like boron or silicon that participate through vacant orbitals, the acquisition of charge must be taken account of automatically. In other words, atoms of this type have a decreased "core" potential when they participate in any $\pi$-system, and their $h_x$ values must be decreased appropriately.

If we now consider nitrogen, first in pyridine IV, it is clear that this nitrogen contributes one electron to the $\pi$-system as does any carbon atom in the system. Therefore, like carbon, the nitrogen in pyridine does not automatically acquire any formal charge through its $\pi$-participation and the value assigned to $h_N$ in this case will be based essentially only on the electronegativity value for nitrogen.

IV    V

However, the nitrogen in pyrrole V formally contributes a doubly occupied $2p_z$ orbital to the $\pi$-system; hence any delocalization automatically results in nitrogen acquiring a formal positive charge (in VB terms) or a decreased electron density. Thus the "core" potential of this nitrogen will be increased over that of the pyridine nitrogen and the $h_N$ values used for nitrogen in pyrrole must be increased appropriately over that used for nitrogen in pyridine.

Similarly, oxygen in a conjugated carbonyl group contributes one electron to the delocalized $\pi$-system and does not automatically acquire any formal charge, whereas oxygen in an ether-type group (such as in phenol or anisole) automatically acquires formal positive charge or decreased electron density. Again the "core" potential will be higher in the latter case and the $h_{\text{O}x}$ values used for ether-type oxygens should be higher than those assigned to carbonyl-type oxygens.

This can be expressed by saying that any atom which contributes a vacant $2p_z$ orbital acquires a *formal* "core" potential of zero since the total number of electrons acquired or shared matches the charge on the nucleus. This formal core potential [CP] can be expressed as

[core potential]
$$= [\text{nuclear charge}] - [\text{number of inner shell electrons}]$$
$$- \begin{bmatrix} \text{number of electrons} \\ \text{in nonconjugated} \\ \text{lone pairs} \end{bmatrix} - \tfrac{1}{2} \begin{bmatrix} \text{number of electrons} \\ \text{in } \sigma\text{-bonds attached} \\ \text{directly to atom} \end{bmatrix}$$

Thus for boron

$$[\text{CP}] = 5 - 2 - 0 - \tfrac{1}{2}(6) = 0$$

For all carbon atoms, or other atoms that in a given structure contribute one electron to the $\pi$-system, the core potential is $+1$. Thus for carbon

$$[\text{CP}] = 6 - 2 - 0 - \tfrac{1}{2}(6) = +1$$

or nitrogen in pyridine

$$[\text{CP}] = 7 - 2 - 2 - \tfrac{1}{2}(4) = +1$$

or oxygen in benzaldehyde

$$[\text{CP}] = 8 - 2 - 4 - \tfrac{1}{2}(2) = +1$$

However, for nitrogen in pyrrole,

$$[\text{CP}] = 7 - 2 - 0 - \tfrac{1}{2}(6) = +2$$

and for oxygen in phenol

$$[\text{CP}] = 8 - 2 - 2 - \tfrac{1}{2}(4) = +2$$

Therefore, other things being equal, the value of $h_x$ for a given nucleus should increase with the formal core potential, that is,

$$h_{\ddot{x}} > h_{\dot{x}} > h_x \quad \text{and} \quad \alpha_{\ddot{x}} < \alpha_{\dot{x}} < \alpha_{\overset{\circ}{x}}$$

## V.1 Systems Involving Heteroatoms

Thus the $2p_z$ electrons of the nitrogen in pyrrole or the oxygen in anisole are less likely to be delocalized into the $\pi$-systems than those in pyridine and benzaldehyde, respectively.

One final point is that if any particular atom initially bears a positive charge (due to protonation for example), it will be even less likely to give up a share of its $2p_z$ electrons to a conjugated $\pi$-system. For example, in an oxygen protonated anisole (VI), the remaining lone

VI

pair is subject to a greatly increased core potential and will not easily be delocalized into the phenyl system. Hence for any atom that is already positively charged in this sense

$$h_x^+ > h_x$$

An accepted set of $h_x$ values[2] based on the above considerations is given in Table V.2. These will be discussed later.

### V.1.2 CHOICE OF BOND INTEGRAL VALUES

In previously considering simple polyene systems, all bonds in the $\sigma$-framework were between carbon atoms. The reasonable simplifying assumption was made that all nearest neighbor interactions would be approximately the same, hence all Coulomb integrals of the type $H_{ij}$ were set equal to some standard value $\beta$ (or $\beta_0$); all nonnearest neighbor terms were set equal to zero. However, when heteroatoms are considered the properties of C—X (or X—X or X—Y) bonds can no longer be considered to be the same as those for C—C bonds. The bond integral value will depend, among other things, on internuclear distance. For example, as the distance between two $\sigma$-bonded atoms increases, the effective orbital overlap will decrease, and hence bonding will decrease. Thus bond integral terms are scaled empirically (like Coulomb integral terms) with reference to a standard C—C bond integral (as in benzene) according to the formula

$$\beta_{rs} = k_{rs}\beta_0$$

Thus if $k_{rs}$ is taken as unity for all normal C—C bonds in polyenes $k_{rs}$ is set at some value less than unity for longer bonds, and greater

TABLE V.1

| r-s distance Å | $k_{rs}$ |
|---|---|
| 1.20 | 1.71 |
| 1.33 | 1.23 |
| 1.397 | 1.00 |
| 1.45 | 0.83 |
| 1.54 | 0.57 |

TABLE V.2

Heteroatom Parameters

| Element | Coulomb integral parameter | Bond integral parameter |
|---|---|---|
| Boron | $h_B = -1$ | $k_{C-B} = 0.7$ <br> $k_{B-N} = 0.8$ |
| Carbon | $h_C = 0$ | $k_{C-C} = 1.0$ |
| Nitrogen | $h_N = 0.5$ <br> $h_{\ddot{N}} = 1.5$ <br> $h_{\overset{\oplus}{N}} = 2.0$ | $k_{C-N} = 0.8$ <br> $k_{C=N} = 1.0$ <br> $k_{N-O} = 0.7$ |
| Oxygen | $h_{\ddot{O}} = 1.0$ <br> $h_{\ddot{O}} = 2.0$ <br> $h_{\dot{O}} = 2.5$ | $k_{C-O} = 0.8$ <br> $k_{C=O} = 1.0$ |
| Fluorine | $h_F = 3.0$ | $k_{C-F} = 0.7$ |
| Chlorine | $h_{Cl} = 2.0$ | $k_{C-Cl} = 0.4$ |
| Bromine | $h_{Br} = 1.5$ | $k_{C-Br} = 0.3$ |
| Silicon[a] | $h_{Si} = -1.2$ | $k_{C-Si} = 0.45$ |
| Germanium[a] | $h_{Ge} = -1.05$ | $k_{C-Ge} = 0.3$ |

[a] N.B. these are for use only with *terminal* Si or Ge, and are for

d-orbital overlap

than unity for shorter bonds. The dependence of $k_{rs}$ on internuclear distance is shown in Table V.1, for which the values have been estimated from calculated $2p_z$-type overlaps.[3]

Thus even for C—C bonds, the bond integral value should be allowed to vary from system to system. However, since most C—C

## V.1 Systems Involving Heteroatoms

bond distances in sp² hybridized systems fall between 1.34 and 1.45 Å, $k_{C-C}$ is normally set equal to unity for all bonds. (A method of taking account of the variation in C—C bonds will be described later.)

The accepted values of $k_{rs}$ for typical bonds involving heteroatoms are listed in Table V.2.[2] Note that apart from the two changes where heteroatoms are involved, namely,

$$\alpha_x = \alpha_0 + h_x\beta_0, \qquad \beta_{c-x} = k_{c-x}\beta_0$$

nothing else is changed in the simple HMO calculation of energies, coefficients, or other derived quantities. Two points to note are that (i) inclusion of heteroatoms generally reduces the symmetry of the system so that calculations may be somewhat more difficult and give less simple roots, and (ii) the generalizations made on p. 78 concerning the values obtained for the roots and energies are no longer valid when heteroatoms are included.

**Notes on the heteroatom parameters in Table V.2** (i) Note that only boron, silicon, and germanium have negative $h_x$ values; most commonly encountered elements in organic molecules are more electronegative than carbon and have positive $h_x$ values.

(ii) Values for silicon and germanium are not for $p_z$ orbitals, which are normally completely involved in σ-bonding in these atoms. The assumption is made that *one* 3d or 4d orbital on these types of atom can participate in a π-system in a similar way to $p_z$ AOs.

(iii) Note that $k_{c-x}$ decreases markedly as second and third row elements are involved in bonding to carbon. This is because the $p_z$ orbitals of these atoms are larger and more diffuse, and overlap less effectively with carbon $2p_z$ AOs.

(iv) Note the very high values of $h_x$ for positively charged atoms such as oxygen or nitrogen. This means that such atoms will not readily participate in π-systems by giving up electron density into a delocalized system.

(v) Comparing the halogens listed, fluorine has a very large and unfavorable Coulomb integral because of its high value of $h_x$. However, it has a much more favorable bond integral term because it is in the same row of the periodic table as carbon. Hence, of the three halogens F, Cl, and Br, it might be expected that fluorine would participate most strongly in a delocalized π-system.

The following examples illustrate the selection of parameters for calculations involving heteroatoms. Great care should be used in properly assigning $h_x$ and $k_{c-x}$ values in each molecule before beginning any HMO calculation.

*2-methoxypyridine VII*

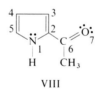

VII

**Bond integrals**

$$H_{12} = H_{16} = \beta$$
$$H_{23} = H_{34} = \cdots = H_{56} = \beta$$
$$H_{27} = 0.8\beta$$

**Coulomb integrals**

$$H_{11} = \alpha + 0.5\beta$$
$$H_{22} = H_{33} = \cdots = H_{66} = \alpha$$
$$H_{77} = \alpha + 2.0\beta$$

*2-acetylpyrrole VIII*

VIII

**Bond integrals**

$$H_{12} = H_{15} = 0.8\beta$$
$$H_{23}, \text{etc.} = \beta$$
$$H_{67} = \beta$$

**Coulomb integrals**

$$H_{11} = \alpha + 1.5\beta$$
$$H_{22} = H_{33} = \cdots = H_{66} = \alpha$$
$$H_{77} = \alpha + \beta$$

(all nonnearest neighbor $\beta$ terms are zero).

## V.1 Systems Involving Heteroatoms

It is instructive to illustrate the inclusion of heteroatoms in $\pi$-systems by carrying out a complete calculation on a simple molecule, and consider if the results obtained are reasonable. This will be done for urea (IX).

*Urea*

$$\underset{\text{IX}}{\underset{H_2\ddot{N}\quad\quad\ddot{N}H_2}{\overset{\overset{\displaystyle O}{\|}}{C}}} \qquad \overset{\overset{\displaystyle D}{|}}{\underset{\underset{B\quad\quad C}{O\quad\quad O}}{A}}$$

**Coulomb integrals**

$$H_{AA} = \alpha, \quad H_{BB} = H_{CC} = \alpha + 1.5\beta, \quad H_{DD} = \alpha + 1.0\beta$$

**Bond integrals**

$$H_{AB} = H_{AC} = 0.8\beta, \quad H_{AD} = \beta, \quad H_{BC} = H_{BD} = H_{CD} = 0$$

Although urea has $C_{2v}$ symmetry, one of the elements of symmetry involved is the common nodal plane of the $\pi$-system. Since all orbitals transform in the same way with respect to this plane, adequate and in fact equivalent simplification is achieved by placing the molecule in the $C_2$ group rather than $C_{2v}$. Thus the transformation table is as shown in Table V.3. and from the character table (Table V.4), the

TABLE V.3

|        | $E$      | $C_2$    |
|--------|----------|----------|
| $\phi_A$ | $\phi_A$ | $\phi_A$ |
| $\phi_B$ | $\phi_B$ | $\phi_C$ |
| $\phi_C$ | $\phi_C$ | $\phi_B$ |
| $\phi_D$ | $\phi_D$ | $\phi_D$ |
|        | 4        | 2        |

TABLE V.4

|   | $E$ | $C_2$ |
|---|-----|-------|
| $A$ | 1   | 1     |
| $B$ | 1   | $-1$  |

symmetry orbitals are, in $\Gamma_A$,

$$N_A = \frac{4 \cdot 1 + 2 \cdot 1}{2} = 3$$

$$S_1 = \phi_A$$

$$S_2 = \frac{1}{\sqrt{2}}(\phi_B + \phi_C)$$

$$S_3 = \phi_D$$

and, in $\Gamma_B$,

$$N_B = \frac{4 \cdot 1 + 2 \cdot (-1)}{2} = 1$$

$$S_4 = \frac{1}{\sqrt{2}}(\phi_B - \phi_C)$$

The simplified secular determinant then becomes

$$\left| \begin{array}{ccc|c} H_{11} - \varepsilon & H_{12} & H_{13} & 0 \\ H_{21} & H_{22} - \varepsilon & H_{23} & 0 \\ H_{31} & H_{32} & H_{33} - \varepsilon & 0 \\ \hline 0 & 0 & 0 & H_{44} - \varepsilon \end{array} \right| = 0$$

The matrix elements can easily be evaluated as before using appropriate values for each heteroatom component. Thus

$$H_{11} = \int S_1 \mathcal{H} S_1 \, d\tau = H_{AA} = \alpha$$

$$H_{22} = \tfrac{1}{2}[H_{BB} + H_{CC} + 2H_{BC}] = \alpha + 1.5\beta$$

$$H_{33} = H_{DD} = \alpha + \beta$$

$$H_{12} = \frac{1}{\sqrt{2}}[H_{AB} + H_{AC}] = \frac{1}{\sqrt{2}}[1.6\beta] = 1.131\beta$$

$$H_{13} = H_{AD} = \beta$$

$$H_{23} = \frac{1}{\sqrt{2}}[H_{BD} + H_{CD}] = 0$$

$$H_{44} = \tfrac{1}{2}[H_{BB} + H_{CC} - 2H_{BC}] = \alpha + 1.5\beta$$

## V.1 Systems Involving Heteroatoms

The roots can then be obtained from

$$\begin{vmatrix} x & 1.131 & 1 \\ 1.131 & x+1.5 & 0 \\ 1 & 0 & x+1 \end{vmatrix} \begin{vmatrix} x+1.5 \end{vmatrix} = 0$$

Expansion of the third-order determinant gives the polynomial

$$P_A = x^3 + 2.5x^2 - 0.78x - 2.78$$

which yields roots $x_i = -2.320, -1.188, +1.008$ and the first-order determinant gives

$$P_B = x + 1.5 = 0$$

with root $x = -1.500$.

The four energy levels for urea are as shown in the energy level diagram of Fig. V.2. The $\pi$-electronic energy is calculated as before,

$$E_\pi = 2(\alpha + 2.320\beta) + 2(\alpha + 1.500\beta) + 2(\alpha + 1.188\beta)$$
$$= 6\alpha + 10.016\beta$$

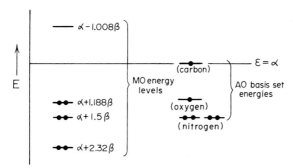

FIGURE V.2

However, the net $\pi$-bonding energy is no longer just the $\beta$ term of this expression, which at first appears to be a very large quantity. Account must be taken of the fact that the starting point of the calculation was not $\varepsilon = \alpha$ for all basis orbitals. It is logical to assume that the $\pi$-system of the molecule was built up from one carbon $2p_z$ electron, one oxygen $2p_z$ electron, and four nitrogen $2p_z$ electrons.[†] Thus the bonding energy

---

[†] For this reason, also, the sum of the roots of the HMO calculation does not vanish. Instead, it is equal to the sum of the $\beta$ portions of the four basis orbitals, that is, $\sum x_i = 4.00$.

is given by

$$B_\pi = (6\alpha + 10.016\beta) - (6\alpha + 7.000\beta) = 3.016\beta$$

This is still a quite significant bonding energy, indicating that urea has a very stable $\pi$-system.

To calculate $DE$ for urea, the most reasonable localized VB structure for calculation of $E_\pi^{loc}$ is XI. This entails a calculation of the energy

$$\underset{\text{XI}}{\overset{\overset{\overset{\text{O}\cdot}{\|}}{\underset{H_2\ddot{N}}{\overset{C\cdot}{\diagup}}\underset{\ddot{N}H_2}{\diagdown}}}{}}$$

levels in an isolated carbonyl fragment, using the same heteroatom values for oxygen as before. (Thus formaldehyde is taken as the localized model rather than ethylene.) This simple calculation involves the determinant

$$\begin{vmatrix} x & 1 \\ 1 & x+1 \end{vmatrix} = 0$$

which gives

$$P_S = x^2 + x - 1 = 0$$

yielding the roots $-1.618$, $+0.618$. Therefore, $E_\pi$ for the localized C=O fragment is $2\alpha + 3.236\beta$. Thus for urea

$$E_\pi^{loc} = (2\alpha + 3.236\beta) + 4(\alpha + 1.5\beta) = 6\alpha + 9.236\beta$$

Therefore,

$$DE = (6\alpha + 10.016\beta) - (6\alpha + 9.236\beta) = 0.78\beta$$

This is quite a large delocalization energy for such a simple system, and indicates that urea is significantly resonance stabilized. On the crude basis that $2\beta \approx 36$ kcal in resonance energy terms (with the reservation discussed previously), one might conclude that urea has about 14 kcal of resonance energy. It is interesting that urea is known to be strongly resonance stabilized and estimates as high as 37 kcal have been obtained for its stabilization energy.[4]

The coefficients can easily be calculated from the simplified secular equations using the above roots. These are listed in Table V.5 for the three occupied orbitals.

## V.1 System Involving Heteroatoms

TABLE V.5

| $\psi_i$ | $x_i$ | $C_C$ | $C_{N_1}$ | $C_{N_2}$ | $C_{Ox}$ | $\varepsilon_i$ |
|---|---|---|---|---|---|---|
| $\psi_1$ | −2.320 | 0.536 | 0.523 | 0.523 | 0.406 | $\alpha + 2.32\beta$ |
| $\psi_2$ | −1.500 | 0 | 0.707 | −0.707 | 0 | $\alpha + 1.50\beta$ |
| $\psi_3$ | −1.188 | 0.153 | −0.394 | −0.394 | 0.816 | $\alpha + 1.19\beta$ |

The bonding MOs can be shown schematically (Fig. V.3) with only the upper lobes indicated. As expected $\psi_1$ is bonding everywhere and is by far the most stable MO. $\psi_2$ is clearly a nonbonding orbital, but since the wave function only has magnitude on the nitrogen centers, it has the same energy as electrons in isolated nitrogen pairs. The third orbital $\psi_3$ is mainly a carbonyl bonding orbital, but with substantial antibonding character in the C—N regions; overall, electrons in this orbital are more stable than in either oxygen or carbon $2p_z$ AOs, thus it can be classified as a weakly bonding MO, although it is higher in energy than $\psi_2$.

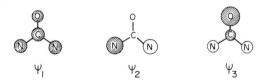

FIGURE V.3

The associated quantities $q$, $\xi$, and $\rho$ are as follows:

$$q_{Ox} = 1.661, \quad \xi_{Ox} = -0.661$$
$$q_C = 0.621, \quad \xi_C = +0.379$$
$$q_N = 1.857, \quad \xi_N = +0.143$$
$$\rho_{CO} = 0.685, \quad \rho_{CN} = 0.440$$

These give us an overall picture of the $\pi$-electron distribution in urea (XII). Thus according to this simple HMO calculation, urea should be

$$\begin{array}{c} ^{-0.661}\text{O} \\ ^{0.685}| \\ ^{+0.143}\text{N}\diagdown\overset{+0.379}{\underset{0.440}{\text{C}}}\diagup\text{N}^{+0.143} \end{array}$$

XII

a strongly dipolar molecule with significant C—N bonding and a lower carbonyl bond order than in simple aldehydes and ketones. These results are in good accord with the well-known properties of urea.

## V.2 INCLUSION OF DIFFERENTIAL OVERLAP

All HMO calculations carried out so far have involved use of the zero differential overlap (ZDO) approximation; namely, that

$$\int \phi_i \phi_j \, d\tau = S_{ij} = \delta_{ij}$$

where $\delta_{ij} = 0$ if the AOs $i$ and $j$ are on different atoms. This is clearly the most drastic approximation involved in the simple HMO treatment.

In the original secular determinants before any approximations were made, each diagonal element was of the form $[H_{ii} - \varepsilon S_{ii}]$ and since the basis set AOs were normalized, all $S_{ii} = 1$. Thus each of these elements simplified, without further approximation, to $[H_{ii} - \varepsilon]$, and using the Hückel definitions, became $(\alpha - \varepsilon)$ for carbon-only systems.

However, every off-diagonal element in the original determinant is also of the above form, namely, $[H_{ij} - \varepsilon S_{ij}]$, and since typical values of $S_{ij}$ for 2p$_z$ orbitals on neighboring carbon atoms are known to be of the order of 0.25, the term $\varepsilon S_{ij}$ is certainly not negligible in these cases compared to $H_{ij}$. Therefore, although the general simplification $[H_{ij} - \varepsilon S_{ij}] \to H_{ij} \to 0$ might be justified in cases involving non-nearest neighbor $i$ and $j$, where $S_{ij} \ll 1$ in most cases, it is certainly not justified for neighboring $i$ and $j$.

Instead of making the above general approximation, we could make the less drastic approximation that

$$[H_{ij} - \varepsilon S_{ij}] = \beta - S\varepsilon \quad \begin{pmatrix} \text{where } i \text{ and } j \text{ are directly} \\ \text{attached in the } \sigma\text{-framework} \end{pmatrix}$$

$$= 0 \quad \begin{pmatrix} \text{where } i \text{ and } j \text{ are not so} \\ \text{attached} \end{pmatrix}$$

and then solve the secular problem, thus including overlap in a significant way. It is of considerable interest to determine in what way the final results (energies, coefficients, etc.) are affected by this inclusion of differential overlap.

## V.2 Inclusion of Differential Overlap

In treating the changed secular determinants, it is convenient to introduce several arithmetical manipulations. If we write

$$[H_{ij} - \varepsilon S_{ij}] = \beta - S\varepsilon$$

then add and subtract the quantity $S\alpha$ on the right-hand side, we obtain

$$\beta - S\varepsilon = \beta - S\varepsilon + S\alpha - S\alpha = (\beta - S\alpha) + S(\alpha - \varepsilon)$$

If we then define

$$\gamma = \beta - S\alpha$$

we have

$$\beta - S\varepsilon = \gamma + S(\alpha - \varepsilon)$$

If we now take the original secular determinant for ethylene,

$$\begin{vmatrix} H_{11} - \varepsilon S_{11} & H_{12} - \varepsilon S_{12} \\ H_{21} - \varepsilon S_{21} & H_{22} - \varepsilon S_{22} \end{vmatrix} = 0$$

and introduce the newly-defined terms based on inclusion of neighboring overlap, this becomes

$$\begin{vmatrix} \alpha - \varepsilon & \beta - S\varepsilon \\ \beta - S\varepsilon & \alpha - \varepsilon \end{vmatrix} = 0$$

and with the substitution $\gamma = \beta - S\alpha$ we obtain

$$\begin{vmatrix} \alpha - \varepsilon & \gamma + S(\alpha - \varepsilon) \\ \gamma + S(\alpha - \varepsilon) & \alpha - \varepsilon \end{vmatrix} = 0$$

Thus all diagonal elements are now expressed in terms of $(\alpha - \varepsilon)$ as before, and all off-diagonal elements also contain this term plus the quantity $\gamma$. If we now divide each element by $\gamma$ (instead of by $\beta$ as was done previously) and let $(\alpha - \varepsilon)/\gamma = x$, we obtain

$$\begin{vmatrix} x & 1 + Sx \\ 1 + Sx & x \end{vmatrix} = 0$$

This is very similar to the previously obtained determinant, except that $x$ is now defined in terms of $\gamma$ and not $\beta$, and the off-diagonal elements have become $(1 + Sx)$ instead of 1.

Suppose we now let $(1 + Sx) = b$, the determinant is changed to

$$\begin{vmatrix} x & b \\ b & x \end{vmatrix} = 0$$

Now division of each element by $b$ does not change the equality, and reduces the determinant to exactly the same form as the original determinant (with complete neglect of overlap):

$$\begin{vmatrix} x/b & 1 \\ 1 & x/b \end{vmatrix} = 0$$

Thus the roots of the secular polynomial will be numerically exactly the same as before, except that now each root $m_j$ corresponds to

$$m_j = x_j/b, \qquad \text{where} \quad x_j = (\alpha - \varepsilon_j)/\gamma$$

whereas before the same numerical values of the individual roots were given by

$$m_j = x_j, \qquad \text{where} \quad x_j = (\alpha - \varepsilon_j)/\beta$$

The new eigenvalues (including overlap) can be obtained simply from the old roots (obtained by using the ZDO approximation), since

$$x_j = m_j b$$
$$= m_j(1 + Sx_j)$$

Thus

$$x_j = \frac{m_j}{1 - m_j S} = \frac{\alpha - \varepsilon_j}{\gamma}$$

and the new eigenvalues have the form

$$\varepsilon_j = \alpha - \frac{m_j}{1 - m_j S}\gamma$$

instead of

$$\varepsilon_j = \alpha - m_j \beta$$

as before.

If we let $m_j/(1 - m_j S) = n_j$, the new eigenvalues take on a very similar pattern to the old values:

$$\varepsilon_j = \alpha - n_j\gamma \qquad \text{(including overlap)}$$
$$\varepsilon_j = \alpha - m_j\beta \qquad \text{(excluding overlap)}$$

## V.2 Inclusion of Differential Overlap

Although this result was derived from a consideration of the simple system ethylene, it is easy to show that it is equally valid for any $\pi$-system, because the secular determinant obtained including overlap can be similarly reduced to the same form as that obtained excluding overlap. Therefore, the only differences brought about by the inclusion of overlap in the above way are:

(i) the bonding term of each eigenvalue will now be smaller in magnitude than the old bond integral $\beta$, which is clearly the case since $\gamma = \beta - S\alpha$ and $S > 0$, and both $\beta$ and $\alpha$ are negative energy quantities;

(ii) the new coefficients of the bonding term $n_j$ can easily be obtained from the previously calculated roots $m_j$, using an appropriate value of the overlap integral, which is taken to be $S = 0.25$.

So far as the Hückel energies and associated quantities are concerned, the overall energy level pattern is not greatly changed by including nearest neighbor overlap, although the individual levels will be shifted relative to each other. What this means is that any general conclusions based on simple Hückel theory, concerning stabilities or delocalization, are not seriously affected, whether overlap is included or neglected. This may seem a very surprising result, but consideration of the following simple $\pi$-systems shows that their energies are only changed in a relative sense, and comparisons of their stabilities give essentially the same predictions as before.

Fortunately, therefore, the basic success of the simple HMO approach is not seriously dependent on the exclusion of overlap approximation. Although nonnearest neighbor overlap is still neglected in the following examples, it is reasonable that its inclusion would affect the results even less in any comparative sense than did the present inclusion of nearest neighbor overlap.

*Ethylene*

Roots: $m_1 = 1.0$
$m_2 = 2.0$

Eigenvalues

| (excluding overlap) | (including overlap) |
|---|---|
| $\varepsilon_1 = \alpha + \beta$ | $\varepsilon_1 = \alpha + 0.8\gamma$ |
| $\varepsilon_2 = \alpha - \beta$ | $\varepsilon_2 = \alpha - 1.33\gamma$ |
| $E_\pi = 2\alpha + 2\beta$ | $= 2\alpha + 1.6\gamma$ |
| $DE = 0$ | $= 0$ |

*Butadiene*

Roots: $m_1 = -1.618$,  $m_3 = +0.618$
$m_2 = -0.618$,  $m_4 = +1.618$

Eigenvalues

(excluding overlap)

$\varepsilon_1 = +1.62\beta$
$\varepsilon_2 = +0.62\beta$
$\varepsilon_3 = -0.62\beta$
$\varepsilon_4 = -1.62\beta$
$E_\pi = 4\alpha + 4.48\beta$
$DE = (4\alpha + 4.48\beta) - (4\alpha + 4\beta)$
$= 0.48\beta$

(including overlap)

$\varepsilon_1 = +1.15\gamma$
$\varepsilon_2 = +0.54\gamma$
$\varepsilon_3 = -0.73\gamma$
$\varepsilon_4 = -2.72\gamma$
$= 4\alpha + 3.38\gamma$
$= (4\alpha + 3.38\gamma) - (4\alpha + 3.2\gamma)$
$= 0.18\gamma$

*Benzene*

Roots: $m_1 = -2$,  $m_4 = +1$
$m_2 = -1$,  $m_5 = +1$
$m_3 = -1$,  $m_6 = +2$

Eigenvalues

(excluding overlap)

$\varepsilon_1 = \alpha + 2\beta$
$\varepsilon_2 = \alpha + \beta$
$\varepsilon_3 = \alpha + \beta$
$\varepsilon_4 = \alpha - \beta$
$\varepsilon_5 = \alpha - \beta$
$\varepsilon_6 = \alpha - 2\beta$
$E_\pi = 6\alpha + 8\beta$
$DE = 2\beta$

(including overlap)

$\varepsilon_1 = \alpha + 1.33\gamma$
$\varepsilon_2 = \alpha + 0.80\gamma$
$\varepsilon_3 = \alpha + 0.80\gamma$
$\varepsilon_4 = \alpha + 1.33\gamma$
$\varepsilon_5 = \alpha + 1.33\gamma$
$\varepsilon_6 = \alpha + 4.00\gamma$
$= 6\alpha + 5.87\gamma$
$= 1.07\gamma$

The energy levels for these three systems can be compared schematically as before (Fig. V.4), but they cannot be compared directly with the previous results since the energies are now scaled in terms of $\gamma$ units and not $\beta$ units. Nevertheless, the patterns are similar in all three cases to those obtained before, and on any comparative basis the same general conclusions would be drawn as previously. It can be seen that

## V.2 Inclusion of Different Overlap

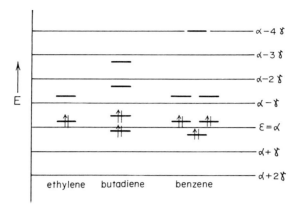

FIGURE V.4

the largest changes occur in the antibonding levels, which are displaced to higher energy considerably more than are the bonding levels, on inclusion of overlap.

Comparing ethylene, butadiene, and benzene, it is evident that the bonding energy terms and $DE$ values are reduced in absolute magnitude, but that benzene is still predicted to be very stable and significantly delocalized, and that butadiene does not have a large delocalization energy or bonding energy per electron compared to ethylene. This kind of comparison can be extended to other systems, as illustrated in Table V.6, where $DE$ and $DEPE$ values based on exclusion and inclusion of overlap are compared. It can be seen that the molecules fall in roughly the same relative order of stabilization, in either set. Systems that were previously predicted to have large stabilization energies due to delocalization, such as the cyclopropenyl cation, still have the highest $DE$ values when overlap is included. Similarly, systems that had low $DE$ values in the previous simple calculations also have low values when overlap is included.

The only notable difference on including overlap is that $DE$ values in some cases turn out to be negative, particularly for systems that would be classified as antiaromatic. Where ABMOs are occupied, the $DE$ values are decreased markedly for the reason noted above. Thus the cyclopropenyl anion is predicted to be very unstable when overlap is included, and also the radical of this system is the only case that is strongly out of line when $DEPE$ values are compared in columns 2 and 4 of Table V.6.

TABLE V.6

|  | Exclusion of overlap | | Inclusion of overlap | |
| --- | --- | --- | --- | --- |
| System | DE ($\beta$ units) | DEPE ($\beta$ units) | DE ($\gamma$ units) | DEPE ($\gamma$ units) |
| cyclopropenyl anion | 0 | 0 | −1.60 | −0.40 |
| cyclobutadiene | 0 | 0 | −0.53 | −0.13 |
| cyclopropenyl radical | 1.00 | 0.33 | −0.27 | −0.09 |
| barrelene[a] | 0 | 0 | −0.43 | −0.07 |
| norbornadienyl anion[a] | 0.56 | 0.09 | −0.32 | −0.05 |
| ethylene | 0 | 0 | 0 | 0 |
| butadiene | 0.47 | 0.12 | 0.17 | 0.04 |
| cyclooctatetraene | 1.60 | 0.20 | 0.41 | 0.05 |
| allyl anion | 0.83 | 0.21 | 0.49 | 0.12 |
| cyclopentadienyl cation | 1.24 | 0.31 | 0.54 | 0.14 |
| allyl radical | 0.83 | 0.28 | 0.49 | 0.16 |
| benzene | 2.00 | 0.33 | 1.07 | 0.18 |
| trimethylenemethane | 1.46 | 0.36 | 0.82 | 0.21 |
| cyclopentadienyl radical | 1.85 | 0.37 | 1.07 | 0.21 |
| norbornadienyl cation[a] | 1.56 | 0.39 | 0.92 | 0.23 |
| allyl cation | 0.83 | 0.41 | 0.49 | 0.25 |
| cyclopentadienyl anion | 2.47 | 0.41 | 1.61 | 0.27 |
| cyclopropenyl cation | 2.00 | 1.00 | 1.07 | 0.53 |

[a] HMO calculations on nonplanar polycylic $\pi$-systems such as these will be described later.

A similar relationship exists between the bonding energy values $B_\pi$ when overlap is either excluded from or included in the calculations. This is illustrated in Fig. V.5, where the correspondence between the bonding energies per electron is quite striking. Therefore, since the results of HMO calculations of energies, or rather their predictions, are not greatly affected by including overlap in the calculations, the simpler approach using the ZDO approximation is most widely used.

It is nonetheless of interest to consider briefly the effect of inclusion of overlap on the calculation of the MO coefficients and derived quantities. This will be done for a simple case, ethylene. It has been shown that the secular determinant, including overlap becomes

$$\begin{vmatrix} x & 1 + Sx \\ 1 + Sx & x \end{vmatrix} = 0$$

## V.2 Inclusion of Differential Overlap

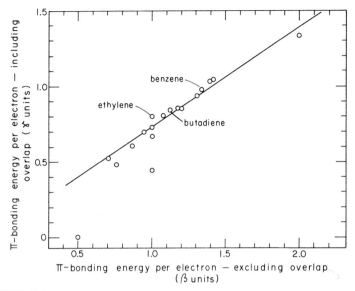

FIGURE V.5

once the appropriate substitutions are made. Instead of making further substitutions to reduce this determinant to the same form as that obtained using the ZDO approximation, as was done previously, the determinant can easily be expanded to obtain the roots directly, in this simple case. This gives the polynomial expression

$$P_s = x^2 - (1 + Sx)^2 = 0$$

and if $S$ is set equal to 0.25 as before,

$$P_s = x^2 - \left(1 + \frac{x}{4}\right)^2$$

$$= x^2 - \left(1 + \frac{x}{2} + \frac{x^2}{16}\right)$$

$$= \frac{15x^2}{16} - \frac{x}{2} - 1 = 0$$

or on multiplying through by a common factor

$$P_s = 15x^2 - 8x - 16 = 0$$

from which the roots are

$$x = \frac{+8 \pm \sqrt{64 + 960}}{30}$$

$$= \frac{+40}{30}, \frac{-24}{30}$$

$$= +1.333, -0.800$$

which verifies the substitution formula used previously.

The secular equations associated with the above determinant are then

$$C_1 x + C_2(1 + x/4) = 0$$
$$C_1(1 + x/4) + C_2 = 0$$

Thus for the root $x_1 = -0.800$ substitution yields

$$-0.8C_1 + 0.8C_2 = 0$$

or

$$C_1 = C_2$$

However, to use the normalization equation to obtain absolute values of $C_1$ and $C_2$, we must recall that the previous condition, namely,

$$\sum_r C_r^2 = 1$$

for all MOs, was based on the ZDO approximation. It will be recalled that this simplified expression arose because all cross-terms in the general expression

$$\sum_r \sum_s C_r C_s \int \phi_r \phi_s \, d\tau = 1$$

vanished, since $S_{rs}$ was taken to be zero unless $r = s$. Now that overlap is being included explicitly, the equation needed to apply the normalization requirement is

$$\sum_r \sum_s C_r C_s S_{rs} = 1$$

which for ethylene is

$$C_1{}^2 S_{11} + 2C_1 C_2 S_{12} + C_2{}^2 S_{22} = 1$$

## V.2 Inclusion of Differential Overlap

or

$$C_1^2 + 2C_1C_2S + C_2^2 = 1$$

If $S$ is taken to be 0.25, this becomes

$$C_1^2 + \frac{C_1 C_2}{2} + C_2^2 = 1$$

and since $C_1 = C_2$ for the root $x_1$, we have

$$C_1^2 + \frac{C_1^2}{2} + C_1^2 = 1$$

or

$$C_1^2 = \tfrac{2}{5}$$

Hence $C_1 = \sqrt{\tfrac{2}{5}} = C_2$ for the MO $\psi_1$, that is,

$$\psi_1 = 0.632\phi_1 + 0.632\phi_2$$

Following the same procedure for $\psi_2$, substitution of the root $x_2 = +1.333$ in the secular equations followed by normalization yields

$$C_1 = \sqrt{\tfrac{2}{3}}, \quad C_2 = -\sqrt{\tfrac{2}{3}}$$

Thus

$$\psi_2 = 0.816\phi_1 - 0.816\phi_2$$

Now the MO coefficients for $\psi_1$ and $\psi_2$ are quite different in magnitude once overlap is included. Naturally the coefficients of each AO in both MOs of ethylene are equal to each other in magnitude, as before, because of the simple symmetry of the system, but the ABMO is more extensive and diffuse than the BMO, which is reasonable.

Although the MO coefficients are changed in magnitude when overlap is taken into account, this does not mean that the electron density or charge distribution is changed. The electron densities $q_r$ can no longer be defined as simply as previously where

$$q_r = \sum_j n_j c_{jr}^2$$

since it is necessary that

$$\sum_r q_r = n_{\text{tot}}$$

however electron density is defined. When overlap is included, the electron density must be redefined to meet this requirement. This can be done by defining $q_r$ as follows:

$$q_r = \sum_j n_j c_{jr}^2 + \sum_s \sum_j n_j c_{jr} c_{js} S_{rs}$$

In the case of ethylene $c_{jr} = c_{js} = \sqrt{\frac{2}{5}}$, thus numerically

$$q_r = (2)(\sqrt{\tfrac{2}{5}})^2 + (2)(\sqrt{\tfrac{2}{5}})(\sqrt{\tfrac{2}{5}})\tfrac{1}{4}$$
$$= 0.8 + 0.2 = 1.0$$

This is the same value for the ethylene carbons as obtained previously. Thus overall the $q_r$ value is unchanged and is still the total electron population in the region of nucleus $r$; however, it is now made up of two terms, the second of which is called an overlap population.

Bond orders also can no longer be defined as simply as before once overlap is included. However, these can be redefined in terms of these overlap populations $\sum_s \sum_j n_j c_{jr} c_{js} S_{rs}$ so that bond orders are unchanged in a relative sense, although they will no longer correspond numerically to simple ideas based on a $\pi$-bond order of unity in ethylene as a standard.

## V.3  SELF-CONSISTENT HMO METHODS

### V.3.1  THE $\omega$-TECHNIQUE

A number of the $\pi$-systems already discussed have been described as having self-consistent charge distributions. This term applies to any system where $q_i$ (or $\xi_i$) at every position has the same value. For example, all neutral AH systems (even-AH molecules and odd-AH radicals) have this property since every $q_i = 1.0$ and every $\xi_i = 0$. In such cases, the initial assumption that the Coulomb integrals $H_{ii}$ for every carbon can be set equal to some constant value $\alpha_0$ is a reasonable one. If each carbon is assumed to have equal Coulombic attraction for the $\pi$-electrons in the system, the calculation should result in an even distribution of electron density over all $\pi$-carbons. In other words the final $q_i$ values should be consistent with the initial $H_{ii}$ values.

However, there are many $\pi$-lattices that are calculated to have uneven charge distributions. This applies to neutral molecules as well as ions, and many neutral systems have $q_i \neq 1$, in general, and hence cer-

## V.3 Self-Consistent HMO Methods

tain carbons bear a positive or negative charge. This is typical of many non-AH systems. In these cases, the Coulomb integrals at certain positions cannot really be taken as being equal to Coulomb integrals at other positions, nor equal to the standard value $\alpha_0$. For example, if $H_{ii}$ for a certain carbon calculated to have $\xi_i = 0$ has a value of $\alpha_0$, the $H_{ii}$ for a carbon bearing positive charge ($\xi_i > 0$) must be larger than $\alpha_0$ (i.e., more positive) since this carbon is unable, because of its position in the $\pi$-system, to attract the same electron density as the first carbon. Similarly, carbons bearing negative charge or high electron density should really have Coulomb integrals that are more negative than $\alpha_0$.

The question is, how to modify the calculations to take into account that certain positions, because of their particular location in the $\pi$-framework, will have either greater or lesser ability to attract electrons than the carbons in self-consistent systems like benzene, for example. The suggestion has been made by Streitwieser[5] and others[6] that the value of $H_{ii}$ for a given carbon could be related to the net charge at that position, obtained from a simple HMO calculation, in the following way.

First, each $H_{ii}$ is set equal to $\alpha_0$ and the charges (or electron densities) are calculated for each position in a zeroth-order Hückel calculation. Then each $H_{ii}$ is modified according to the relation

$$\alpha_i = \alpha_0 + \omega(n_i - q_i)\beta_0$$

where $\alpha_0$ is the standard value of carbon Coulomb integrals; $\omega$ is a dimensionless parameter; $n_i$ is the number of $2p_z$-type electrons that a neutral atom of the type $i$ contributes to the $\pi$-system ($n_i = 1$ for all carbon atoms); $q_i$ is the calculated electron density at position $i$ in the zeroth-order calculation; $\beta_0$ is the value of the standard bond integral, which is used here simply as a scaling unit of energy, since all $E_\pi$ and $DE$ values are expressed in terms of this quantity.

Thus all $H_{ii}$ values are scaled with respect to $\alpha_0$ in terms of some fraction of a $\beta_0$ unit, which may be positive or negative. The constant $\omega$ is a parameter whose value is chosen to give the best agreement with experimental results for a wide range of $\pi$-systems and experimental quantities. (However, appropriate values of $\omega$ can be estimated theoretically from more sophisticated calculations,[7] which take explicit account of interelectronic repulsions.) The accepted value of $\omega$ used in most calculations is 1.4,[8] but a value of 1.0 is also used in some cases. The precise value of $\omega$ is not of critical importance.

The Hückel calculation is then repeated using these modified Coulomb integrals. The term $\omega(n_i - q_i)\beta_0$ introduces some measure of interelectronic repulsion into the calculations, since if $q_i > 1.0$ in the zeroth-order calculation, $(n_i - q_i)$ is negative and the value of $\alpha_i$ is thereby reduced. Thus in the first iteration using the $\omega$-expression, positions of high electron density in the zeroth-order calculation will be less able to attract electrons. Conversely, positions with $q_i < 1.0$ in the zeroth-order calculation will have their Coulomb integrals modified to make them more electron attracting, since $(n_i - q_i)$ will be positive.

Once this procedure has been carried out, it must be reiterated using the same expression

$$\alpha_i'' = \alpha_0 + \omega(n_i - q_i')\beta_0$$

except that the value of $\alpha_i$ used in the second iteration incorporates the electron density $q_i'$ recalculated in the first iteration. This is because the simple $\omega$-expression overestimates the effect of build-up of electron density, or positive charge, at the various positions. The calculations are then reiterated until successively calculated values of $q_i$ at each position, and hence successive $\alpha_i$ values cease to vary.

Convergence usually requires ten or more iterations, and the $q_i$ values oscillate significantly, particularly in the first few iterations. However, a simple method of estimating the finally converged values will be described later. Once convergence is reached, the calculated $q_i$ values at each position become fully consistent with the $\alpha_i$ values assigned to calculate them. Thus a self-consistent charge distribution is achieved.

The application of this method, which is called the $\omega$-technique, will be illustrated for a simple case, the allyl cation. We have already seen that a simple HMO calculation on this system involves the determinant

$$\begin{vmatrix} x & 1 & 0 \\ 1 & x & 1 \\ 0 & 1 & x \end{vmatrix} = 0$$

which yields the polynomial

$$P_s = x^3 - 2x = 0$$

with roots $x_i = 0, \pm\sqrt{2}$.

Substitution of the root $-\sqrt{2}$ into the secular equations yields the following coefficients for the first MO, which is the only one occupied

## V.3 Self-Consistent HMO Methods

in the allyl cation:
$$\psi_1 = 0.500\phi_1 + 0.707\phi_2 + 0.500\phi_3$$
From these coefficients the calculated electron densities are
$$q_1 = 0.500 = q_3, \qquad q_2 = 1.000$$
which gives a simple picture of the electron distribution in the cation whereby the terminal carbons each bear half the cationic charge and the central atom is neutral.

$$\begin{array}{ccc} +0.5 & 0 & +0.5 \\ CH_2 \!=\!\!=\! CH \!=\!\!=\! CH_2 \end{array}$$

Application of the $\omega$-expression now modifies the Coulomb integrals as follows:
$$\alpha_1 = \alpha_0 + 1.4(1 - 0.5)\beta_0 = \alpha_3$$
$$\alpha_2 = \alpha_0 + 1.4(1 - 1.0)\beta_0 = \alpha_0$$

Thus the modified secular determinant becomes
$$\begin{vmatrix} x + 0.7 & 1 & 0 \\ 1 & x & 1 \\ 0 & 1 & x + 0.7 \end{vmatrix} = 0$$

Solution of this determinant gives a new set of eigenvalues[†]:
$$\varepsilon_1 = \alpha + 1.81\beta$$
$$\varepsilon_2 = \alpha + 0.7\beta$$
$$\varepsilon_3 = \alpha - 1.11\beta$$

(Note that the energy of the nonbonding MO $\psi_2$ is displaced to $\alpha + 0.7\beta$, which is the lowest energy Coulomb integral introduced into the secular determinant.)

Substitution of the new eigenvalue for $\varepsilon_1$ into the modified secular equations gives a new set of coefficients for $\omega_1$. These in turn yield the new set of electron densities
$$q_1 = 0.621 = q_3, \qquad q_2 = 0.757$$

[†] The eigenvalues obtained in successive iterations by the $\omega$-technique do not usually oscillate, as do the $q_i$ values, but are generally displaced monotonically toward lower, or sometimes higher energies.

Thus after the first iteration the electron density has been increased from 0.500 to 0.621 at the terminal positions as a result of the above introduction of modified Coulomb integrals for $H_{11}$ and $H_{33}$. This increase takes place at the expense of the electron density at position 2, where the value of $H_{22}$ was not modified; hence this position was less able to attract electron density relative to positions 1 and 3.

A second iteration is now carried out, again modifying the Coulomb integrals, as follows:

$$\alpha_1 = \alpha_0 + 1.4(1 - 0.621)\beta_0 = \alpha_3$$
$$\alpha_2 = \alpha_0 + 1.4(1 - 0.757)\beta_0$$

This will result in displacement of some electron density from positions 1 and 3 back to position 2. Table V.7 shows how the electron densities

TABLE V.7

| No. of iterations | $q_1 (=q_3)$ | $q_2$ |
|---|---|---|
| 0 | 0.500 | 1.000 |
| 1 | 0.621 | 0.757 |
| 2 | 0.534 | 0.934 |
| 3 | 0.597 | 0.806 |
| 4 | 0.552 | 0.896 |
| 5 | 0.584 | 0.830 |
| 6 | 0.560 | 0.880 |
| $\infty$ | 0.571 | 0.858 |

vary with successive iterations until the final convergence values are reached. Thus the final and self-consistent charge distribution in the allyl cation XIII, according to the $\omega$-technique, is

$$\begin{array}{ccc} +0.429 & +0.142 & +0.429 \\ CH_2 === CH === CH_2 \end{array}$$

XIII

which probably represents a much more realistic picture than that obtained by the simple HMO technique.

A useful labor-saving approximation that can be used to arrive at the convergence values is as follows. If a simple HMO calculation and just three iterations are carried out, the three sets of electron densities $q_1'$, $q_i''$, and $q_i'''$ for each position can be combined to give a good

## V.3 Self-Consistent HMO Methods

estimate of the final values $q_i^\infty$ from the relation

$$q_i^\infty \cong \frac{1}{2}\left[\frac{q_i' + q_i''}{2} + \frac{q_i'' + q_i'''}{2}\right] \cong \tfrac{1}{4}(q_i' + 2q_i'' + q_i''')$$

This is shown graphically in Fig. V.6, where the means of the first and second iterated values and the means of the second and third iterations bracket the convergence values very closely. For example, in the allyl cation,

$$q_2^\infty \cong \tfrac{1}{4}(0.757 + 2(0.934) + 0.806) = 0.8577 \simeq 0.858$$

This figure also shows the typical oscillation of electron density values in an $\omega$-type calculation.

The principal use of the $\omega$-technique has been to obtain more realistic charge distributions in organic molecules than are available from the simple Hückel technique.[6,9] These have been used more successfully for predictions of both physical properties and chemical reactivities. In addition, the energy values, such as $E_\pi$, are also changed by the $\omega$-type iteration procedure. The difficulty here however is how to calculate such quantities as $\pi$-bonding energies ($B_\pi$) and the delocalization energies ($DE$). The problem is that once convergence is reached in an $\omega$-calculation, the starting Coulomb integrals are no longer equal to $\alpha_0$, nor are they in general equal to each other. Therefore, $B_\pi$ is no longer simply equal to the $\beta$ term of $E_\pi$ as it would be in a simple

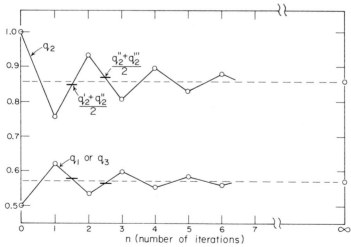

FIGURE V.6

Hückel calculation, where the bonding energy can be equated to the change in energy of the electrons between the starting $2p_z$ orbitals (all of energy $\alpha_0$) and the total energy they have in the final molecule (which is $E_\pi$). This difficulty is clearly illustrated in the case of the allyl cation, as shown diagramatically in Fig. V.7.

From the simple Hückel calculation, the initial electronic energy is based on the starting AOs, each at $\alpha$. The final energy after formation of the allyl $\pi$-system is based on the occupied MO, $\psi_1{}^0$ at $\alpha + 1.414\beta$. Thus the net gain in energy on molecule formation:

$$B_\pi = 2.828\beta$$

In the case of the $\omega$-result, two of the starting AOs were at $\alpha + 0.6\beta$ (which incidentally is the final energy of the nonbonding allyl MO in the $\omega$-calculations) and one was at $\alpha + 0.2\beta$. Since the calculation began with two electrons in AOs, then resulted in formation of a molecule with these electrons in an MO, the question is what energy to take as the starting point for $E_{AO}$. If we take it that it is reasonable to place both electrons initially in the lower AO level ($\alpha + 0.6\beta$), the bonding energy becomes

$$B_\pi = (2\alpha + 3.66\beta) - (2\alpha + 1.2\beta) = 2.46\beta$$

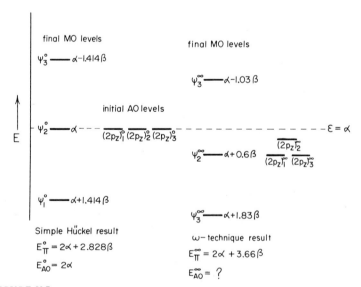

FIGURE V.7

## V.3 Self-Consistent HMO Methods

which is a significantly worse energy than that obtained in an unmodified HMO calculation. However, if we assume that we started with the electrons in AOs on positions 1 and 2 as in XIV, the bonding energy would become XIV

$$CH_2 - CH - CH_2$$

XIV

$$B_\pi = (2\alpha + 3.66\beta) - (2\alpha + 0.8\beta) = 2.86\beta$$

which is slightly better than the simple Hückel result. The appropriate choice is not clear, even in this very simple case, so that it is advisable in general not to draw conclusions based on $E_\pi$ values obtained from use of the $\omega$-technique.

The situation becomes even more difficult when $DE$ values are considered. The problem is that the standard value of $E_\pi$ for ethylene can no longer be used as a reference for a localized double bond, since the Coulomb integrals involved in calculating this value are not the same as those used in calculating the delocalized $E_\pi$ value based on the $\omega$-technique. Furthermore, it is unrealistic to use any of the particular $H_{ii}$ values involved in the final iteration of an $\omega$-calculation to arrive at a value of $E_\pi$ for a model localized system, since these $H_{ii}$ values were based entirely on a delocalized model. This would be true even if a reasonable choice could be made of which centers to include in the localized model, which, as shown previously, is not always an easy question to answer. Thus questions concerning $DE$ values are best dealt with on the basis of the simple Hückel technique without applying the $\omega$-modification.

There is an additional difficulty involved in calculating energy values using the $\omega$-technique. As pointed out by Harris,[10] the sets of coefficients calculated for each MO in the simple Hückel technique correspond to minimum energy values for each MO, since they were arrived at using the variation method. However, once the parameters are varied using the $\omega$-technique the resulting coefficients (and MOs) no longer necessarily correspond to minimum energy values for the system in question. Although the difference between the calculated $\omega$-technique energies and minimized values is not in general expected to be large,[11] it is nonetheless a source of concern in using this technique.

## V.3.2 USE OF THE $\omega$-TECHNIQUE WITH HETEROATOMS

The application of iterative $\omega$-calculations to systems containing heteroatoms is basically no different to that described above for the allyl system. The only difference is that in the basic $\omega$-equation account must be taken of intrinsic differences in the electron attracting ability of various atoms, as well as those that arise because of the nature of the $\pi$-system and the particular location of the atoms in it. The $\omega$-expression for carbon centers has been given as

$$\alpha_c = \alpha_0 + \omega(n_i - q_i)\beta_0$$

where $n_i = 1$. However, the analogous expression for a heteroatom X must be modified to take into account the different intrinsic electron attracting abilities of X and carbon. This is done in a similar way to that described for the inclusion of heteroatoms in simple HMO calculations. Thus for heteroatoms in an $\omega$-calculation

$$\alpha_x = (\alpha_0 + h_x\beta_0) + \omega(n_i - q_i)\beta_0$$

where $h_x$ is the heteroatom parameter described previously and $n_i$ can be equal to 0, 1, or 2, depending on the particular atom and structure involved (i.e., $n_i$ is equal to the number of electrons in $2p_z$ orbitals which atom X can contribute to the delocalized $\pi$-system). The first part of this expression can be thought of as arising from the intrinsic electron attracting ability of X, and the second part as arising from the particular location of X in the non-self-consistent $\pi$-system. It should be noted that the full expression for heteroatoms must be retained in every iteration of the $\omega$-treatment. For example, in a calculation involving vinylamine (XV),

$$\text{CH}_2\!=\!\text{CH}\!-\!\ddot{\text{N}}\text{H}_2$$
$$\phantom{\text{CH}_2\!=}1\phantom{\text{CH}\!-\!\ddot{\text{N}}}2\phantom{\text{H}}3$$

XV

the Coulomb integrals for carbons 1 and 2 in the simple HMO calculation (or zeroth iteration) would be set equal to $\alpha_0$ and that for nitrogen set equal to $\alpha_0 + h_N\beta_0$, with $h_N = 1.5$. However, in the first and all subsequent iterations, these terms would be changed to

$$\alpha_1 = \alpha_0 + \omega(1 - q_1)\beta_0 = \alpha_2$$
$$\alpha_3 = (\alpha_0 + 1.5\beta_0) + \omega(2 - q_3)\beta_0$$

### V.3 Self-Consistent HMO Methods

It is a common error to use the full expression for the heteroatom in the first iteration, then omit the heteroatom part of the first term in subsequent iterations. If this were done in the present case, the $\omega$-calculations would converge on the isoelectronic carbon system, namely, the allyl anion. Since the intrinsic electron attracting ability of nitrogen in this system does not change, it is incorrect to allow the Coulomb integral for the nitrogen to become

$$\alpha_3{}^n = \alpha_0 + \omega(2 - q_3{}^{n-1})\beta_0$$

in any iteration, independent of what values $q_3$ might take on.

### V.3.3 BOND ORDER–BOND INTEGRAL CONSISTENCY

To be fully self-consistent, all matrix elements in a Hückel calculation should be consistent with the final properties calculated for the system, and not just the diagonal terms $H_{ii}$. Therefore, in addition to use of the $\omega$-approach, consistency should strictly be sought between the bond integral terms $H_{ij}$ and the calculated bond orders. Originally all these integrals were either set equal to $\beta_0$, or taken to be zero for nonnearest neighbors. Thus if the nonzero bond integrals are all the same in any system, then one might logically expect that all bond orders would be uniform, as in benzene (where all $\rho_{ij} = 0.667$). This is clearly not the result in general; for example, in butadiene the 1,2 and 2,3 bond orders are 0.894 and 0.447, respectively, which is clearly not consistent with the assignment $H_{12} = H_{23} = \beta$.

Therefore, to achieve consistency between bond integrals and bond orders, the $H_{ij}$ values are made a function of the bond orders obtained in a simple HMO calculation, and an iterative procedure is followed, as in the $\omega$-technique, until convergence is obtained. Following a suggestion by Coulson and Golebiewski,[12] Boyd and Singer have used an exponential expression[13] to relate $H_{ij}$ and $\rho_{ij}$ as follows:

$$H_{ij} = \beta_{ij} = \beta_0 e^{-[k\rho_{ij} + a]}$$

where $\beta_{ij}$ is the bond integral used in the first and all successive iterations, $\beta_0$ is the standard bond integral, and $k$ and $a$ are dimensionless constants. Values of these constants that have been used for conjugated polyene systems are $k = 0.55$ and $a = -0.37$. Calculations performed on typical hydrocarbon systems show that convergence is achieved more rapidly than in the $\omega$-technique, since usually after four or five iterations there are no further significant changes in the calculated

$\rho_{ij}$ (and hence $\beta_{ij}$) values. These final $\rho_{ij}$ values have been found to show very precise correlations with experimental bond lengths for a wide variety of hydrocarbons.[12]

It should be pointed out that calculations of the above type have been carried out so far only on AH-hydrocarbons, which already have self-consistent charge distributions; thus the $\omega$-technique was not necessary. In general, to achieve fully self-consistent parametrization, both types of modification should be employed simultaneously until convergence with respect to both $\rho_{ij}$ and $q_i$ values is obtained. Little work has been done so far in this area.

## V.4 THE EXTENDED HÜCKEL (EHMO) METHOD

This method, developed principally by Hoffmann,[14] is a widely employed, and in many cases very useful extension of the simple Hückel approach. It has been mainly applied to hydrocarbon systems and is very useful for estimating energies and conformational stabilities; however, it is less useful for heteroatom systems and for estimating charge distributions. Like the simple Hückel (SHMO) approach, the EHMO treatment is an electron-free type of LCAO–MO calculation. That is to say, the electrons are not inserted into the MOs until the calculation of their energy levels is complete. Thus the electron occupancy pattern of the various MOs has no effect whatever on their energies, so that in such methods electron–electron repulsion terms are totally neglected in any explicit way (although iterative methods such as the $\omega$-technique can be applied later in an attempt to correct for the neglect of interelectronic repulsion). Other progessively more sophisticated LCAO methods do attempt to take account of such electron repulsions in either an empirical or ab initio way, but such methods are beyond the scope or intent of this book. Nonetheless, it is worth considering the EHMO method in some detail to illustrate the increase in complexity of molecular orbital calculation that occurs whenever any more sophisticated LCAO–MO treatment than the SHMO method is used.

The basic approach used in the EHMO method is as follows:

(1)   All atoms in the molecular system are taken into account.

(2)   The basis set is formed from Slater-type[15] atomic orbitals (STOs) that are hydrogenlike AOs and are defined explicitly in the treatment (see Section V.4.1).

### V.4 The Extended Hückel (EHMO) Method

(3) All orbitals on each atom that are part of its valence shell are included in the calculations, whether these AOs are occupied in the free atom or not. Thus for hydrogen the orbital used is 1s; for carbon, 2s, $2p_x$, $2p_y$, $2p_z$; for oxygen, 2s, $2p_x$, $2p_y$, $2p_z$; and so forth.

(4) The secular determinant is set up as before, but all off-diagonal elements are retained (i.e., no assumptions are made about any of these being zero). Thus each matrix element is of the form $[H_{ij} - \varepsilon S_{ij}]$, whether $i = j$ or not.

(5) Overlap integrals $S_{ij}$ are calculated between all pairs of orbitals $\phi_i$ and $\phi_j$ on all pairs of atoms, not just nearest neighbors. Thus no integrals $S_{ij}$ are assumed to be zero, although some in fact may turn out to be zero. To calculate all $S_{ij}$ values it is necessary to know the molecular geometry in terms of bond lengths and bond angles. Thus the coordinates of every atom must be specified. This can either be done by taking available data on molecular geometries from the literature, or by optimizing the final energies with respect to geometrical changes.

(6) Coulomb integrals are not calculated explicitly, but for all normally occupied atomic orbitals (or subshells) they are taken to be equal to the appropriate valence state ionization potentials (VSIP) that are available in the literature in many cases.[16] For AOs in the basis set that are normally vacant, the Coulomb integrals are taken to be equal to the appropriate electron affinity values. The assumption involved here is that the energies of electrons in isolated AOs, which correspond to the terms $H_{ii}$ in the calculations, can be approximated on a scale relative to an electron at infinity by using suitable ionization potentials and electron affinities. This is shown schematically in Fig. V.8.

(7) The bond integrals are not calculated explicitly, but are assigned numerical values based on the Wolfsberg–Helmholtz approximation,[17] where

$$H_{ij} = K \cdot \frac{[H_{ii} + H_{jj}]}{2} \cdot S_{ij}$$

This is not an unreasonable approximation, since the Coulomb integral for an interaction between two orbitals $i$ and $j$ should be related to the energies of electrons in the individual AOs, as well as to the degree of overlap between them. The best value of the constant $K$, which is used as a scaling factor, is 1.75, although the final results are not very sensitive to the precise value of $K$ used.

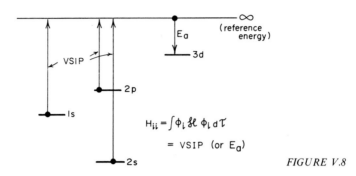

FIGURE V.8

On the above basis, the complete secular determinant is then solved to obtain the eigenvalues (in terms of eV or other energy units) and the coefficients (or eigenvectors). Solution of the secular problem requires the use of a matrix diagonalization procedure and a computer.[†]

The major differences between this method and the SHMO method are that all valence orbitals on all atoms are included, not just $2p_z$ orbitals on $sp^2$ hybridized atoms of the $\pi$-framework; all off-diagonal elements are retained and overlap is fully included; and finally energies are obtained as numerical quantities and not in terms of $\alpha$ and $\beta$ units.

The EHMO method will be illustrated for a simple molecule, in terms of the input data needed and the type of results obtained. Since ethylene is the simplest organic molecule containing a $\pi$-system it is of interest to compare the EHMO approach on this system with the SHMO treatment. It will be recalled that this is a trivial problem in the simpler Hückel treatment, since it involves the solution of only a second-order secular determinant, with various simplifying approximations. However, in the EHMO treatment a twelfth-order determinant is involved with all terms included.

## V.4.1  EHMO APPROACH TO ETHYLENE

First of all, the atoms are designated as shown (XVI) and the basis

$$\begin{array}{cc} H_2 & H_3 \\ \diagdown & \diagdown \\ C_1 \!\!-\!\! C_2 \\ \diagup & \diagup \\ H_1 & H_4 \end{array}$$

XVI

[†] Computer programs are readily available for EHMO calculations.[18]

## V.4 The Extended Hückel (EHMO) Method

set is chosen as:

one 1s orbital on each hydrogen
one 2s, $2p_x$, $2p_y$, and $2p_z$ orbital on each carbon

Thus there are 12 AOs altogether in the basis set: that is 4 (1s), 2 (2s), and 6 (2p).

As mentioned, these orbitals are taken as the hydrogenlike AOs devised by Slater[15] for use in LCAO calculations, which have the general form

$$\phi_{STO} = N \cdot \underbrace{r^{n*-1} e^{-(z-\sigma)r/n*}}_{\text{radial function}} \cdot \underbrace{\Theta_{lm} \cdot \Phi_m}_{\text{angular function}}$$

In this expression $N$ is a normalization constant. In the radial part of the function, $n*$ is the effective principal quantum number, which for first row elements is equal to the actual principal quantum number; $Z$ is the nuclear charge; $\sigma$ is a shielding constant that can be calculated from a set of rules given by Slater. The whole term $(Z - \sigma)/n*$ is referred to as the Slater exponent and designated as $\alpha$. This term $\alpha$ gives the dependence of the wave function $\phi_{STO}$ on the distance $r$ from the nucleus. The angular part of the wave function consists of the usual spherical harmonic terms that are functions of the angles $\Theta$ and $\phi$ in a spherical polar coordinate system and involve the quantum numbers $l$ and $m$.

For the AOs involved in the ethylene basis set, the appropriate expressions are

$$\phi_{1s} = \left(\frac{\alpha^3}{\pi}\right)^{1/2} e^{-\alpha r}$$

$$\phi_{2s} = \left(\frac{\alpha^5}{3\pi}\right)^{1/2} r e^{-\alpha r}$$

$$\phi_{2p_x} = \left(\frac{\alpha^5}{\pi}\right)^{1/2} r \sin\theta \cos\phi \, e^{-\alpha r}$$

$$\phi_{2p_y} = \left(\frac{\alpha^5}{\pi}\right)^{1/2} r \sin\theta \sin\phi \, e^{-\alpha r}$$

$$\phi_{2p_z} = \left(\frac{\alpha^5}{\pi}\right)^{1/2} r \cos\theta \, e^{-\alpha r}$$

Note that the s functions have no angular dependence, and that the p functions have nodal planes (i.e., $2p_x \to 0$ everywhere in the $yz$-plane since $\sin\theta = 0$ at $\theta = 0$ or $180°$ on the $z$-axis, and $\cos\phi = 0$ at $\phi = 90°$ or $270°$ on the $y$-axis). These orbitals, or functions, are very similar in appearance to the ordinary 1s, 2s, and 2p hydrogen AOs obtained by solution of the Schrödinger equation for the hydrogen atom. The major difference is that the 2s Slater orbital has no inner nodal surface[†] unlike the hydrogen 2s orbital.

Contour diagrams for these STOs are very similar to those obtained for hydrogen AOs, as shown for example for $2p_x$ STO (Fig. V.9).

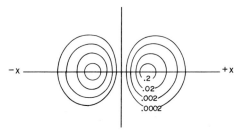

FIGURE V.9  Contour diagram for $Zp_x$ STO. Contours represent regions of constant electron density ($\psi^2$).

To consider ethylene specifically, the coordinate system is chosen so that the origin lies at the midpoint of the C—C bond, and this bond is placed along the $x$-axis, as shown in Fig. V.10. The coordinates[‡] of each atom are precisely specified according to the known geometry of ethylene. The complete basis set, in arbitrary order, is then

$\phi_1$—2s AO on $C_1$     $\phi_7$—$2p_x$ AO on $C_2$
$\phi_2$—$2p_z$ AO on $C_1$     $\phi_8$—$2p_y$ AO on $C_2$
$\phi_3$—$2p_x$ AO on $C_1$     $\phi_9$—1s AO on $H_1$
$\phi_4$—$2p_y$ AO on $C_1$     $\phi_{10}$—1s AO on $H_2$
$\phi_5$—2s AO on $C_2$     $\phi_{11}$—1s AO on $H_3$
$\phi_6$—$2p_z$ AO on $C_2$     $\phi_{12}$—1s AO on $H_4$

---

[†] This is characteristic of all STOs.

[‡] Although the STOs are normally expressed in terms of spherical polar coordinates for convenience, the coordinates of the atoms and the AO types (e.g., $2 0_x$, $2p_y$, $2p_z$) are usually designated in terms of Cartesian coordinates. Any necessary coordinate transformations can be incorporated into the EHMO program.

## V.4 The Extended Hückel (EHMO) Method

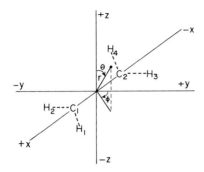

FIGURE V.10

Since the coordinates and AO functions are now specified, $S_{ij}$ values can be calculated between all pairs of orbitals. This is done by means of a subroutine incorporated into the EHMO program. Next are needed approximate energies for the $H_{ii}$ and $H_{ij}$ matrix elements. For the $H_{ii}$ values, the appropriate VSIP values are

- $-13.60$ eV for hydrogen 1s AOs
- $-21.40$ eV for carbon 2s AOs
- $-11.40$ eV for carbon 2p AOs

From these and the appropriate $S_{ij}$, the off-diagonal elements involving $H_{ij}$ can easily be calculated using the Wolfsberg–Helmholtz formula.

All the necessary input data is now available and the program calculates all required quantities, such as eigenvalue and electron occupancy distribution, total energy, MO coefficients, electron and charge densities, and overlap populations. In the computed results the MOs $\psi_i$ are expressed in the form shown in Fig. V.11. The energies obtained for the first six (doubly occupied) MOs of ethylene are (in electron volts)

$\varepsilon_1 = -26.981$ corresponding to an electron in $\psi_1$
$\varepsilon_2 = -20.605$ corresponding to an electron in $\psi_2$
$\varepsilon_3 = -16.215$ corresponding to an electron in $\psi_3$
$\varepsilon_4 = -14.448$ corresponding to an electron in $\psi_4$
$\varepsilon_5 = -13.776$ corresponding to an electron in $\psi_5$
$\varepsilon_6 = -13.217$ corresponding to an electron in $\psi_6$

and the MO coefficients (for the occupied MOs only) are given in Table V.8. These MOs are shown schematically in Fig. V.12, from

**196**                              V    *Extensions and Improvements*

$$\begin{pmatrix}\psi_1\\\psi_2\\\psi_3\\\psi_4\\\psi_5\\\psi_6\\\psi_7\\\psi_8\\\psi_9\\\psi_{10}\\\psi_{11}\\\psi_{12}\end{pmatrix} = \begin{pmatrix}C_{11} & C_{12} & C_{13} & \cdots & C_{1,12}\\C_{21} & C_{22} & C_{23} & \cdots & C_{2,12}\\ & & & & \\ & & & & \\ & & & & \\ & & & & \\ & & & & \\ & & & & \\ & & & & \\ & & & & \\ & & & & \\ C_{12,1} & \cdots & & & C_{12,12}\end{pmatrix}\begin{pmatrix}\phi_1\\\phi_2\\\phi_3\\\phi_4\\\phi_5\\\phi_6\\\phi_7\\\phi_8\\\phi_8\\\phi_{10}\\\phi_{11}\\\phi_{12}\end{pmatrix}$$

                   ortho-                  coefficient or            normalized
         normalized         eigenvector matrix        basis set
         set of MO's                                             AO's

*FIGURE V.11*

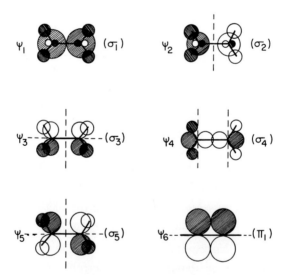

*FIGURE V.12*    MOs $\psi_1$ and $\psi_2$ are represented as projections in the xy-plane. MO $\psi_6$ is taken as a projection in the xy-plane. Nodal planes are shown as dashed lines.

TABLE V.8

| $\psi$ | $C_1(2s)$ | $C_2(2p_z)$ | $C_3(2p_x)$ | $C_4(2p_y)$ | $C_5(2s)$ | $C_6(2p_z)$ | $C_7(2p_x)$ | $C_8(2p_y)$ | $C_9(1s)$ | $C_{10}(1s)$ | $C_{11}(1s)$ | $C_{12}(1s)$ |
|---|---|---|---|---|---|---|---|---|---|---|---|---|
| $\psi_1$ | 0.4871 | 0 | 0 | −0.0240 | 0.4871 | 0 | 0 | 0.0240 | 0.0888 | 0.0888 | 0.0888 | 0.0888 |
| $\psi_2$ | 0.3860 | 0 | 0 | 0.1768 | −0.3860 | 0 | 0 | 0.1768 | 0.2396 | 0.2396 | −0.2396 | −0.2396 |
| $\psi_3$ | 0 | 0 | 0.3821 | 0 | 0 | 0 | −0.3821 | 0 | −0.2668 | 0.2668 | −0.2668 | 0.2668 |
| $\psi_4$ | −0.0741 | 0 | 0 | 0.5265 | −0.0741 | 0 | 0 | −0.5265 | 0.1904 | 0.1904 | 0.1904 | 0.1904 |
| $\psi_5$ | 0 | 0 | 0.4428 | 0 | 0 | 0 | −0.4428 | 0 | 0.3436 | −0.3436 | −0.3436 | −0.3436 |
| $\psi_6$ | 0 | 0.6275 | 0 | 0 | 0 | 0.6275 | 0 | 0 | 0 | 0 | 0 | 0 |

which it is seen that the first five MOs are $\sigma$-type orbitals since their electron distribution is centered in the $xy$-plane, and they have zero contribution from the $2p_z$ AOs that are orthogonal to this plane. These MOs can be designated as $\sigma_1$ to $\sigma_5$. The sixth MO conversely has zero contribution from the 1s, 2s, $2p_x$, and $2p_y$ AOs and is made up of equal contributions from the two $2p_z$ AOs on $C_1$ and $C_2$. Hence $\psi_6$ is a simple $\pi$-orbital, and can be designated as $\pi_1$.

A consideration of the other six (vacant) MOs would show that $\psi_7$ is an antibonding $\pi$-type MO since it has only equal and opposite contributions from the $2p_z$ AOs and zero contribution from all other AOs. This can be designated as $\pi_2^*$. The remaining MOs, $\psi_8$ to $\psi_{12}$, are all $\sigma$-type, but are antibonding and can be designated as $\sigma_6^*$ to $\sigma_{10}^*$. This ordering of MOs is illustrated schematically in Fig. V.13, showing the pattern and occupancy of all 12 ethylene MOs. Note that the final MOs are designated as bonding or antibonding (*), not according to whether their final energy is positive or negative, but according to whether or not they are more or less stable than any of the starting AO levels.

An estimate of the net bonding energy that results from the EHMO calculation on ethylene can be arrived at by comparing the total energy

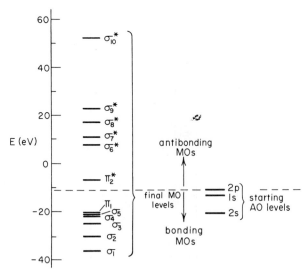

FIGURE V.13

### V.4 The Extended Hückel (EHMO) Method

of all 12 electrons occupying the six bonding MOs, with the sum of the energies assigned to these electrons in the starting AOs.[†] Thus

$$E_{bond} = E_{tot} - E_{atom}$$

$$= \sum_{j=1}^{n_{occ}} n_j \varepsilon_j - \sum_{i=1}^{n_{occ}} n_i \varepsilon_{AO_i}$$

$$= [2(-26.981) + 2(-20.605) + 2(-16.215)$$
$$+ 2(-14.448) + 2(-13.776) + 2(-13.217)]$$
$$- [4(-21.40) + 4(-11.40) + 4(-13.60)]$$

$$= -24.884 \text{ eV} \simeq -574 \text{ kcal}$$

Since a carbon–carbon double bond has a bond energy of about 142 kcal and each C—H bond has about 99 kcal, the total bond energy in ethylene is about 538 kcal.[19] It is thus interesting that the EHMO estimate, although very approximate, is of the correct order of magnitude.

However, since all the MOs in the EHMO treatment are completely delocalized, as expected in any LCAO treatment, the calculated $\sigma$-bond energies or MOs cannot be associated with any particular C—C or C—H bond in the valence bond formula. This is shown in the schematic representations of the ethylene MOs in Fig. V.12. The lowest energy MO, $\sigma_1$, is mainly a C—C $\sigma$ bonding orbital through overlap of the 2s AOs on each carbon, but also has significant C—H bonding in each region, arising through 1s–2s overlap. Since all overlaps in this MO are favorable, $\sigma_1$ is clearly a strongly bonding MO. The next MO, $\sigma_2$, is mainly a C—H bonding orbital, but is strongly antibonding in the C—C region due to unfavorable overlap of both the 2s–2s and $2p_x$–$2p_x$ type. The next orbital $\sigma_3$ is mainly a C—H bonding orbital through strong and favorable $2p_y$–1s overlap, but is also bonding in the C—C region through sideways $2p_y$–$2p_y$ overlap. The orbital $\sigma_4$ is mainly C—C bonding through end on $2p_x$–$2_x$ overlap, but also has bonding contributions in the C—H regions from $2p_x$–1s overlap. The final $\sigma$-type bonding orbital, $\sigma_5$, is a strongly bonding orbital in the C—H region due to favorable $2p_y$–1s overlap, but is also strongly antibonding in the C—C region due to unfavorable $2p_y$–$2p_y$ overlap. The only $\pi$-type bonding MO, $\pi_1$, has only favorable overlap of the $2p_z$–$2p_z$ type and corresponds to the bonding MO in the simple Hückel picture.

---

[†] It is assumed that the carbons initially have a $(2s)^2 (2p)^2$ electron distribution.

Although it is possible to make a clear distinction between σ- and π-type orbitals in the case of ethylene, this will not be possible in general unless the system possesses a plane of symmetry containing all the atoms. If such a plane is absent, the EHMO orbitals will not fall into two distinct types, where the $2p_z$ contributions to each MO are either zero or finite and thus they cannot be classified simply as σ- or π-orbitals.

In this simple case it is seen that the nodal properties of MOs can be clearly designated, and that for the σ-type MOs the number of nodes again increases with the energy. The π-type MO is of a different type and is of higher energy than $\sigma_4$ or $\sigma_5$, although it has less nodes. This is because the $2p_z$–$2p_z$ type of overlap is less effective in a bonding sense than the various overlaps involved in the σ-type MOs. This is illustrated by the following values of $S_{ij}$ for typical orbital overlaps (Fig. V.14). These figures are based on the particular geometry of ethylene, and since $S_{ij}$ varies with internuclear distance, they are not

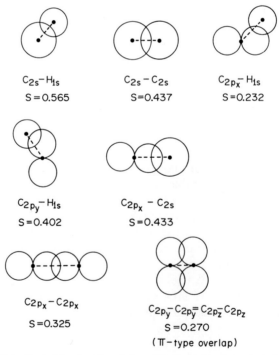

FIGURE V.14  Typical S values for ethylene. Bond axes are shown as dashed lines.

directly transferable to other molecules. However, these values of $S_{ij}$ give a good idea of the magnitudes of typical overlap integrals and indicate the degree of approximation involved in the simple Hückel neglect of overlap.

One final point concerning the EHMO results for ethylene is that the electron densities on each atom can be calculated from the coefficient matrix using a similar expression to that involved in the SHMO technique, with inclusion of overlap. These give the charge distribution for ethylene (XVII). As expected the highest electron density is in the

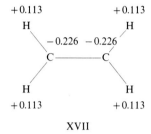

XVII

C—C double bond region, and each C—H bond is polarized in the sense

$$\overset{\delta^-\phantom{xx}\delta^+}{C\text{—}H}$$

It should be pointed out that the EHMO method is still a very naive LCAO method in comparison with such methods as the CNDO-type and the various LCAO–SCF approaches that attempt to take explicit account of electron repulsion terms. However, explanations and discussions of such advanced methods tend to obscure the descriptive and pictorial aspects of MO theory, since the mathematical treatments involved are considerably more complex than in the SHMO or EHMO approaches. It will be recognized that even with the EHMO approach the computed output (energy levels, coefficients, etc.) becomes very unwieldy as the size of the molecule, and hence the size of the basis set, increases.

## PROBLEMS

1. Set up but do not solve the simplified secular determinant for the diazacyclobutanedione system (I) using appropriate symmetry

elements. Express each of the nonzero matrix elements in terms of $\alpha$ and $\beta$ units. Depict the set of symmetry orbitals you use, indicating nodes.

I

2. Calculate the $\pi$-energy levels, $E_\pi$, the $\pi$-bonding energy, and $DE$ for the diazacyclobutadiene system (II). Obtain the coefficients for the occupied MOs and calculate the bond orders and charge densities on carbon and nitrogen. (In your answer show the energy level distribution, electron occupancy, and depict the MOs schematically.)

II

3. Set up and simplify the secular determinant for the pyrazine (IV) system using its symmetry properties (III). Solve this to obtain the

III

eigenvalues for pyrazine (IV) and 1,4-dihydropyrazine (V). Draw the energy level diagram for each molecule showing the occupancy of the MOs

IV  V

Calculate $E_\pi$ and the π-bonding energy in each case. Also calculate $DE$ using the above diagrams as the basis for your localized π-energy (you may need to calculate $E_\pi$ for the C=N system). Compare these energy values for (IV) and (V) and state whether Hückel's rule is valid for these heterosystems or not. Also predict whether you would expect each molecule to be *completely* planar.

4. Calculate $E_\pi$ and $DE$ for the π-isoelectronic systems furan (VI), pyrrole (VII), and thiophene VIII, and compare their electronic stabilities with that of the cyclopentadienyl anion (IX). To what extent is Hückel's rule valid for these $(4n + 2)$ heteromolecules?

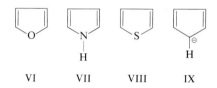

5. Carry out simple HMO calculations on the π-systems of the molecules formamide (X) and urea (XI) (use $C_2$ symmetry)

Calculate $E_\pi$, the π-bonding energy, and $DE$ value[†] for each system and compare their stabilities. [Note: The amide group has been estimated to have an internal resonance energy (nonvertical) of about 15 kcal.] Use the heteroatom parameters given.

6. Use the eigenvalues and secular equations from Problem 5 to calculate the MO coefficients for the four MOs of urea and three MOs of formamide. Depict these schematically, showing their nodal properties. Calculate the charge density on each atom and the C—O and C—N bond orders. On the basis of these values, rank the resonance (VB) contributors that can be written for each of these systems in order of importance.

7. (a) Calculate the HMO energies for the formaldehyde system (which you will need to calculate $DE$ values in Problem 5). Use these to obtain the coefficients for the bonding orbital only, and obtain $q$, $\xi$, and $\rho$ values. (Compare the electron

[†] See Problem 7.

distribution and bond order of the carbonyl group in formaldehyde with that in the amido compounds in Problems 1 and 5.)

(b) Carry out two iterations on formaldehyde using the $\omega$-technique, that is,

$$\alpha_x'' = \alpha_x + \omega(n - q_x')\beta$$

setting $\omega = 1.00$. Take the average charge densities from these two iterations as being close to the final convergence values and use the resulting charges to calculate the dipole moment[†] of formaldehyde. (Note that your calculation will give only the $\pi$-contribution to the total dipole moment, i.e., $\mu_{obs} = \mu_\pi + \mu_\sigma$. Estimated C—O bond moments are of the order of 0.9 D; therefore, take the $\mu_\sigma$ contribution from the C—O bond as being equal to this value.) Many simple carbonyl compounds have $\mu_{obs} \approx 2.85$ D in the gas phase. Compare this with the value you obtain.

8. Solve the allyl secular problem, including nearest neighbor overlap. (Use a value of $S_{12} = 0.25$.) Verify that the calculated energy levels agree with the formula given in this chapter, which relates these eigenvalues to the simple Hückel results obtained with the zero differential overlap approximation. Obtain the MO coefficients for all three MOs, based on inclusion of overlap. Compare the resulting picture of the MOs with that obtained previously (in Chapter II).

9. Borazole ($B_3N_3H_6$) (XII) is an interesting analog of an organic aromatic system, and in fact has been called "inorganic benzene." Calculate the MO energies for this system using the available heteroatom parameters. From these calculate $E_\pi$ and $DE$ and compare these values with those available for benzene itself. What can you conclude about the possible aromatic character of borazole?

XII

[†] Take 1.22 Å as being a typical C—O bond length. The electronic charge is 4.803 × $10^{-10}$ esu and 1 Debye unit = $10^{-18}$ esu-cm.

## REFERENCES

1. G. W. Wheland and L. Pauling, *J. Am. Chem. Soc.* **57**, 2086 (1935); C. A. Coulson, "Valence," p. 242. Oxford Univ. Press, London and New York, 1952.
2. A. Streitwieser, Jr., "Molecular Orbital Theory for Organic Chemists," p. 135. Wiley, New York, 1962.
3. C. Sandorfy, *Bull. Soc. Chim. Fr.* p. 615 (1949); G. W. Wheland, *J. Am. Chem. Soc.* **64**, 900 (1942).
4. L. Pauling, "The Nature of the Chemical Bond," 2nd Ed., p. 286. Cornell Univ. Press, Ithaca, New York, 1960.
5. A. Streitwieser, Jr., *J. Am. Chem. Soc.* **82**, 4123 (1960).
6. G. W. Wheland and D. E. Mann, *J. Chem. Phys.* **17**, 264 (1949).
7. N. C. Baird and M. A. Whitehead, *Can. J. Chem.* **44**, 1933 (1966).
8. N. Muller, L. W. Pickett, and R. S. Mulliken, *J. Am. Chem. Soc.* **76**, 4770 (1954); N. Muller and R. S. Mulliken, *J. Am. Chem. Soc.* **80**, 3489 (1958); A. Streitwieser, Jr. and P. M. Nair, *Tetrahedron* **5**, 149 (1959).
9. G. Berthier and A. Pullman, *C. R. Acad. Sci.* **229**, 761 (1949).
10. F. E. Harris, *J. Chem. Phys.* **48**, 4027 (1968).
11. K. Jug, *J. Chem. Phys.* **51**, 2779 (1969).
12. C. A. Coulson and A. Golebiewski, *Proc. Phys. Soc.* **78**, 1310 (1961).
13. G. V. Boyd and N. Singer, *Tetrahedron* **22**, 3383 (1966).
14. R. Hoffmann, *J. Chem. Phys.* **39**, 1397 (1963).
15. J. C. Slater, *Phys. Rev.* **36**, 57 (1930).
16. H. A. Skinner and H. O. Pritchard, *Chem. Rev.* **55**, 745 (1955); C. J. Ballhausen and H. B. Gray, "Molecular Orbital Theory," p. 122. Benjamin, New York, 1964.
17. M. Wolfsberg and L. Helmholtz, *J. Chem. Phys.* **20**, 837 (1952).
18. *Quantum Chem. Program Exchange* **3**, Program No. 30 (1967).
19. E. S. Gould, "Mechanism and Structure in Organic Chemistry," p. 36. Holt, New York, 1959.

## SUPPLEMENTARY READING

*Hoffmann, R., *J. Chem. Phys.* **39**, 1397 (1963).
*Streitwieser, A., Jr., "Molecular Orbital Theory for Organic Chemists." Wiley, New York, 1961

# VI | *THE QUANTITATIVE SIGNIFICANCE OF HMO RESULTS*

The calculation of $\pi$-electronic energies by the SHMO technique in terms of $\alpha$ and $\beta$ units is relatively straightforward. We have seen that the calculated MOs, and in particular the energy level distributions and occupancies, give a reasonable qualitative account of why formation of some systems leads to strongly bonding, or significantly delocalized situations, and in other cases why less favorable situations result. However, the questions still remain: How valid are the calculated energies in a quantitative sense and how applicable are they in a predictive or correlative sense to the properties of molecules of the size of interest to organic chemists? If we examine the approximations and assumptions inherent in the method of calculation, it is amazing that the Hückel technique should work at all, even in a rough qualitative way. However, we can attempt to answer the above questions by examining the relationship of various HMO quantities to available experimental results. Thus the two main areas we should examine are the physical properties and chemical reactivities of organic molecules containing $\pi$-electrons.

## VI.1 THE RELATIONSHIP OF HMO RESULTS TO MOLECULAR PROPERTIES

This is a very large topic, and the considerable literature on this subject has already been extensively reviewed elsewhere.[1] Therefore,

# VI.1 HMO Results and Molecular Properties

only a survey of some of the relationships will be given here to indicate the extent to which HMO results are successful in explaining or correlating various types of experimental data.

## VI.1.1 PHYSICAL PROPERTIES

In general, we would expect the various calculated values obtained from HMO calculations such as total energies, energy level spacings, electron densities, and bond orders to correlate with one or possibly more of a variety of physical properties. It can be stated at the outset that this expectation is fulfilled in a remarkably consistent way, with one notable exception, namely, the correlation of calculated and observed dipole moments.

### Dipole Moments

The contribution to total molecular dipole moment arising from the $\pi$-electron distribution only can be calculated from HMO electron density values and known molecular geometries. No information is available from HMO results on the $\sigma$-contribution, so that if the total (observed) moment is represented as

$$\mu_{tot} = \mu_\pi + \mu_\sigma$$

the $\sigma$-contribution can either be assumed to be effectively zero, as it would be for most polyene systems, or else it can be estimated from the observed moments of analogous saturated systems.

If this is done, it usually turns out that the calculated $\pi$-moment is far too high, and is generally about a factor of five greater than the observed dipole moment. For example, azulene (I) has been calculated to have a $\pi$-moment of 6.9 D in the direction shown, whereas the observed value is only 1.0 D.[2] Similarly, fulvene (II) is calculated to have a $\mu_\pi$

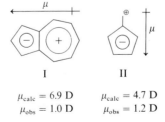

I

II

$\mu_{calc} = 6.9$ D       $\mu_{calc} = 4.7$ D
$\mu_{obs} = 1.0$ D        $\mu_{obs} = 1.2$ D

contribution of 4.7 D whereas the observed value is much lower than this. Thus, even if the reasonable assumption is made that the $\sigma$-frameworks of these two molecules contribute little to the observed moments, the $\pi$-contributions are grossly overestimated by the HMO calculations. Although application of iterative methods such as the $\omega$-technique helps the HMO method arrive at more reasonable final electron distributions, resulting in lower calculated moments, these are still much too high compared with $\mu_{obs}$. For example, the $\omega$-technique only reduces $\mu_{calc}$ for azulene to 3.8 D.

Despite this lack of quantitative agreement, HMO calculations are nonetheless quite successful in predicting dipole moments in a rough sort of way, since many hydrocarbons such as azulene are predicted to have significant moments and, in fact, such molecules do have values of $\mu$ that are very high for typical hydrocarbon systems. Similarly, other polyene systems that are predicted by HMO theory to have zero dipole moment do, in fact, have zero or negligibly small moments. In addition, HMO theory is successful in predicting the directions of these moments. Although $\mu$ is a scalar quantity experimentally, its direction can be determined by substitution experiments, and where the direction is known for polyene systems it is in good agreement with the HMO prediction. (It should be pointed out that the direction of dipole moments cannot in general be predicted by valence bond arguments.)

Better agreement between $\mu_{calc}$ and $\mu_{exp}$ has been obtained for heteroatom systems, but this is largely because more adjustable parameters are involved in the calculations, such as $h_x$ values, which have been adjusted in most cases to give a reasonable fit with experimental quantities. For example, an HMO calculation of the charge distribution in formaldehyde (III) gives the following result, using the heteroatom parameters in Table V.2.

$$\begin{array}{c} H \\ \phantom{H}\diagdown \\ \phantom{HH}C=O \\ \phantom{H}\diagup \\ H \end{array} \qquad \begin{array}{l} q_C = 0.553 \\ q_{OX} = 1.447 \end{array}$$

III

Taking the C—O bond length to be 1.22 Å, the $\pi$-moment can be calculated from

$$\mu_\pi = e \cdot d$$

where $e$ is the net charge on either atom and $d$ is the interatomic separation. This gives a value

$$(\mu_\pi)_{calc} = (0.447)(1.22 \times 10^{-8})(4.85 \times 10^{-10}) \quad \text{esu-cm}$$

where the factor $4.85 \times 10^{-10}$ converts electronic charge to electrostatic units. Thus

$$(\mu_\pi)_{\text{calc}} = 2.6 \times 10^{-18} \quad \text{esu-cm}$$
$$= 2.6 \quad \text{D}$$

Estimating the $\sigma$-bond contribution to the total moment from typical values for C—O bond moments give

$$(\mu_\sigma)_{\text{est}} \simeq 0.9 \quad \text{D}$$

Thus the total calculated moment becomes

$$(\mu)_{\text{calc}} = (\mu_\pi)_{\text{calc}} + (\mu_\sigma)_{\text{est}}$$
$$= 3.5 \quad \text{D}$$

This is quite a reasonable value since most simple carbonyl compounds have $\mu$ in the range 2.3–3.0 D and formaldehyde itself has $\mu = 2.33$ D.[3]

## Bond Order–Bond Length Correlations

Since the calculated HMO bond orders give a measure of bond strength arising from the $\pi$-electron system, it would be expected that $\rho_{rs}$ values would show some kind of inverse correlation with experimental bond lengths. Such a correlation can never be absolute since the $\sigma$-bond order is neglected in the calculations. However, if it is assumed that each $\sigma$-bond contribution to the total bond order is unity, a correlation on a relative basis might be expected when different $\pi$-systems are compared.

Such correlations have been observed, and these are remarkably good, as illustrated in Fig. VI.1.[4] In this figure $\rho_{rs}$ for typical bonds in polyenes are plotted against the experimental bond lengths. If ethane is taken as a model for a $\pi$-bond order of zero, and ethylene as a model for a $\pi$-bond order of unity, these two systems provide fixed reference points. It can be seen that most of the values of $\rho_{rs}$ for other typical bonds fall very closely on a line joining these two reasonable limiting values. Similar correlations have been found for the C—C bonds in a variety of polycyclic hydrocarbons, and also for $\rho_{CO}$ values in carbonyl compounds.[5] The remarkable precision of these relationships is illustrated by the fact that standard deviations between experimental and calculated values of bond lengths, based on empirical correlations such as in Fig. VI.1 are usually not much greater than the experimental errors involved in the bond length measurements.

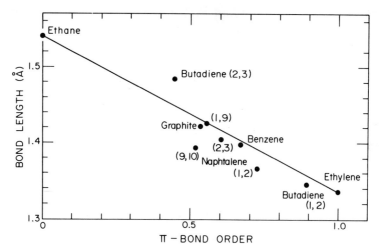

*FIGURE VI.1*

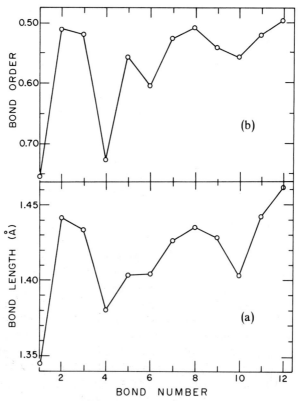

*FIGURE VI.2*

## VI.1 HMO Results and Molecular Properties

A particularly striking example of the relative agreement between HMO bond orders and bond lengths is illustrated by the case of ovalene. This hydrocarbon has 12 different C—C bonds for which the lengths are known experimentally.[6] The pattern of variation of bond lengths in ovalene IV is shown in Fig. VI.2a. The corresponding HMO bond order variations are also shown (Fig. VI.2b). The general similarity in pattern of the variations from bond to bond is quite remarkable.

Application of the technique described earlier to achieve bond integral-bond order consistency in HMO calculations, has in some cases refined the above type of correlation even further, so that deviations between calculated and observed bond lengths actually become smaller than the experimental errors in the latter values.

IV

### Infrared Spectra

Since force constants reflect the strengths of bonds, it is reasonable to expect correlations of infrared stretching frequencies with calculated bond orders. Again, since the $\sigma$-contribution to bond order is neglected in the HMO method, only relative correlations can be expected. Thus, for example, C=C and C=O $\pi$-bond orders could not be compared directly in the same correlation.

For a series of structurally diverse carbonyl compounds a good linear correspondence has been found to exist between $\nu_{CO}$ and $p_{CO}$ values.[7] Unlike the bond length correlations, the dependence of $\nu_{CO}$ on $p_{CO}$ is direct, since the higher the bond order the greater the force constant is expected to be. Similar correlations have not been reported for $\nu_{CC}$ values, probably because of the difficulty in definitely assigning bands in the C—C stretching region of the infrared spectrum.

Correlations of infrared spectral data with other HMO quantities have also been found. For example, C—H and N—H stretching frequencies cannot be correlated with calculated bond orders, since these bonds are not taken into account in the calculations. However, $\nu_{N-H}$ values have been found to correlate well with the calculated electron densities on nitrogen, and $\nu_{C-H}$ has been found to correlate with the free valence indices on carbon.[8]

*Electronic Spectra*

More interesting correlations are to be expected between electronic transition energies and HMO energies. If the HMO orbital energy level pattern is considered, it could be assumed that the gap between the highest occupied filled $\pi$-orbital (HOMO) and lowest unoccupied orbital (LUMO) should be associated with the $\pi \to \pi^*$ transition energy. This supposes that the energy level spacings will not be significantly affected by the excitation process (this crude approximation is frequently made in considering electronic transition energies), as shown schematically in Fig. VI.3. Thus $\Delta E = \varepsilon_{LUMO} - \varepsilon_{HOMO}$ is taken to be approximately equal to the transition energy $h\nu$. If this is done, surprisingly good correlations are found between $\Delta E_{calc}$ and $\Delta E_{exp}$ for various classes of absorbing systems, such as linear polyenes, $\alpha$-$\omega$-diphenylpolyenes, polyene aldehydes, and polycyclic aromatic compounds.[9] Each class of compounds is found to follow a somewhat different correlation, since it is too much to expect that $\pi \to \pi^*$-type excitations in a variety of chromophoric systems will not also depend strongly on factors extraneous to the simple energy level distribution obtained in an HMO calculation.

The most interesting aspect of such correlations is that because of the definition of the transition energy in simple Hückel terms, it is possible to obtain numerical estimates of the important quantity $\beta$. Since

$$\Delta E_{calc} = E_{LUMO} - E_{HOMO}$$
$$= (n\alpha + q\beta) - (n\alpha + m\beta)$$
$$= (q - m)\beta$$

the slope of a plot of $\Delta E_{exp}$ (in eV or other appropriate energy units) versus $\Delta E_{calc}$ (in $\beta$ units) will give a direct estimate of $\beta$ in eV units if a

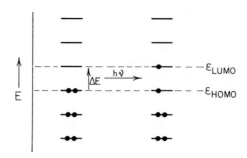

FIGURE VI.3

## VI.1 HMO Results and Molecular Properties

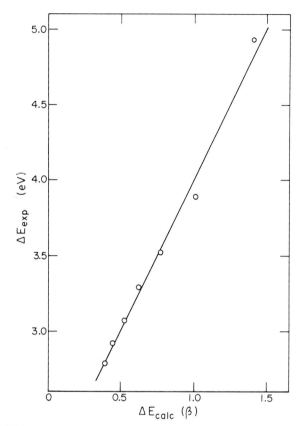

*FIGURE VI.4*

one-to-one correspondence exists between the two sets of $\Delta E$ values. An example of such a plot is given in Fig. VI.4. Since, as pointed out already, each class of compound generates its own correlation line, various estimates of $\beta$ are obtained. These will be summarized later, along with other independent estimates, but it is sufficient to say that although various uv-spectral correlations do give different values for $\beta$, they all fall in the range 2–3 eV per $\beta$ unit (45–70 kcal). This further indicates that the "thermochemical" value of $\beta = 18$ kcal based on the resonance stabilization energy of benzene is much too low.

### Charge-Transfer Spectra

Solutions of aromatic hydrocarbons and iodine are found to give intense new spectral bands (in either the uv or visible regions) that are not characteristic of either the polyene or iodine. These bands are believed to arise from complex formation in which one species (ArH) acts as an electron donor (D) and the other (I$_2$) acts as an electron acceptor (A). These complexes are called charge-transfer (CT) complexes, and their electronic absorptions are called CT-bands.

In the case of a benzene–iodine complex, this can be viewed in valence-bond terms as in (V) and the ground-state complex can be

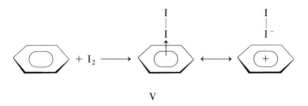

V

represented as

$$\psi_{complex} = \psi_{D \to A} + b\psi_{D^+ A^-}$$

where the weighting factor $b$ gives the amount of CT character in the complex. Spectroscopic excitation can then be viewed as taking place from a ground state with a certain value of $b$ to an excited state with a larger value $b^*$, that is,

$$\psi_{complex}{}^* = \psi_{D \to A} + b^*\psi_{D^+ A^-} \qquad (b^* > b)$$

Thus the energy involved in this excitation

$$\Delta E_{CT} = E_{complex}{}^* - E_{complex}$$

will be proportional in some way to the difference in contribution of the charge-transfer structure to the total complex. Thus in terms of simple Hückel results, it might be expected that $\Delta E_{CT}$ would correlate with $E_{HOMO}$ since it is presumably from this level that charge-transfer takes place. In fact, a good correlation has been found between experimental $\Delta E_{CT}$ values for a series of aromatic hydrocarbon–iodine complexes and calculated $E_{HOMO}$ of the hydrocarbons.[10] Again, since $E_{HOMO}$ is expressed in $\beta$ units (the $\alpha$ term is a constant factor), the slope of this correlation gives a further, independent estimate of $\beta$. This turns out to be in the same range as the estimate based on $\pi \to \pi^*$ transition energies, although a quite different property is being considered here and a different HMO quantity is being used.

## VI.1 HMO Results and Molecular Properties

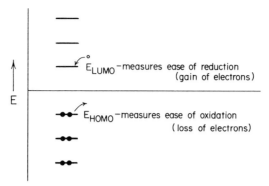

FIGURE VI.5

*Redox Potentials*

The standard reduction or oxidation potentials ($E°$) of organic molecules in solution in different solvents can be measured by means of their polarographic half-wave potentials. These can be determined using either a dropping mercury electrode (reduction) or a rotating platinum electrode (oxidation). A large number of redox potentials have been measured in this way and have been found to give remarkably good correlations with HMO energy levels. It is found that $E_{red}°$ correlates with $E_{LUMO}$ values[11] and $E_{ox}°$ correlates equally well with $E_{HOMO}$ values.[12] This is quite reasonable, as shown schematically in Fig. VI.5. What is even more remarkable about these correlations, shown in Fig. VI.6(a) and (b), is that their slopes also give estimates of $\beta$ that fall in the same range (2–3 eV) as those obtained from the electronic- and CT-spectral correlations described earlier.

*Ionization Potentials*

Molecular ionization potentials are usually determined mass-spectrometrically either by photoionization or by electron impact. These processes can be represented as

$$R \xrightarrow{h\nu} R^+ + e^- \quad \text{(photoionization)}$$
$$R + e^- \longrightarrow R^+ + 2e^- \quad \text{(electron impact)}$$

In the first case the frequency of the exciting radiation is varied until an ion current appears in the mass-spectrometric detection system. The electronic energy required to achieve the threshold of this ion

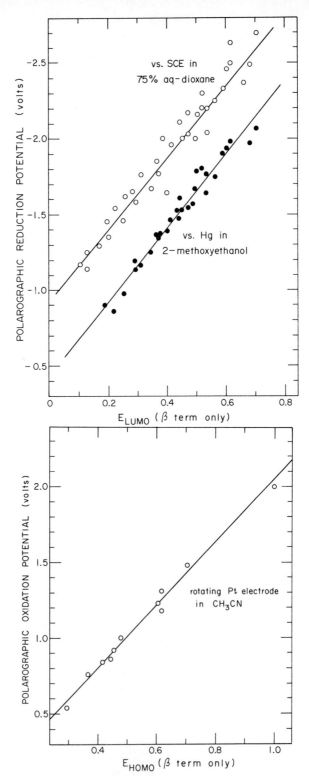

FIGURE VI.6

## VI.1 HMO Results and Molecular Properties

current is called the appearance potential and gives a measure of the molecular ionization potential ($I_p$). In the second case electrons of known energy (i.e., passed through a known potential difference) are used to bombard molecules in the ionization chamber of the mass spectrometer. Again the ion current (from $R^+$) is measured as a function of the energy of the bombarding electrons, and the appearance potential is obtained in a similar way to above. These ionization processes are reasonably assumed to involve the highest energy electrons in the organic molecule, which in the case of a polyene would be the $\pi$-electrons occupying $E_{HOMO}$. Values of $I_p$ measured in the above ways do in fact give good correlations with $E_{HOMO}$,[13] and again yield estimates of $\beta$ from slopes based on different classes of organic molecule. Once more these are in the same range as obtained from the other types of correlation.

*Summary*

There are a number of other physical properties [such as electron spin resonance (ESR) and NMR spectra] that give reasonably good correlations with HMO results, and when one considers the simplicity of the calculations and the assumptions involved, the nature of the agreement between the calculated quantities and such a wide range of physical properties is truly amazing (apart from dipole moments where the agreement is only qualitative). From the semiquantitative explanations of the properties of molecules arising from their $\pi$-electronic structure, one must conclude that the Hückel approach, although simple and approximate, is basically sound. Particularly noteworthy is the consistency of the values of $\beta$ estimated from correlations of HMO energies with the results of vastly different experimental techniques. A range of such values is listed in Table VI.1. The mean value based on the different estimates is 2.42 eV with an average deviation of 0.3 eV. Thus in more conventional units $\beta$ is estimated to have a value of about 56 kcal ($\pm 7$ kcal). This illustrates clearly that vertical resonance stabilization energies (based on *DE* values) are probably much greater in general than thermochemically estimated resonance energies.

It is perhaps not surprising that of all the types of experimental data, the correlation of HMO quantities with dipole moment shows the poorest agreement. It should be recalled that the HMO results (and in fact all LCAO–MO results) are based on application of the variation method. Since this approach optimizes the approximate electronic

TABLE VI.1

| Estimate of $\beta$ (in eV) | Experimental source | State properties involved |
| --- | --- | --- |
| 2.62 | uv-spectra | ground state→electronically excited state |
| 3.08 | uv-spectra | ground state→electronically excited state |
| 2.02 | uv-spectra | ground state→electronically excited state |
| 2.36 | uv-spectra | ground state→electronically excited state |
| 2.06 | uv-spectra | ground state→electronically excited state |
| 2.37 | reduction potentials | ground state→anion in solution |
| 2.41 | reduction potentials | ground state→anion in solution |
| 2.05 | oxidation potentials | ground state→cation in solution |
| 2.48 | ionization potentials | ground state→cation in gas phase |
| 3.14 | ionization potentials | ground state→cation in gas phase |
| 2.00 | charge-transfer bands | complex of ground state with second molecule |

wave functions with respect to their energies and not their charge distributions, it is reasonable that the calculated HMO results should show better agreement with experiment when energy quantities are considered than when only charge distributions are involved.

### VI.1.2  CHEMICAL REACTIVITIES

The problem of predicting chemical reactivities from MO theory on any quantitative basis is much more difficult to deal with than questions concerning physical properties. In the case of physical properties, it is only necessary to consider one, or at most two, states of the system at a time, and one usually has a reasonable knowledge of the structure of one or both of these states. In the case of chemical reactivity one would really need to have a complete picture of the energy profile in going from reactants to products, including any important reaction intermediates and transition states. Such a complete knowledge of the reaction profile, and particularly the nature of the transition state, is not available to us except for the most simple reacting systems, and it is necessary to use one of various approximate approaches. It is probably fair to say that at present the quantitative treatment of reaction rates is far too difficult a problem for MO calculations. However, it is interesting to consider various qualitative or semiquantitative approaches in the light of calculated HMO quantities. Considering the idealized reaction profile for a two-step reaction involving a reactive intermediate (Fig. VI.7), we could attempt to

### VI.1 HMO Results and Molecular Properties

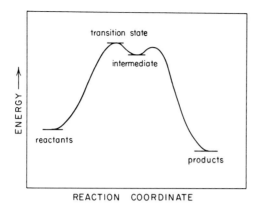

FIGURE VI.7

make predictions based on one or more of the states involved. Since the nature of the reactants is best known, the simplest approach is to consider how their properties, such as charge distributions, might affect subsequent progress along the reaction coordinate. This could be called the initial-state approach. Alternatively, if the reactions were known to be thermodynamically controlled, the relative stabilities of the products could be assessed. This could be called the final-state approach. It is unlikely that sufficient information will be known about the structure of transition states to make MO calculations possible on these states, but in certain reactions characteristic intermediates are involved whose structures are reasonably well known. Thus a reaction–intermediate approach can be used in some cases whereby the stabilities of different intermediates are compared. An alternative to the initial-state approach is to consider how the MO energies change as the system is perturbed from the reactant state during approach of a reagent. This could be called the perturbation approach. All these approaches have been used, and the following brief descriptions give an idea of the type of success achieved in predicting reactivities.

*Initial-state calculations*

In this approach, the assumption is made that an attacking reagent will have a preference for a particular site or sites in the molecule before it is close enough to perturb the molecular structure significantly by partial bond formation. Roberts has considered this approach,[14] using azulene VI as an example of a substrate that could be attacked by three types of reagent: electrophiles, nucleophiles, and free radicals. If we consider the molecular diagram in terms of the HMO quantities $q_r$, $\xi_r$, and $\mathscr{F}_r$, certain predictions can be made.

$$
\begin{array}{c}
\text{structure VI with labels:} \\
-0.17 \text{ at C8}, +0.15 \text{ at C7}, -0.03, +0.01, +0.13 \text{ at C6}, \\
-0.05 \text{ at C1}, 0.42, 0.45 \text{ at C5}, \\
0.15, 0.48, 0.48, 0.43
\end{array}
$$

VI

(i) It is reasonable that in the case of electrophilic attack a reagent $E^+$ would be expected to react most favorably with the $\pi$-system at positions of highest electron density. Clearly positions 1 and 3 would be favored in this sense over any others, because they are the only positions bearing substantial negative charge.

(ii) Conversely, in the case of nucleophilic attack, an attacking reagent $Nu$: would be expected to prefer centers in the $\pi$-system where electron density is lowest. Thus on this simple basis nucleophiles should prefer to attack at the 4,8 positions, closely followed by attack at the 6 position.

Certainly in the above two cases electrophilic reactions would be expected to occur preferentially in the five-membered ring system and nucleophilic reactions in the seven-membered ring.

(iii) In the case of free radical attack the situation is less clear, because the reagent is neutral; thus attack would not be expected to depend simply on electron density or charge. Since the free valence index gives an indication of the extent to which a given carbon is already $\pi$-bonded to adjoining atoms, positions of high free valence might be expected to be favorable reaction sites for a free radical. In this case no strong positional preference should be exhibited by azulene for free radical attack, because all positions except the bridgehead carbons have roughly equivalent values of $\mathscr{F}_r$.

From the limited evidence available so far, the above predictions are in good accord with experiment, particularly for electrophilic attack. For example, azulene is a remarkably strong base for a hydrocarbon and is protonated even in moderate concentrations of acid.[15] Protonation is known to occur exclusively at the 1 (or 3) position, in excellent agreement with the above prediction.

## VI.1 HMO Results and Molecular Properties

This kind of approach works quite well in general for aromatic substitution reactions. Using the heteroatom parameters in Table V.2, calculations on the two model systems aniline VII and benzaldehyde VIII give the following electronic charge distributions.

$$
\begin{array}{cc}
+0.18 & \text{H} \quad \text{O}\ -0.513 \\
\text{NH}_2 & \diagdown \text{C} \diagup \\
+0.062 & +0.354 \\
-0.088 & -0.023 \\
+0.003 & +0.064 \\
-0.072 & -0.002 \\
 & +0.057 \\
\end{array}
$$

Total charge on ring carbons:  $-0.180$        $+0.158$

VII                                          VIII

Thus in terms of electrophilic aromatic substitution reactions, aniline would be predicted to be more reactive than benzene since its ring carbons bear a significant net negative charge whereas benzaldehyde should be less susceptible to attack by electrophilic reagents because its ring carbons bear a significant net positive charge. In terms of positional preferences, aniline should be attacked most readily at the ortho and para positions, while benzaldehyde should react most readily at the meta position, since this is the only available position with reasonably high electron density. Thus, according to the HMO results, $-\text{NH}_2$ should be an activating, ortho–para directing substituent and $-\text{CHO}$ should be deactivating and meta directing, in excellent agreement with the well-known pattern of electrophilic substitution effects of these substituents.

This kind of approach also works well for nonbenzenoid aromatic systems such as pyridines,[16] but clearly will not work for any AH system, where, because of the uniformity of $q_r$ values, no reactivity preference would be predicted. For example, naphthalene, being an AH-system, would be predicted to be equally reactive towards electrophiles (or nucleophiles) at the $\alpha$- and $\beta$-positions. This is in clear disagreement with experiment since the $\alpha$-position is known to be much more susceptible to electrophilic substitution than the $\beta$-position.

In such cases, where all $q_r = 1.00$ (and all $\xi_r = 0$) the free valence index $\mathscr{F}_r$ has been suggested as a secondary indicator of reactivity.[17] In naphthalene $\mathscr{F}_r$ is much higher at the $\alpha$-position (IX), indicating that this carbon is significantly less strongly bonded to its neighbors in the $\pi$-system, and should therefore be more reactive to the incoming reagent.

0.453
0.106 ↑   ↗ 0.404

[structure IX: naphthalene-like bicyclic]

IX

A somewhat different initial-state approach, which has been used with success, mainly by Fukui et al.[18] is the *frontier electron approach*. It is particularly useful in cases where all $q_r$ values are the same, or negligibly different, and thus give no clear indication of preference for electrophilic or nucleophilic attack at any position. In this approach an electrophile $E^+$ is considered to be attacking the $\pi$-electrons of the highest occupied MO, which are assumed to be the most available or reactive pair in the system—or the "frontier electrons." Thus the positions in this MO with the highest electron density (largest value of $C_r^2$) are predicted to be where bonding to the vacant orbital of $E^+$ will occur preferentially. Conversely, a nucleophilic reagent $Nu:$ is assumed to be attacking the lowest unoccupied MO, since it is seeking to form a bond using its own pair of electrons. Thus the LUMO is taken to be the "frontier orbital" since it is the lowest energy vacant MO. Attack is predicted to occur preferentially where this MO has the highest value of $C_r^2$, or the position of greatest stability of an electron in this normally vacant MO. This approach is based on the criterion of maximizing overlap in the developing bond as reaction proceeds.

The frontier electron approach is illustrated (Table VI.2) for butadiene: The frontier electrons in HOMO have $C_1^2 = C_4^2 = 0.361$,

TABLE VI.2

|            | $C_1$ | $C_2$  | $C_3$  | $C_4$  |      |
|------------|-------|--------|--------|--------|------|
| $\psi_1(2)$ | 0.372 | 0.601  | 0.601  | 0.372  |      |
| $\psi_2(2)$ | 0.601 | 0.372  | −0.372 | −0.601 | HOMO |
| $\psi_3(0)$ | 0.601 | −0.372 | −0.372 | 0.601  | LUMO |
| $\psi_4(0)$ | 0.372 | −0.601 | 0.601  | −0.372 |      |

$C_2^2 = C_3^2 = 0.138$. Thus $E^+$ has a clear preference for 1,4 attach. The frontier orbital (LUMO) also has $C_1^2 = C_4^2 = 0.361$, $C_2^2 = C_3^2 = 0.138$. Therefore $Nu:$ also has a clear preference for 1,4 attack. Alternatively, positions 1 and 4 can be thought of as having the largest orbital lobes in both the HOMO and the LUMO. Hence these positions would provide most effective initial overlap with the orbitals of incoming reagents. The orbital $\psi_2$ would naturally interact in a more favorable way with a vacant orbital on an attacking reagent, such as $E^+$, whereas $\psi_3$ would interact more favorably with a filled orbital on an attacking reagent such as $Nu:$.

If we consider the case of electrophilic substitution in naphthalene (Fig. VI.8) using the frontier electron approach, then $\psi_5$ is the frontier orbital for attack by $E^+$. The MO coefficients in this orbital are as follows:

$$C_1, C_4, C_5, C_8, \quad C_2 C_3 C_6 C_7 \quad \{C_9 C_{10}$$
$$\pm 0.425 \qquad \pm 0.263 \quad 0.000$$

Thus $C_r^2$ is clearly greatest at the α-positions. Hence the frontier electron approach is in agreement with the earlier prediction, and with experiment.

The simple frontier electron approach does not allow comparisons of reactivity between different molecules, but Fukui and his co-workers have proposed an alternative method[19] related to frontier electron densities which is described in the next section.

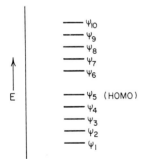

FIGURE VI.8  *Naphthalene energy level distribution.*

*Perturbation Methods*

These resemble the initial-state approaches, but are based on changes in the initial state that occur as bonding with an attacking reagent takes place. Thus instead of considering just the electron distribution at a point on the reaction profile corresponding to the initial reactant, a further point on the profile is considered where bond formation with a second reagent has occurred to such an extent that the initial-state electron distribution has been appreciably perturbed. At this point, changes in energy resulting from perturbations corresponding to different reaction paths are compared. For example, in an electrophilic substitution reaction such as nitration, the effect of an $NO_2^+$ ion attacking the $i$th position of an aromatic ring would probably be mainly to change the Coulomb integral at that position as significant bonding begins to occur. Thus one approach would be to consider the effect of changing $\alpha_i$ at various positions of attack on the total $\pi$-electronic energy of the aromatic system. Coulson and Longuet–Higgins[20] have shown that

$$\left(\frac{\partial E_\pi}{\partial \alpha_i}\right)_{\beta_{ij}} = k \cdot q_i$$

Thus if all other $\alpha_j$ remain effectively constant during attack at position $i$, and the bond integral terms also remain essentially constant, the changes in $E_\pi$ caused by attack of an electrophile such as $NO_2^+$ at various ring positions are simply proportional to the electron densities at those positions in the initial state. This approach then gives the same predictions as the initial-state approach based on $q$ values. This lends confidence to the use of $q$ values for predicting reactivity, for even when a significant departure from the initial electron distribution occurs as bonding takes place, the same answer is obtained.

If we now consider free radical attack on the same type of system, then as bonding occurs at the $i$th position little change would be expected in the Coulomb integral $\alpha_i$, because the attacking reagent is neutral. However, it might be expected that the principal effect would be to change $\beta_{ij}$ values involving position $i$ and its neighbors, as this position is on its way to becoming a saturated carbon incapable of $\pi$-bonding. Thus it might be expected that $\beta_{ij} \to 0$ as the free radical bonds more and more strongly. It has also been shown by Coulson and Longuet–Higgins[20] that

$$\left(\frac{\partial E_\pi}{\partial \beta_{ij}}\right)_{\alpha_i} = k - k' \cdot \mathscr{F}_i$$

Thus, at constant $\alpha_i$ and with all other bond integrals remaining effectively constant, $E_\pi$ changes with $\beta_{ij}$ in a way that is proportional to the free valence at position $i$; the larger the free valence at any given position, the less the energy increases.

Again this result lends confidence to the use of initial-state parameters for predicting relative reactivities.

## VI.1 HMO Results and Molecular Properties

Fukui has applied perturbation theory to a model in which an attacking reagent (electrophile, radical, or nucleophile) forms a weak $\pi$-bond to an atom $r$ of the $\pi$-system as the transition state is approached. The remainder of the $\pi$-system is assumed to be otherwise unchanged. The $\pi$-activation energy involved is related to a quantity referred to as the superdelocalizability index,[19] which is defined by

$$S_r = \sum_{j=i}^{N} (v_j - v) C_{jr}^2 / \lambda_j$$

where $N$ is the total number of $\pi$-orbitals in the reacting molecule, $v_j$ is the number of electrons in the $j$th MO, and $C_{jr}$ is the coefficient of the $r$th AO in the $j$th MO that is of energy $\varepsilon_j = \alpha + \lambda_j \beta$. The quantity $v$ is 0, 1, or 2 depending on whether the attacking reagent is an electrophile, radical, or nucleophile, respectively.

Where the substrate is an even electron molecule, with energy levels that can be denoted by $1, 2, \ldots, m$ for the occupied levels and $m + 1, m + 2, \ldots, N$ for the unoccupied levels the superdelocalizability index can be defined for each of the three above cases by

$$S_r(E) = 2 \sum_{j=1}^{m} \frac{C_{jr}^2}{\lambda_j}$$

$$S_r(R) = \sum_{j=1}^{m} \frac{C_{jr}^2}{\lambda_j} + \sum_{j=m+1}^{N} \frac{C_{jr}^2}{(-\lambda_j)}$$

$$S_r(Nu) = 2 \sum_{j=m+1}^{N} \frac{C_{jr}^2}{(-\lambda_j)}$$

In many compounds, especially large molecules, the terms $\lambda_m$ and $\lambda_{m+1}$ are much smaller than the other $\lambda_j$, and the magnitude of $S_r$ is frequently dominated by the term in the summation whose $j$ is $m$ or $m + 1$. This clearly relates the frontier electron density approach to the model used by Fukui et al.[19] in deriving superdelocalizabilities. This quantity has been found to correlate reasonably well with experimental reactivities for a variety of hydrocarbon systems.

Closely related to frontier electron theory is the perturbation approach of Brown.[21] This is based on a charge-transfer model of the transition state in which stabilization is achieved by transfer of electron density from an aromatic substrate to an electrophilic reagent, without otherwise modifying the $\pi$-system. Brown's perturbation treatment of this model (for electron transfer from the HOMO only) leads to a reactivity index $Z_r$ that is defined by

$$Z_r = Y_e - \lambda_m + \frac{2g_e^\ddagger C_{mr}^2}{Y_e - \lambda_m}$$

where $Y_e$ is the HMO energy coefficient of LUMO of the electrophile and $\lambda_m$ is the corresponding term for the HOMO of the substrate, and $g_e^\ddagger$ is a measure of the extent of reaction at the transition state. By assigning reasonable values to $Y_e$ and $g_e^\ddagger$ Brown obtained good correlation with experimental data based on aromatic nitration.[22]

The resemblance between the functional form of $Z_r$ and $S_r(E)$ is apparent, and in fact these two reactivity indices correlate reasonably well against each other.

## Intermediate Stabilities

In cases where the nature of the reaction intermediate is well established, comparisons of their stabilities for different paths of the same reaction or of different reactions can sometimes be made based on MO calculations. The most interesting and applicable of these is the localization energy approach to aromatic substitution.[23] This is based on calculations involving the well-known Wheland intermediate in electrophilic substitution, or the corresponding intermediate in nucleophilic substitution. The approach used is to consider the $\pi$-electronic energy of the pentadienate cation that is formed when an electrophile bonds to an aromatic substrate, and to compare this with the energy of the starting aromatic molecule, as in the simple case of

$$E_\pi = 6\alpha + 8\beta \qquad E_\pi^+ = 4\alpha + 5.46\beta$$

$$\text{X} \qquad\qquad \text{XI}$$

benzene. The question is: How much $\pi$-bonding energy or delocalization energy is lost in going from the fully delocalized aromatic substrate X to the partly localized pentadienate cation XI? Since the two systems contain different numbers of $\pi$-electrons, they are not strictly comparable†; but if it is assumed that the carbon at which reaction occurs

---

† This difficulty could be avoided by comparing $DE$ values for reactant and intermediate, since both are expressed solely in $\beta$ units.

## VI.1 HMO Results and Molecular Properties

still has the same value of the Coulomb integral as before, the two electrons formerly part of the $\pi$-system can be assigned an energy of $2\alpha$, although they are now involved in the $\sigma$-bond to $E^+$. This is not really a very sound assumption, but if the same approximation is made for all substrates being compared, it should not be very serious. Thus the quantity $L_r^+$ or *localization energy*† is defined as

$$L_r^+ = E_{\text{benzene}} - E_{\text{pentadienate}}^+$$
$$= (6\alpha + 8\beta) - [(4\alpha + 5.46\beta) + (2\alpha)]$$
$$= 2.54\beta$$

This is taken to be equal to the loss in $\pi$-stabilization in forming the Wheland intermediate, and should provide a relative measure of the reactivity of benzene compared to other similar systems. This type of definition of $L_r^+$ is used in general for other substrates:

$$L_r^+ = (E_\pi)_{\text{substr}} - (E_\pi)_{\text{intermed}} - 2\alpha$$

If we next consider aniline as a substrate, we can not only compare this with benzene using $L_r^+$ values, but also compare positional reactivities in aniline itself. Aniline (XII) could react at the ortho, meta, or para positions to give three different intermediates (XIII–XV). The $E_\pi$

† A similar quantity can be defined for anionic reaction intermediates.

values of each of these can easily be calculated, and hence their $L_r^+$ values can be calculated by comparing them with $E_\pi$ for aniline. This yields

$$L_r^+(\text{ortho}) = 2.371\beta$$
$$L_r^+(\text{meta}) = 2.541\beta$$
$$L_r^+(\text{para}) = 2.401\beta$$

Therefore the loss of delocalization energy is least for reaction at the ortho position, closely followed by the para position, and is significantly worse for the meta position. Therefore, by the $L_r^+$ approach, aniline is predicted to react preferentially in the order $o \gtrsim p \gg m$ in reasonable agreement with experiment. (Note that this simple approach totally neglects steric effects on reaction at the ortho position.) In addition, these results for aniline can be compared with the result for benzene. Since $L_r^+$ for the most favorable position of attack on aniline is significantly less serious than for a corresponding attack on benzene ($L_r^+ = 2.536\beta$), aniline would be predicted to be much more susceptible to electrophilic substitution. This is also in good agreement with experiment.

This type of approach can be extended to other systems if the nature of the reactive intermediate is reasonably well established. The method is also particularly useful for predicting relative reactivities in AH-systems where the $q_r$ values cannot be used.

It should be pointed out that although predictions based on localization methods are frequently in agreement with those based on $q_r$ values, as illustrated above, this is not always so. The problem that can arise to cause disagreement between the two approaches is illustrated by the following reaction profiles. If two competing reaction pathways are compared, the free energy curves leading to the corresponding intermediates could either cross or not cross before reaching the transition state, as shown in Fig. VI.9b and a, respectively. If the curves do not cross (Fig. VI.9a), then the initial-state approach (or the perturbation approach) would predict that path (b) would be favored over path (a) since the curve rises less steeply for this pathway. The reactive intermediate approach would also predict the same preference since $I_b$ (and presumably the corresponding transition state) is lower in energy than $I_a$. In such a case the prediction based on $q_r$ values would agree with the prediction based on $L_r^+$ values.

However, if the curves do cross before the transition state (Fig. VI.9b) the initial-state approach would predict path (b) to be favored because

## VI.1 HMO Results and Molecular Properties

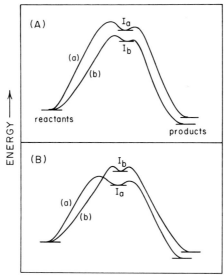

FIGURE VI.9

its curve initially rises less steeply, but the reactive intermediate approach would favor path (a) since $I_a$ is lower in energy than $I_b$. Therefore, in such cases $q_r$ and $L_r^+$ values would disagree in their predictions. It can be envisaged that such a crossing of curves might easily occur if one pathway involved an "early" transition state and the other a "late" transition state. It is advisable in general to use both approaches to predicting relative reactivity. If the two methods give consistent results, one can feel some confidence that a noncrossing system is involved and that the predictions are reliable. On the other hand, if they disagree, it is possible that a crossing-system is involved, and in the absence of other information, no safe choice between the two approaches can be made.

### Delocalization Approaches

In contrast with the above localization energy approach where the intermediate is less delocalized than the reactants, other approaches have been applied to reactions where the intermediate (or product) is more delocalized than the reactants.

Two simple examples can be used to illustrate qualitatively the idea involved. The solvolyses of allyl (XVI) and benzyl (XVII) chlorides each involve formation of intermediate cations that have $\pi$-systems which are more delocalized than those of the parent chlorides. The ionization of allyl chloride involves a change in $\pi$-energy of $0.828\beta$, and that of benzyl chloride involves a change of $0.720\beta$.

$$CH_2=CH-Cl \longrightarrow [CH_2\text{---}CH\text{---}CH_2]^+ + Cl^-$$
$$E_\pi = 2\alpha + 2\beta \qquad\qquad E_\pi = 2\alpha + 2.828\beta$$

XVI

$$\text{Ph-CH}_2\text{Cl} \longrightarrow [\text{Ph-CH}_2]^+ + Cl^-$$
$$E_\pi = 6\alpha + 8\beta \qquad\qquad E_\pi = 6\alpha + 8.720\beta^2$$

XVII

If we assume that these changes take place in addition to the other energy changes involved in breaking the $-CH_2-Cl$ bond, and that the latter are roughly comparable in both saturated and unsaturated systems, then the ionizations of the two unsaturated systems should be considerably more favorable than those of their saturated analogs. Taking $\beta$ to be about 18 kcal in resonance energy terms, then both allyl and benzyl chloride should solvolyse about $10^9-10^{10}$ times as fast as n-propyl or cyclohexylmethyl chlorides.

Although the rate differences are not known experimentally, the well-known inertness of saturated primary alkyl halides to $S_N I$ solvolysis and the known rates of solvolysis of allyl and benzyl chlorides[24] suggest that the predicted rate differences are not unreasonably high.

A more quantitative treatment is that of Streitwieser,[25] who calculated the difference in delocalization energy ($\Delta DE$) for a number of p-substituted triarythylmethyl chlorides and their corresponding triaryl-carbonium ions. He found that these calculated values

$$Ar_3CCl \rightleftharpoons Ar_3C^+ + Cl^-$$
$$\Delta DE = DE_{(Ar_3C^+)} - DE_{(Ar_3CCl)}$$

correlated remarkably well with experimental values of $\Delta F_{ion}$ for these systems. Most points fitted on a smooth curve and showed that as $\Delta DE$ increased with substitution, $\Delta F_{ion}$ decreased proportionately. In other words, as the stability of the extended conjugated $\pi$-system of the carbonium ion is increased, the extent of ionization of the chloride is also increased.

## VI.1 HMO Results and Molecular Properties

Equally interesting was the observation that for several *m*-substituted triarylmethyl chlorides, $DE$ of the carbonium ion is lower than the value for the parent chloride, and in these cases $\Delta F_{\text{ion}}$ is actually higher than that for the unsubstituted triphenylmethyl chloride. As pointed out by Streitwieser, this decrease in stabilization is reasonable in terms of the valence bond structures for the *m*-substituted chlorides and carbonium ions.

### Product Stabilities

Predictions of relative reactivity based on the stability of the final product have been made in a few cases. These predictions will only be valid in cases where the rate profiles of the reactions being compared do not cross (as illustrated in Fig. VI.9a).[26]

In the Diels–Alder reactions of aromatic compounds acting as dienes, such as the addition of maleic anhydride to benzene, two of the $\pi$-electrons of the aromatic system can be envisaged as becoming

$E_\pi = 6\alpha + 8\beta$ $\qquad$ $E_\pi = 4\alpha + 4\beta$

XVIII

more and more localized as the new $\sigma$-bonds are being formed. Although the extent of this localization at the transition state cannot be calculated, if it can be assumed that this type of process continues smoothly to give the product molecule XVIII without crossing of the energy curves for reactions of different aromatic compounds, or for reactions at different positions of the same compound, the relative $\pi$-stabilization energy of the resulting adduct may give information as to which reactions are preferred.

In the simple case of benzene, the reactant diene has a $\pi$-energy of $6\alpha + 8\beta$ and the adduct has a $\pi$-energy of $4\alpha + 4\beta$. Therefore, the change in $\pi$-energy due to adduct formation is $2\alpha + 4\beta$. This loss of $\pi$-stabilization can then be compared with that resulting in other Diels–Alder reactions. Assuming that the other bonding energy terms associated with the formation of the $\sigma$-bonds are about the same for a given dienophile, an aromatic that gives a product $\pi$-system with a lower loss in $\pi$-energy should be more reactive than benzene.

If we consider anthracene, for example, values of $\Delta E_\pi$ can easily be calculated for reaction at the 1,4 and 9,10 positions. (Fig. VI.10) Thus 9,10 addition with $\Delta E_\pi = 2\alpha + 3.32\beta$ is predicted to be favored over 1,4 addition with $\Delta E_\pi = 2\alpha + 3.64\beta$, in agreement with experiment.[27]

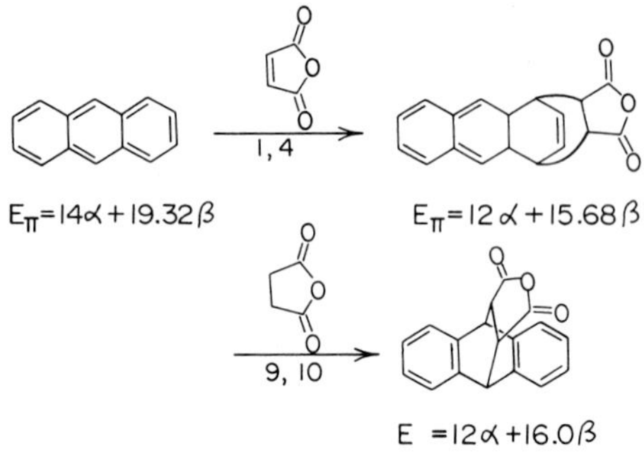

$E_\pi = 14\alpha + 19.32\beta$

$E_\pi = 12\alpha + 15.68\beta$

$E = 12\alpha + 16.0\beta$

FIGURE VI.10

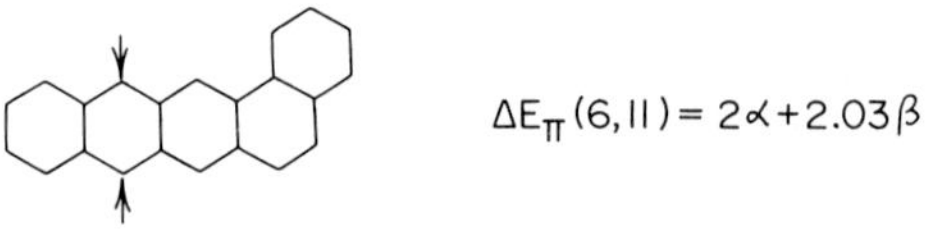

I  1,2-benzonaphthacene

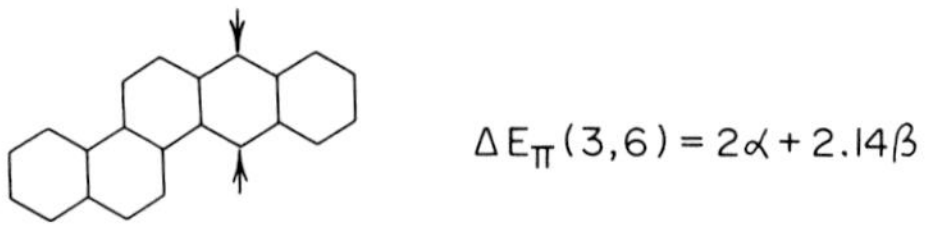

$\Delta E_\pi (6,11) = 2\alpha + 2.03\beta$

II  4,5-benzochrysene

$\Delta E_\pi (3,6) = 2\alpha + 2.14\beta$

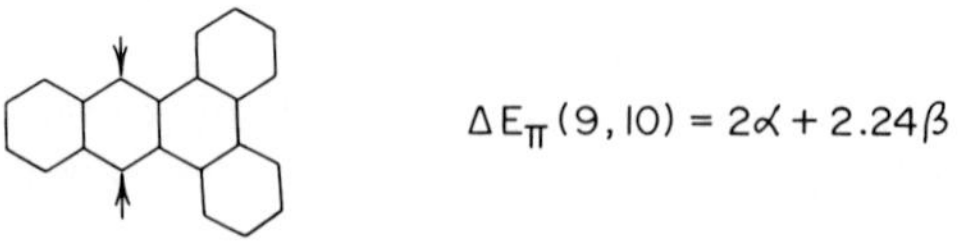

III  1,2-3,4-dibenzanthracene

$\Delta E_\pi (9,10) = 2\alpha + 2.24\beta$

FIGURE VI.11

## VI.2 Conclusion

The same kind of approach can be used to compare reactivities of different molecules. Values of $\Delta E_\pi$ can easily be calculated for the possible Diels–Alder reactions of polynuclear aromatics at different positions to decide on the favored mode of addition in each case. These preferred reactivities (indicated by arrows) can then be compared for different molecules as shown in Fig. VI.11. Thus the order of reactivity with a typical dienophile is predicted to be I > II > III, in agreement with experimental results[28] for the reactions of maleic anhydride with these three hydrocarbons. In addition, the preferred position of reaction in 1,2-benzonaphthacene is correctly predicted to be (6,11) and not (5,12) ($\Delta E_\pi = 2\alpha + 2.11\beta$).

The same kind of product stability approach has been used with more limited success for other reactions of hydrocarbons, such as osmium tetroxide oxidation and ozonolysis.[28] Basically the approach is very similar to the localization energy method, except that products rather than reaction intermediates are considered.

### VI.2 CONCLUSION

The foregoing sections have attempted to give some idea of the kind of considerations involved in predicting reactivities from numerical HMO results, and indicate the limited success achieved by the various methods. There are other equally good approaches that have been used and the discussion of the above types is meant to be illustrative rather than exhaustive. However, it should be clear that the whole problem of treating chemical reactivity in a numerical or quantitative way is much too complex and difficult for the simple HMO approach.

Nevertheless, simple LCAO–MO approaches have been extremely useful in assisting highly successful qualitative treatments of chemical reactivity. Probably the most outstanding predictive achievements of theoretical chemistry are based on application of the principle of conservation of orbital symmetry, which has leaned heavily on both simple and extended Hückel results for its theoretical basis. The Woodward–Hoffmann rules, which have been derived for concerted reactions of certain types, may give only qualitative predictions of reactivity, but they have nonetheless been remarkably successful in explaining reactivity patterns and in making real predictions (and not ex post facto rationalizations) that have subsequently been verified experimentally for a very large number of systems. This, and related topics, will be treated in the following chapters.

## PROBLEMS

1. The coefficients for the first five MOs of pentalene (I) are given in the accompanying tabulation.

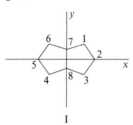

I

| $\psi_i$ | MO symmetry | $C_{1,3,4,6}$ | $C_{2,5}$ | $C_{7,8}$ |
|---|---|---|---|---|
| $\psi_1$ | $S_x S_y$ | 0.318 | 0.271 | 0.473 |
| $\psi_2$ | $S_x A_y$ | ±0.354 | ±0.500 | 0.000 |
| $\psi_3$ | $A_x S_y$ | ±0.408 | 0.000 | ±0.408 |
| $\psi_4$ | $S_x S_y$ | −0.121 | −0.513 | 0.456 |
| $\psi_5$ | $A_x A_y$ | ±0.500 | 0.000 | 0.000 |

(a) Calculate the charge (or electron) distribution in pentalene, and the bond order of the 1—2, 1—7, and 7—8 bonds. Give the significant contributing (resonance) structures for I and compare the conclusions reached on the basis of qualitative VB (resonance) theory with the above MO charges and bond orders.

(b) Using calculated electron densities and free valence indices, predict which positions in (I) would be most reactive toward (i) electrophilic reagents, (ii) nucleophilic reagents, (iii) free radicals. Determine whether these reactivities [in (i) and (ii) only] are in agreement with those predicted using the frontier electron approach. If any of these are not in agreement, what does this suggest about the relative reactivities towards certain reagents.

2. The coefficients for the MOs of fulvene (II) are given in the accompanying tabulation.

II

| $\psi$ | $C_1$ | $C_2$ | $C_3$ | $C_4$ | $C_5$ | $C_6$ |
|---|---|---|---|---|---|---|
| 1 | 0.429 | 0.385 | 0.384 | 0.429 | 0.523 | 0.247 |
| 2 | 0 | −0.500 | −0.500 | 0 | 0.500 | 0.500 |
| 3 | 0.602 | 0.372 | −0.372 | −0.602 | 0 | 0 |
| 4 | −0.351 | 0.280 | 0.280 | −0.351 | −0.190 | 0.750 |
| 5 | 0.372 | −0.602 | 0.602 | −0.372 | 0 | 0 |
| 6 | 0.439 | −0.153 | −0.153 | 0.439 | −0.664 | 0.357 |

(a) Explain the significance of these coefficients and show how they can be used to depict the electron distribution schematically (indicate nodes).

(b) Show roughly how the six energy levels of fulvene are arranged and how the electrons occupy them.

(c) Calculate the charge densities at all positions, the bond orders of the 5—6, 1—2, 2—3, and 1—5 bonds, and the free valence indices.

(d) Give the significant contributing structures for II and compare the conclusions reached on the basis of qualitative VB (resonance) theory with the above MO picture. How might the MO and VB pictures of the charge distribution be tested experimentally?

(e) Predict from the values obtained in (c) which positions in fulvene should be most susceptible to attack by electrophiles, nucleophiles, and free radicals, respectively.

(f) Compare the predictions in (e) (for nucleophilic and electrophilic attack only) with those obtained using the frontier electron approach.

(g) Explain (*but do not do any calculations*) how you could use the cation localization energy ($L_r^+$) approach to predict the preferred position of electrophilic attack.

(h) Use the electron densities to calculate an approximate dipole moment for fulvene (in Debye units) showing its direction. (Assume regular pentagonal geometry and that all bond lengths are equal at 1.4 Å.)

3. The MO coefficients for the ring carbons of acetophenone (III) are given in the accompanying tabulation.

III

|  | $X_i$ | $C_1$ | $C_2$ | $C_3$ | $C_4$ | $C_5$ | $C_6$ |
|---|---|---|---|---|---|---|---|
| $\psi_1$ | −2.150 | 0.514 | 0.386 | 0.316 | 0.294 | 0.316 | 0.386 |
| $\psi_2$ | −1.504 | −0.251 | 0.059 | 0.341 | 0.453 | 0.341 | 0.059 |
| $\psi_3$ | −1.000 | 0 | −0.500 | −0.500 | 0 | 0.500 | 0.500 |
| $\psi_4$ | −0.845 | −0.424 | −0.282 | 0.185 | 0.439 | 0.185 | −0.282 |
| $\psi_5$ | +0.503 | 0.261 | −0.329 | −0.095 | 0.377 | −0.095 | −0.329 |

(a) Calculate $q_i$ and $\xi_i$ values and predict from these the preferred positions of electrophilic and nucleophilic aromatic substitution. Also predict whether acetophenone should be more or less reactive than benzene towards each type of attack.

(b) Use the frontier electron approach to predict the preferred positions of attack, first by electrophiles, then by nucleophiles. Do these agree with the results in (a)?

(c) Use the localization energy ($L_r^+$) approach to make similar predictions of preferred mode of attack and overall reactivity towards electrophiles only. How well do these agree with the predictions made in (a)?
(See data in tabulation following Problem 5 for relevant $\pi$-energies.)

(d) Predict which should be oxidized (and reduced) more easily, acetophenone or benzene.

(e) Predict which would absorb at longer wavelength in the ultraviolet region.

4. (a) From the coefficients for the occupied MOs of heptafulvene (IV) (which are *not* listed in order of increasing energy) make predictions based on electron distribution and bond order concerning:
   (i) the most probable site of protonation in acid solution;
   (ii) the positions most susceptible to nucleophilic attack;
   (iii) the preferred sites of free radical attack.
   (b) Draw schematic representations of the occupied MOs and rank these in order of increasing energy. Explain the basis of your ranking.
   (c) Check the prediction in (a)(i) above by using the frontier electron approach.

IV

| MO | $C_1$ | $C_2$ | $C_3$ | $C_4$ | $C_5$ | $C_6$ | $C_7$ | $C_8$ |
|---|---|---|---|---|---|---|---|---|
| $\psi_i$ | 0.417 | 0.521 | 0.232 | −0.232 | −0.521 | −0.417 | 0 | 0 |
| $\psi_j$ | −0.180 | 0.221 | 0.498 | 0.498 | 0.221 | −0.180 | −0.480 | −0.333 |
| $\psi_k$ | −0.334 | −0.223 | 0.285 | 0.285 | −0.223 | −0.334 | 0.151 | 0.700 |
| $\psi_l$ | 0.391 | 0.332 | 0.305 | 0.305 | 0.332 | 0.391 | 0.484 | 0.232 |

5. Compare the $\pi$-electron stabilities of the Diels–Alder adducts of phenanthrene, naphthacene, 3,4-benzphenanthrene, and triphenylene. Predict the order of reactivity of these four hydrocarbons towards maleic anhydride.

| $\pi$-energies | |
|---|---|
| benzene, molecule | $6\alpha + 8.00\beta$ |
| pentadienyl cation | $5\alpha + 5.46$ |
| 1-acetylpentadienyl cation | $6\alpha + 7.78\beta$ |
| 2-acetylpentadienyl cation | $6\alpha + 8.05\beta$ |
| 3-acetylpentadienyl cation | $6\alpha + 7.69\beta$ |

## REFERENCES

1. A. Streitwieser, Jr., "Molecular Orbital Theory for Organic Chemists," pp. 139–306. Wiley, New York, 1962.
2. G. Wheland and D. E. Mann, *J. Chem. Phys.* **17**, 264 (1949).

3. A. L. McClellan, "Tables of Experimental Dipole Moments," pp. 41, 580. Freeman, San Francisco, California, 1963.
4. C. A. Coulson, *Proc. R. Soc., Ser. A* **169**, 413 (1939).
5. J. Trotter, "Crystal Structure Studies of Aromatic Hydrocarbons," *R. Inst. Chem., Lect. Ser.* No. 2. Roy. Inst. Chem., London, 1964.
6. A. J. Buzeman, *Proc. Phys. Soc. London, Ser. A* **63**, 827 (1950).
7. G. Berthier, B. Pullman, and J. Pontes, *J. Chim. Phys.* **49**, 367 (1952).
8. S. F. Mason, *J. Chem. Soc.* p. 3619 (1958); G. M. Badger and A. G. Moritz, *Spectrochim. Acta* 672 (1959).
9. Reference 1, Sects. 8.2 and 8.3.
10. R. Bhattacharya and S. Basu, *Trans. Faraday Soc.* **54**, 1286 (1958).
11. G. J. Hojtink, *Rec. Trav. Chim.* **74**, 1525 (1955); I. Bergman, *Trans. Faraday Soc.* **50**, 829 (1954).
12. G. J. Hojtink, *Rec. Trav. Chim.* **77**, 555 (1958).
13. A. Streitwieser, Jr. and P. M. Nair, *Tetrahedron* **5**, 149 (1949); A. Streitwieser, Jr., *J. Am. Chem. Soc.* **82**, 4123 (1960).
14. J. D. Roberts, "Molecular Orbital Calculations," p. 94. Benjamin, New York, 1962.
15. J. Colapietro and F. A. Long, *Chem. Ind. (London)* p. 1056 (1960); J. Schulze and F. A. Long, *J. Am. Chem. Soc.* **86**, 322 (1964).
16. C. Sandorfy, *Bull. Soc. Chim. France.* **16**, 615 (1949); C. Sandorfy and P. Yvan, *C. R. Acad. Sci.* **229**, 715 (1949).
17. B. Pullman and A. Pullman, *Prog. Org. Chem.* **4**, p. 41. (1958).
18. K. Fukui, T. Yonezawa, and H. Shingu, *J. Chem. Phys.* **20**, 722 (1952); K. Fukui, T. Yonezawa, C. Nagata, and H. Shingu, *J. Chem. Phys.* **22**, 1433 (1954).
19. K. Fukui, T. Yonezawa, and C. Nagata, *Bull. Chem. Soc. Jpn.* **27**, 423 (1954); K. Fukui, T. Yonezawa, and C. Nagata, *J. Chem. Phys.* **27**, 1247 (1957).
20. C. A. Coulson and H. C. Longuet-Higgins, *Proc. R. Soc., Ser. A* **191**, 39 (1947).
21. R. D. Brown, *J. Chem. Soc.* p. 2232 (1959).
22. M. J. S. Dewar, T. Mole, and E. W. T. Warford, *J. Chem. Soc.* p. 3581 (1956).
23. G. W. Wheland, *J. Am. Chem. Soc.* **64**, 900 (1942).
24. A. Streitwieser, Jr., "Solvolytic Displacement Mechanisms." McGraw-Hill, New York, 1962.
25. A. Streitwieser, Jr., *J. Am. Chem. Soc.* **74**, 5288 (1952).
26. R. D. Brown, *J. Chem. Soc.* p. 1612 (1951).
27. E. Clar and L. Lombardi, *Ber. Dtsch. Chem. Ges.* **65**, 1411 (1932).
28. R. D. Brown, *Q. Rev., Chem. Soc.* **6**, 89 (1952).

## SUPPLEMENTARY READING

*Brown, R. D. *Q. Rev., Chem. Soc.* **6** 63 (1952).
Flurry, R. L., "Molecular Orbital Theories of Bonding in Organic Molecules." Dekker, New York, 1968.
Higasi, K., Baba, H., and Rembaum, A., "Quantum Organic Chemistry." Wiley (Interscience), New York, 1965.
Zahradiuk, R., and Carsky, P., "Organic Quantum Chemistry Problems." Plenum, New York, 1973.

# VII | *THE PRINCIPLE OF CONSERVATION OF ORBITAL SYMMETRY*

This principle, which was first clearly stated and generalized by Woodward and Hoffmann,[1] is fundamental to all concerted reactions, whether they take place thermally or photochemically. By concerted reaction is meant any reaction that occurs in one smooth, continuous process from reactants to products, with formation of no discrete intermediates of significant lifetime. The principle can be stated as follows:

> Concerted reactions occur readily when there is congruence between the orbital symmetry characteristics of reactants and products, and only with great difficulty or not at all, when such congruence is absent.

This can be restated briefly by saying that orbital symmetry is conserved in concerted reactions, or that the stereochemistry of concerted reactions is controlled by orbital symmetry relationships.

Thus under a given set of conditions (either thermal or photochemical excitation) reactions that conserve orbital symmetry are designated as "allowed," and reactions that do not are said to be "forbidden." The terms "allowed" and "forbidden" are used here in the spectroscopic sense, where allowed processes are those that take place readily and rapidly under normal conditions, and are vastly preferred over forbidden processes, which either take place very slowly, with great difficulty, or do not occur at all. Hence application of the principle of conservation of orbital symmetry leads to the development of selection rules for concerted processes. These rules are now known as the Woodward–Hoffmann rules and will be described later.

This principle has been applied to a wide variety of organic reactions to predict which modes of reaction will occur most readily, and under what conditions; in every case investigated so far, its predictions are borne out by experiment. There are no clear-cut violations of the principle, where it can be strictly applied. (It should be pointed out that the selection rules cannot always be applied, since, for example, the appropriate symmetry elements may be absent from the reacting system. However, it will be shown later on that there are other approaches based on orbital symmetry properties that can be used in these cases.)

The ideas involved in the application of orbital symmetry relationships are based on quantum mechanical concepts and approaches, but do not in any way depend on *numerical* quantum mechanical calculations. However, the predictions arrived at have been supported by calculated results (of the SHMO and EHMO type as well as more advanced methods) in every case where these have been carried out.

The principle of orbital symmetry conservation has its fundamental basis not in the presence or absence of molecular symmetry in the absolute sense, but is based on the idea that those reactions will be preferred that maximize overlap in a bonding sense as the system proceeds from reactants to products, through the transition state. This can be expressed by saying that reactions in which bonding levels in the reactants have orbital symmetry characteristics that correlate best with bonding levels in the products will be most preferred. Symmetry (of the reacting orbitals, not necessarily of the molecules as a whole) is simply used as a tool to decide which levels correlate, or which orbitals can most effectively maximize developing overlap as bonds are broken and formed during reaction.

What follows is an attempt to illustrate the basic ideas and approaches used in applying orbital symmetry relationships to organic reactions. This chapter is aimed at providing a sound basic understanding of orbital symmetry conservation and the Woodward–Hoffmann rules, rather than at giving a comprehensive and detailed treatment of the very wide application of these rules. Such detailed and excellent treatments are already available.[2]

The three main types of concerted reaction that have been treated most extensively using orbital symmetry ideas are:

(i)  intramolecular cycloadditions and cycloreversions;

## VII Conservation of Orbital Symmetry

(ii) intermolecular cycloadditions and cycloreversions;
(iii) sigmatropic reactions.

The first two types are called electrocyclic reactions and are defined as follows:

(i) The formation of a single ($\sigma$) bond between the termini of a linear, conjugated $\pi$-system containing $k$ $\pi$-electrons, to give a cyclic conjugated $\pi$-system containing $(k-2)$ $\pi$-electrons, and the reverse of this process. Such processes are shown schematically by (1). Examples of this type of interconversion are given by (2) and (3).

(ii) The formation of $n$ single ($\sigma$) bonds between the termini of $n$ linear conjugated $\pi$-systems to give a cyclic system (which will be conjugated if $\pi$-electrons are still present), and the reverse of this process. Examples are given by (4), which is called a $(_\pi 2 + {}_\pi 2)$ reaction, or simply a $(2 + 2)$ reaction, also by (5), which is a $(_\pi 4 + {}_\pi 2)$ or $(4 + 2)$ reaction, and by (6), which is a $(_\pi 2 + {}_\pi 2 + {}_\pi 2)$ or $(2 + 2 + 2)$ reaction.

(6)

The third type, sigmatropic reactions, are defined as:

(iii) The uncatalyzed migrations of a σ-bond, flanked by one or more π-systems to a new position in the molecule, also flanked by one or more π-systems. Example are given by (7) and (8). These reactions

$$\begin{array}{c} H_1 \\ C_1-C_2H=C_3H-C_4H=C_5H_2 \end{array} \xrightarrow{[1,5]} \begin{array}{c} H_1 \\ C_1=C_2H-C_3H=C_4H-C_5 \\ H \end{array}$$

(7)

(8)

are named according to the relationship of the migration termini of the σ-bond to their initial termini in the molecule. Thus there are two basic types, as illustrated above. Where the migrating bond is flanked by only one π-system, the shift is a $[1,j]$ sigmatropic shift; where it is flanked by more than one, it is an $[i,j]$ shift.

These three types of reaction can be discussed from the point of view of orbital symmetry relationships on several bases. Electrocyclic reactions will be treated first, starting with simple molecular orbital diagrams, then by considering orbital energy level correlation diagrams and finally by considering state-correlation diagrams. At the same time, overall selection rules will be derived for each type of reaction.

## VII.1 SELECTION RULES FOR INTRAMOLECULAR CYCLOADDITIONS

In a reaction such as the cyclization of a butadiene derivative to form a cyclobutene, any fixed geometrical isomerism imposed on the open-chain system by the presence of substituents could be related to the

## VII.1 Selection Rules

rigid geometrical isomerism at the tetrahedral carbons in the cyclic structure, as shown by (9).

(9)

A priori, this relationship could be either one of two types. For example, *trans,trans*-2,4-hexadiene could give two possible dimethylcyclobutenes (10). Thus, in general, any electrocyclic reaction of this

(10)

type could give one or other, or both, types of product, as in (11).

(11)

Processes such as (12), in which cyclization involves twisting of the two termini in the same direction as each other are called *conrotatory*, as is the reverse process of ring opening. Thus, both are conrotatory processes.

(12)

Conversely, processes such as (13) that involve rotation of the two

(13)

termini in opposite directions are called *disrotatory*. Thus, both the reactions (13) are disrotatory processes. In practice, many reactions of this type have been carried out, either under thermal ($\Delta$) or photochemical ($hv$) conditions, and in every known case the reaction proceeds in a highly stereospecific manner. For example, the thermal isomerization of cyclobutenes (14) is a cleanly conrotatory process,[3] as illustrated

(14)

by the reaction of *cis*-3,4-dimethylcyclobutene which gives only the *cis-trans*-2,4-hexadiene as the product. (Therefore, by the principle of microscopic reversibility, the reverse process must also be conrotatory.)

In contrast with the above, the thermal cyclization of hexatrienes (15) is completely stereospecific in a disrotatory sense.[4] This reaction

(15)

is even more striking in its stereospecificity since on the basis of angle strain and steric effects alone, it would have been expected that any stereoselectivity factors would have been distinctly in favor of the conrotatory mode of ring closure. Despite these unfavorable steric factors, the reaction nonetheless proceeds thermally only in a disrotatory fashion.

It should be noted at this point that conformational changes in the open-chain partner in these reactions do not affect the nature of the process involved. For example, the *cis-trans*-2,4-hexadiene discussed in the conrotatory processes can easily undergo conformational changes as in (16) by rotations about the central bond. (It is reasonable to assume that unless the termini can approach each other closely in a particular conformation, no cyclization can occur.) The stereochemistry of the final product is unaffected by these conformational changes.

## VII.2 THE WOODWARD–HOFFMANN RULES

Woodward and Hoffmann originally suggested[5] that the course of these and other intramolecular cyclizations is determined by the symmetry characteristics of the highest occupied MO in the open-chain partner in the reaction. Thus in an open-chain system containing $4n$ $\pi$-electrons, the symmetry of the HOMO in the ground state is always such that a bonding interaction between the termini must involve developing overlap between $\pi$-orbital envelopes on opposite faces of the system. This can only be achieved in a conrotatory process, hence the thermal (ground-state) reactions of $4n$ systems are always conrotatory. In butadiene, for example, the HOMO can be represented as either I or II from which it can be seen that conrotatory twisting of the termini will always juxtapose lobes of the same sign, and hence lead to development of a bonding situation at the termini.

Conversely, in open-chain systems containing $(4n + 2)$ $\pi$-electrons, the symmetry of the HOMO is such that a terminal bonding interaction must involve overlap of orbital envelopes on the same face of the system. This can only be achieved in a disrotatory process. For example, the HOMO of hexatriene is as shown by either III or IV. Thus dis-

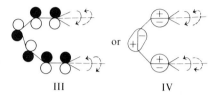

III          IV

rotatory twisting, in either of the two possible directions, will lead to favorable overlap of the terminal lobes. Hence $(4n + 2)$ systems will react thermally (i.e., from the ground state) in a disrotatory fashion. Since there are a priori two possible directions for the termini to twist in either a conrotatory or a disrotatory process, a preference could be exhibited in some cases. For example, steric effects in cis-5,6-di-t-butyl-1,3-cyclohexadiene would certainly be expected to control the thermal ring-opening so that only one of the two possible disrotatory modes (17) would be observed. In conrotatory processes it is usually not

possible to distinguish between the two available modes in this way, since they are frequently equally probable and give the same end result.

The above simple rules apply only to the thermal reactions of these systems, since we have only considered symmetry and bonding interactions involving the HOMO in the lowest electronic or ground states. Let us now consider promotion of an electron from the HOMO to the LUMO by means of uv or visible radiation, and reactions of the first excited state of the molecule. According to Woodward and Hoffmann, the orbital mainly involved in bond redistribution during the electrocyclic reaction will again be the highest occupied one. This now becomes what is normally designated as the LUMO. This leads to a complete reversal of the terminal symmetry relationships just considered for $4n$ and $(4n + 2)$ systems. This can again be illustrated using butadiene (Fig. VII.1) and hexatriene, (Fig. VII.2) indicating the signs of the appropriate MO coefficients. To achieve cyclization from the excited state, developing favorable overlap now requires interaction of orbital envelopes on the same face of the system, or a disrotatory

## VII.2 The Woodward–Hoffmann Rules

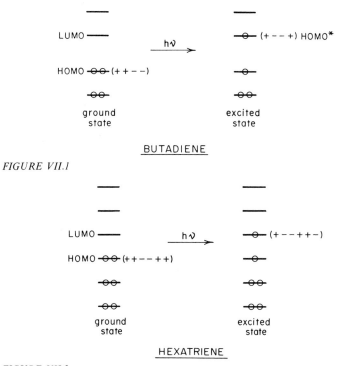

FIGURE VII.1

FIGURE VII.2

process (18). This is characteristic of all $4n$ π-electron systems, since their terminal orbital symmetry characteristics are the same as for the LUMO of butadiene.

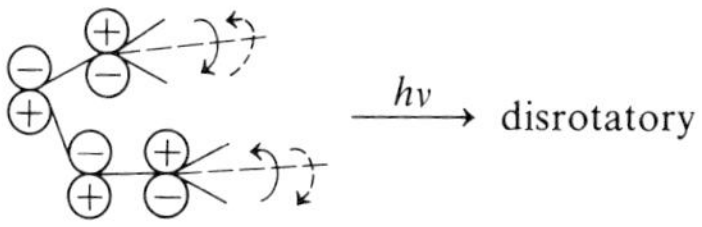

$$\text{(18)} \xrightarrow{h\nu} \text{disrotatory}$$

Cyclization of hexatriene in its excited state now requires a conrotatory process (19). This again is characteristic of all $(4n + 2)$ π-

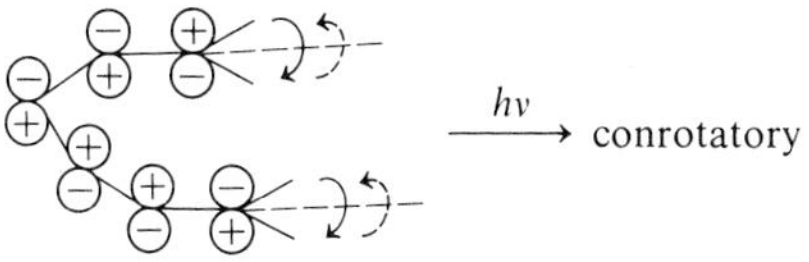

$$\text{(19)} \xrightarrow{h\nu} \text{conrotatory}$$

electron systems, since the terminal symmetry properties of their LUMOs are the same as shown for hexatriene.

Thus, any system that cyclizes or opens in a conrotatory fashion in the ground state should follow a disrotatory mode when the system is photochemically excited, and vice versa. These predictions are in complete agreement with experimental observation, and can be summarized as shown in Table VII.1.

*TABLE VII.1*

*Selection Rules for Intramolecular Cycloadditions and Cycloreversions*

| No. of $\pi$-electrons in cyclic members | Favored process | |
|---|---|---|
| | Thermally | Photochemically |
| $4n$ | | |
| $(4n - 1)$ | con | dis |
| $4n + 2$ | | |
| $(4n + 1)$ | dis | con |

Note that odd-electron systems obey the same rule as that for the even-electron system containing one more electron, since the HOMO–LUMO terminal symmetry characteristics are the same for $(4n - 1)$ $\pi$-systems as for $4n$ systems, and the same for $(4n + 1)$ as for $(4n + 2)$ $\pi$-systems. Ions obey the same rules as the neutral molecules containing the same number of electrons. The Woodward–Hoffmann rules fit all known cases, which are highly stereospecific reactions, and also allow some interesting predictions that have not yet been verified experimentally. For example, in allyl systems the cation should cyclize thermally to give the disrotatory product exclusively, as in (20), whereas

allyl cation (HOMO)

(20)

under the same conditions the closely related radical or anion should each give only the product formed in a conrotatory fashion (21). (It has recently been confirmed by Schleyer that ring opening in a cyclopropyl-X solvolysis occurs thermally in an exclusively disrotatory manner.[6])

## VII.2 The Woodward–Hoffmann Rules

(21)

It should be emphasized at this stage that the above rules only give the favored (or allowed) process, but do not exclude the possibility of operation of the alternative (or forbidden) process under highly energetic conditions. Thus the Woodward–Hoffmann rules are restrictive rather than absolute. For example, although the cyclobutene derivative shown (22) readily undergoes ring opening in a cleanly

(22)

conrotatory process, as predicted by the rules, the corresponding [3.2.0]-bicycloheptene derivative cannot open in an analogous way because of the fused five-membered ring. Such a process would result in formation of a *cis-trans*-cycloheptadiene product, which would be an extremely unfavorable process energetically. Therefore, if ring opening is to occur thermally, it must do so in a forbidden disrotatory mode (23), or not at all. In fact, ring opening does occur in the bicyclic

(23)

system, but slowly and with great difficulty in comparison with the allowed process in the monocyclic analog.[7]

Although the selection rules proposed for intermolecular cycloadditions and cycloreversions work perfectly well, the explanation of why they work in terms of simple orbital pictures of the HOMO and LUMO and their terminal symmetry relationships is incomplete, and not really satisfactory in a theoretical sense. For example, it is not clear from the above description why the properties of only one MO (either the HOMO or the LUMO) should be all that is important

**250**                                        *VII Conservation of Orbital Symmetry*

in determining the steric course of a reaction, and why the properties of the other MOs should not be taken into account as well. Similarly, it is not clear why promotion of one electron from the HOMO to the LUMO should so completely change the overall stereospecificity of the reaction.

In fact, Woodward and Hoffmann have shown that the symmetry properties of all the MOs, and indeed the total state symmetry, should be taken into account to provide a proper theoretical explanation of these and other selection rules. They have done this by developing two types of correlation diagram based on the symmetry properties of both the reactants and products in electrocyclic reactions. These are called energy level correlation diagrams and state-correlation diagrams, and it is necessary to construct such diagrams to arrive at and explain the selection rules for the more complicated cases of intermolecular cycloadditions. The simple MO picture used above is only practical for arriving at selection rules for the simpler intramolecular processes.

## VII.3  ENERGY LEVEL CORRELATION DIAGRAMS

To illustrate the procedure involved in constructing correlation diagrams, let us consider the simple case of a conrotatory cyclization of butadiene to give cyclobutene. This can be represented as in Fig. VII.3, where both reactant and product are placed in a common plane. We next identify any elements of symmetry that are common to reactant and product, and that are thus retained by the system as it moves concertedly from initial to final state. Only those elements of symmetry are considered that bisect bonds made or broken during reaction.[†]

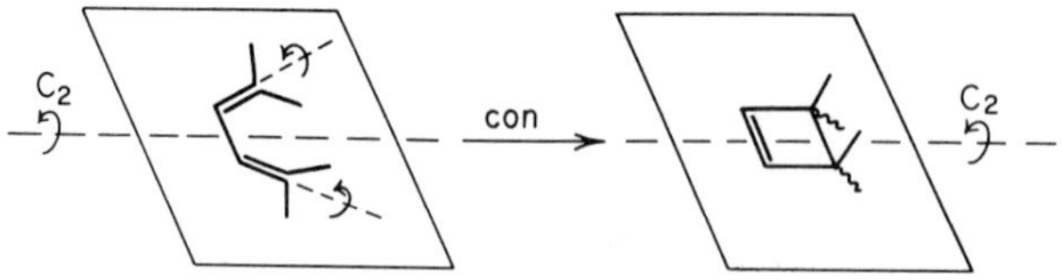

FIGURE VII.3

[†] Consideration of other symmetry elements, which do not bisect any bonds being made or broken, can only lead to the conclusion that the reaction will be symmetry allowed.

## VII.3 Energy Level Correlation Diagrams

In the above case the only such element is the $C_2$ axis passing through the center of the $C_2$—$C_3$ bond of butadiene (as shown). Then the symmetries of each MO in reactant and product with respect to this element are considered.[†] It is simple to write down the MOs for the reactant butadiene, since these are already known from the HMO calculation. These are designated as shown in Fig. VII.4. In the case of the product, it is reasonable that cyclobutene will have a $\pi$-bonding and a $\pi$-antibonding orbital, with similar symmetry properties to those of ethylene. In addition, there will be a $\sigma$-bonding orbital as a result of bond formation at the termini of the reacting diene, and a corresponding $\sigma$-antibonding orbital. These are shown in Fig. VII.4.

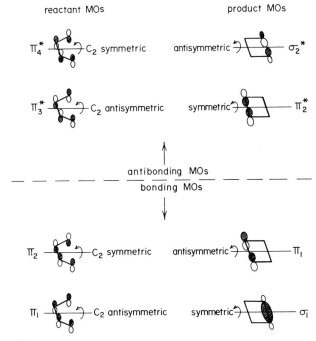

FIGURE VII.4

[†] Only the essential symmetry properties of the electronic system undergoing reaction are taken into account. The presence of substituents such as alkyl groups, which make only trivial changes in MO properties, are ignored. Thus cyclobutene and 1-methyl-cyclobutene would be considered to have equivalent symmetries for the present purposes. Also, any heteroatom orbitals in the system are replaced by isoelectronic carbon orbitals.

There are several points to note about the cyclobutene MOs. First of all the $\sigma$-orbitals are considered to be made up from overlaps (favorable or unfavorable) of some type of $sp^n$ hybridized orbitals. It is not important to specify which type of hybrid is involved, since it is only the symmetry characteristics of the resulting MOs that are of interest. It is reasonable that $\sigma$-orbitals will be ranked in energy according to their nodal properties, as are $\pi$-orbitals, and there is usually no problem in representing these $\sigma$-MOs, even in larger systems. Second, although these $\sigma$-MOs will be effectively delocalized over the whole $\sigma$-framework, it is only really necessary to specify those parts of the MOs that occupy regions where bond making and breaking is taking place. Finally, it is customary to place the $\sigma$-bonding MOs at somewhat lower energies than the $\pi$-bonding MOs, which is reasonable in terms of the known overlap properties of $\sigma$- and $\pi$-type orbitals. Correspondingly, the $\sigma$-antibonding MOs are placed at somewhat higher energies than the $\pi$-antibonding MOs. This ranking of $\sigma$ and $\pi$ (or $\sigma^*$ and $\pi^*$) levels is not essential to the treatment; all that is really important is the separation of the MO energy levels into two sets, bonding and antibonding.

The symmetry characteristics of each MO with respect to the $C_2$ axis are now examined. Each MO of whatever type will transform under the $C_2$ rotation either completely symmetrically (S) or completely antisymmetrically (A), as indicated alongside each MO diagram[†] in

---

[†] It is important to note that each MO considered in any system must be either totally symmetric or totally antisymmetric with respect to each independent symmetry element for that system. Any MO representation that does not have this property, such as V

V

for butadiene, is not an acceptable wave function. This is because the properties of the system arising from electrons occupying such an MO would be different in the $C_1$—$C_2$ region from $C_3$—$C_4$ region, which is not reasonable. Alternatively, it can be seen that the properties of such an MO would change physically on reflection in the central plane of symmetry or on rotation about the $C_2$ axis. This is not possible for any real wave function, since performance of a symmetry operation on any system can have no physical effect on that system. Note that this is not the same as the result obtained for antisymmetric MOs, since multiplication through of any wave function by $-1$ changes nothing in a physical sense.

## VII.3  Energy Level Correlation Diagrams

Fig. VII.4. The energy level diagram Fig. VII.5 for the reacting system can now be constructed showing the symmetry characteristics of each MO level. Levels of like symmetry in reactant and product that are nearest to each other in energy are now connected, and the energy level correlation diagram is complete.

Woodward and Hoffmann make the reasonable postulate that if the energy level correlation diagram contains no correlations between bonding levels in the reactant and antibonding levels in the product, the reaction will be allowed thermally. It can be seen that this is the case for the conrotatory ring closure of butadiene. The correlation diagram for this system shows that there are equal numbers of bonding MOs of each type ($S$ or $A$) in both reactants and products. Thus all the bonding levels in the reactant correlate directly with only bonding levels in the product. If electron occupancies are considered, it can be seen that a ground-state butadiene can proceed directly and smoothly to a ground-state cyclobutene with no symmetry-imposed barrier.

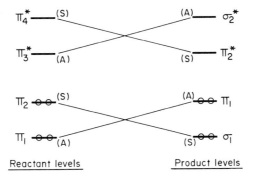

FIGURE VII.5

Using the same approach, a correlation diagram can now be constructed for the analogous disrotatory reaction Fig. VII.6. The MOs and energy levels remain the same, but now the important element of symmetry retained by the system during the reaction is different. Instead of a $C_2$ axis, we must now consider orbital symmetries with respect to the mirror plane $\sigma$ that passes through the $C_2$—$C_3$ bond of butadiene, and which is common to both systems. Examination of the MOs shows that the symmetries are now as designated in Fig. VII.7. Construction of the correlation diagram shows that bonding

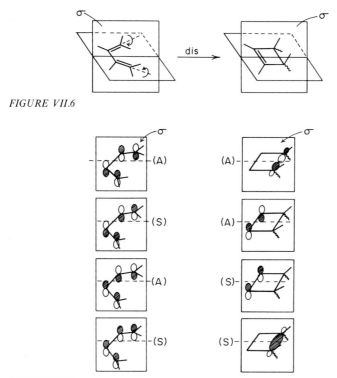

FIGURE VII.6

FIGURE VII.7

levels in the reactant do not correlate well with bonding levels in the product (Fig. VII.8) (This is clear from the fact that there are different numbers of S-type bonding orbitals on each side of the diagram.) The $\pi_2$ level in the reactant correlates with $\pi_2^*$ in the product. Thus the reaction is symmetry forbidden thermally. Considering the electron occupancies of the MOs it is clear that a ground-state butadiene cannot easily proceed directly to a ground-state cyclobutene if orbital symmetry is to be conserved, since this would entail initial formation of a doubly excited cyclobutene. Thus for the thermal reaction (g.s. to g.s.) there would be a large symmetry-imposed barrier that the reacting system would have to overcome because of the intended correlation between bonding and antibonding levels.

If we now consider the same two processes (conrotatory and disrotatory) in a photochemical sense, we can show the correlation diagrams (Fig. VII.9) as before and simply change the electron occu-

## VII.3 Energy Level Correlation Diagrams

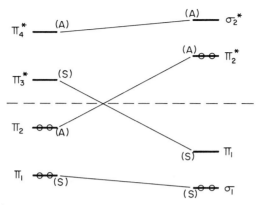

FIGURE VII.8

pancies by exciting an electron from $\pi_2$ in butadiene to $\pi_2^*$. The thermal diagrams are shown for comparison.

It can be seen that whereas the thermal conrotatory process can take place easily from ground state to ground state, the analogous

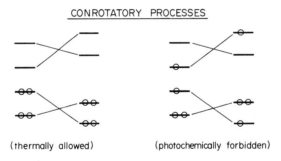

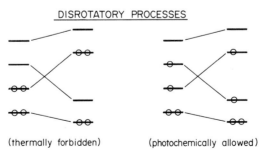

FIGURE VII.9

photochemical process will be much more difficult since the first excited state of butadiene (a $\pi \leftarrow \pi^*$ state) is trying to correlate with a much higher energy and quite different type of excited state (a $\sigma \leftarrow \sigma^*$ state) of cyclobutene, in order to conserve orbital symmetry. Thus the photochemical process is forbidden.

For the disrotatory modes, we have seen that the ground-state to ground-state process has a symmetry-imposed barrier, and that the thermal process is therefore forbidden. However, in a photochemical reaction the first excited state of butadiene ($\pi \leftarrow \pi^*$ type) can now proceed directly to the first excited state of cyclobutene (also a $\pi \leftarrow \pi^*$ type). Hence the reaction is photochemically allowed.

Based on these and similar considerations of other systems, we can make the general statement:

> If the energy level diagram contains no correlations between bonding levels in the reactant and antibonding levels in the product, the reaction will be permitted thermally, but will be very difficult photochemically. If the diagram does contain such correlations, the reaction will be very difficult thermally, but will be permitted photochemically.

It is worthwhile considering briefly the case of a typical $(4n + 2)$ system, hexatriene, and constructing the analogous correlation diagrams for its conrotatory and disrotatory modes of ring closure. The reason for this further treatment is to illustrate that even for such simple cases, it is not always evident why a certain reaction that is allowed thermally should necessarily be forbidden photochemically, or vice versa. However, these conclusions can be justified in general by proceeding one stage further and constructing state-correlation diagrams from the simpler energy level diagrams. These state diagrams are based on the idea that in concerted reactions, each state can be assigned a total symmetry and that only states of like symmetry can easily interconvert during a concerted chemical reaction.

## VII.4 STATE-CORRELATION DIAGRAMS

In the case of the reaction of hexatriene to form cyclohexadiene, we begin by constructing energy level diagrams for the conrotatory and disrotatory modes (Fig. VII.10). The symmetry characteristics of the $\pi$-MOs of hexatriene are easy to assign, since the properties

## VII.4 State-Correlation Diagrams

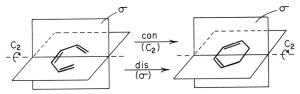

FIGURE VII.10

of these MOs are known from HMO theory. For the product cyclohexadiene, it is reasonable that there will again be one $\sigma$-bonding MO and one $\sigma^*$, antibonding, MO and that these will have similar nodal characteristics in the region of formation of the new $\sigma$-bond to the $\sigma$-MOs of cyclobutene. In addition, the diene system of the product will have two $\pi$-bonding MOs and two $\pi^*$-antibonding MOs similar to those in the open-chain analog, butadiene. Thus the MOs of the reactant and product will be as depicted in Fig. VII.11, and their symmetries with respect to the elements $C_2$ (retained during the conrotatory process) and $\sigma$ (retained during the disrotatory process) will be as shown. From these orbital symmetries, the energy level correlation diagram can be constructed as before (Fig. VII.12) by connecting the nearest levels of like symmetry in reactants and products.

From these diagrams it is clear that the conrotatory mode will be forbidden thermally, and it is not unreasonable that the photochemical process should be much easier energetically. It is also clear that there is no symmetry-imposed barrier to the thermal disrotatory process, and that this should be an allowed process. However, it is not so clear why the photochemical reaction should be forbidden in a disrotatory fashion. This would involve interconversion of a $\pi \leftarrow \pi^*$ excited state of hexatriene and a higher, but still $\pi \leftarrow \pi^*$, excited state of cyclohexadiene. Therefore, it does not seem reasonable at first sight why this photochemical process should be unfavorable energetically, at least in the reverse sense of ring opening.

To provide an answer to this and related questions, we consider the total electronic symmetry of various possible states (of hexatriene or cyclohexadiene) with respect to each stereochemical process, and how these states correlate between reactant and product stages. The states to be considered are the ground state of each molecule, various reasonably accessible singly excited states, and for the sake of completion one doubly excited state, although it is unlikely that such highly energetic states would be involved in most organic photoreactions.

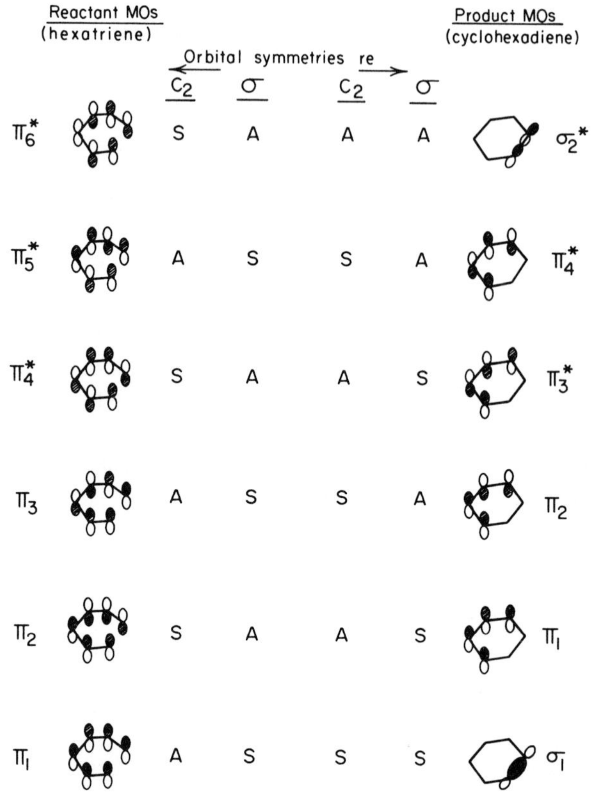

258                                                    VII  *Conservation of Orbital Symmetry*

|   | Reactant MOs (hexatriene) | | Orbital symmetries | | | | Product MOs (cyclohexadiene) | |
|---|---|---|---|---|---|---|---|---|
|   |   | $C_2$ | $\sigma$ | $C_2$ | $\sigma$ |   |   |   |
| $\pi_6^*$ |   | S | A | A | A |   | $\sigma_2^*$ |
| $\pi_5^*$ |   | A | S | S | A |   | $\pi_4^*$ |
| $\pi_4^*$ |   | S | A | A | S |   | $\pi_3^*$ |
| $\pi_3$ |   | A | S | S | A |   | $\pi_2$ |
| $\pi_2$ |   | S | A | A | S |   | $\pi_1$ |
| $\pi_1$ |   | A | S | S | S |   | $\sigma_1$ |

FIGURE VII.11

For hexatriene, reacting in a conrotatory mode, the energy levels and orbital symmetries are as shown in Fig. VII.13. The total symmetries of each of these states can be assigned by taking account of the product of the individual orbital symmetries and their electron occupancies. Taking $(S) \times (S) = (S)$, $(S) \times (A) = (A)$, and $(A) \times (A) = (S)$, we have for the ground-state configuration $(A)^2(S)^2(A)^2 = (S)$. Therefore, the ground state can be classified as a totally symmetric state with respect to the conrotatory process. Similarly, the first excited state (1) has an electronic configuration $(A)^2(S)^2(A)^1(S)^1 = (A)$, which makes it an antisymmetric state, and so forth, for the other excited states, as shown in Table VII.2.

Similar consideration of the energy levels and orbital symmetries for the product cyclohexadiene is given in Fig. VII.14.

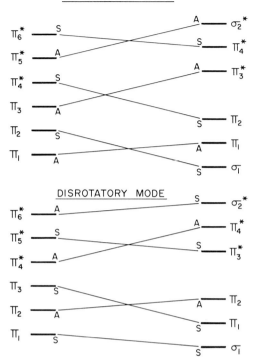

*FIGURE VII.12*

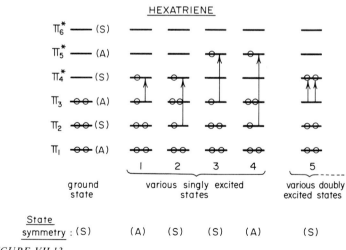

*FIGURE VII.13*

TABLE VII.2.

State Symmetries

| g.s. | $A^2S^2A^2 = S$ | g.s. | $S^2A^2S^2 = S$ |
|---|---|---|---|
| (1) | $A^2S^2AS = A$ | (1) | $S^2A^2SA = A$ |
| (2) | $A^2SA^2S = S$ | (2) | $S^2AS^2A = S$ |
| (3) | $A^2S^2AA = S$ | (3) | $S^2A^2SS = S$ |
| (4) | $A^2SA^2A = A$ | (4) | $S^2AS^2S = A$ |
| (5) | $A^2S^2S^2 = S$ | (5) | $S^2A^2A^2 = S$ |

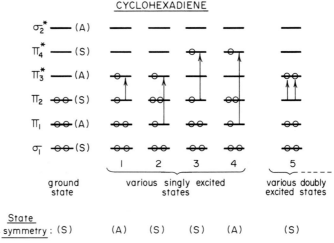

FIGURE VII.14

A state-correlation diagram (Fig. VII.15) can now be constructed by ranking the various states roughly according to energy, and considering which reactant states correlate most directly with product states of like symmetry. However, two things must be considered before connecting any of these states. One is that not only must the two states possess the same total symmetry, but the various electronic components must also correlate with respect to their individual symmetries. Thus for example although two states such as $(S)^2(A)^2(S)(A)$ and $(S)^2(A)(S)^2(S)$ are both totally $(A)$ or antisymmetric, they cannot correlate directly with each other and still conserve orbital symmetry since the individual terms are different. However, two states such as $(S)^2(A)^2(S)(A)$ and $(S^2)(A)(S)(A)^2$ could correlate directly since their

## VII.4 State-Correlation Diagrams

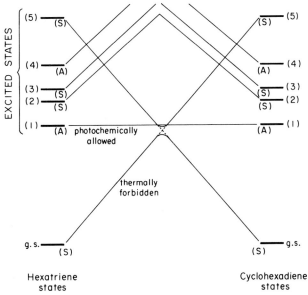

FIGURE VII.15 *State correlation diagram for the conrotatory interconversions of hexatriene and cyclohexadiene.*

components, although not the order of the individual levels, are the same with respect to their symmetry characteristics.

The second point is that although two states of appropriate symmetry do correlate directly, such intended conversions may not actually take place because of the operation of the quantum-mechanical noncrossing rule.[8] This rule states that energy curves of states of like symmetry never cross, so that if a line showing interconversion between state $R_1$ of the reactants and $P_1$ of the products crosses a line joining $R_2$ and $P_2$, where all four states have like symmetry, an interaction will take place at the crossing point, leading to a splitting of the energy levels in such a way that the lowest two reactant and product states will interconvert directly (as will the highest two). This has the consequence, illustrated in the state diagram (Fig. VII.15 and Table VII.2) that an intended crossing may lead to a symmetry-imposed barrier to the process in question. (The two ground states in the diagram have been placed arbitrarily at the same energy level, as have correspondingly excited states, for convenience of representation. The relative placings of the two systems are not essential to the argument.)

From the state diagram in Fig. VII.15 it can be seen that although both ground states are $(S)$ they cannot correlate directly. The symmetric ground-state configuration of hexatriene is $(A)^2(S)^2(A)^2$, and this will try to correlate with the doubly excited state (5) of cyclohexadiene. A similar correlation is intended between the cyclohexadiene ground state and doubly excited state (5) of hexatriene. Because of the noncrossing rule, the curves will split as shown, so that g.s. → g.s. interconversion can occur directly. However, this reaction will be very difficult energetically because of the large symmetry-imposed barrier, as shown in the diagram. Thus the thermal process is forbidden. However, no such barrier exists for the interconversion of the first excited states, $[(1) \to (1)]$ each of which is $(A)$ and has the same individual symmetry characteristics. This one process can take place easily photochemically and the reaction is therefore allowed in this mode. It will be noted that the other excited states [(2), (3), (4), etc.], although reasonably accessible, do not easily lead to photochemical reaction, since they have intended correlations with much higher excited states (not shown on the diagram Fig. VII.15).

An analogous state diagram can easily be constructed for the disrotatory process. The energy levels are the same as before, as are the various states considered; only the symmetries are changed because the $\sigma$ element of symmetry is now the relevant one. The state diagram that results is shown in Fig. VII.16.

This differs from the diagram for the conrotatory mode in that the ground states now correlate directly with no barrier, and the disrotatory process is therefore thermally allowed. Of the various excited states, the first excited states have intended correlations with higher excited states, which are avoided because of the noncrossing rule. Thus there is a significant symmetry imposed barrier to processes involving the most accessible electronically excited states and the reaction will be difficult photochemically. A disrotatory photochemical reaction would be possible from the second excited state configuration, but this state is higher in energy to begin with and also has a small symmetry imposed barrier.

Thus the state-correlation diagrams give the same conclusion as the simpler approaches and provide a more satisfactory explanation of the selection rules. Similar diagrams could be constructed for other $(4n + 2)$ systems, and for typical $(4n)$ systems to provide further illustration of the basis of the Woodward–Hoffmann rules for intramolecular electrocyclic reactions.

## VII.5  Intermolecular Cycloadditions

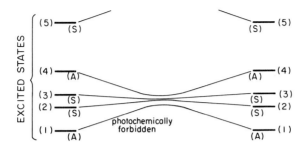

FIGURE VII.16  State correlation diagram for the disrotatory interconversions of hexatriene and cyclohexadiene.

### VII.5  INTERMOLECULAR CYCLOADDITIONS

The more difficult case of an electrocyclic reaction involving cyclization by two or more reactant molecules cannot be treated in terms of simple molecular orbital diagrams, as mentioned earlier. It is therefore necessary in every case to use energy level correlation diagrams, and it is preferable to amplify and justify the explanation of the resulting selection rules by using state-correlation diagrams as well. To illustrate the procedure involved in treating intramolecular processes, let us first consider the simplest possible case, the (2 + 2) cycloaddition of two ethylenes to give cyclobutane (24). It is first of all necessary to

(24)

assume a reasonable geometry of approach of the two ethylenes. Although other stereochemical approaches are feasible, it is most reasonable to assume that the two ethylenes approach each other perpendicularly with respect to the parallel nodal planes of their two

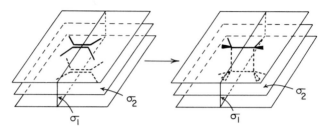

FIGURE VII.17

$\pi$-systems (as shown in Fig. VII.17). In this way the $\pi$-lobes will overlap most directly to form the new $\sigma$-bonds of the cyclic product. It is evident that two essential elements of symmetry are retained by the system as the cyclobutane is formed. These are the plane of symmetry $\sigma_1$ that bisects the two ethylenes, and the plane $\sigma_2$ that lies midway between the parallel nodal planes of the approaching $\pi$-systems. Each of these planes is a symmetry element of the product cyclobutane, and each bisects bonds being made or broken during the reaction.

Without necessarily considering the relative energies of the MOs involved, it is clear that as the four $\pi$-levels of the two ethylenes interact to give cyclobutane, there will arise four corresponding $\sigma$-levels. It is reasonable to depict these levels as shown in Fig. VII.18. The question is: How do the $\pi$-MOs corresponding to these levels interact and what are their symmetries with respect to the two planes $\sigma_1$ and $\sigma_2$? In looking at the reaction in the direction of cyclization, it is necessary to treat the two ethylenes as one system, since once overlap begins to occur they will in effect be one system. Only in this way can the appropriate symmetry elements be considered. It is simplest to project the above diagrams of reactants and products onto the plane that contains

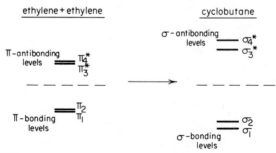

FIGURE VII.18

## VII.5  Intermolecular Cycloadditions

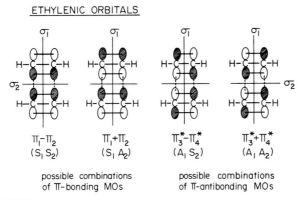

FIGURE VII.19

all four carbons and depict all possible combinations of the two sets of ethylenic $\pi$-orbitals. The planes $\sigma_1$ and $\sigma_2$ then become the orthogonal lines shown in Fig. VII.19. It is seen that only four combinations are possible. These are the symmetric and antisymmetric combinations of each pair of either bonding or antibonding $\pi$-orbitals. Other combinations are not allowed, since they would lead to formation of product MOs that are unacceptable wave functions for the cyclobutane system (see footnote on p. 252).

If these interactions are allowed to continue to form the final $\sigma$-type orbitals of cyclobutane, they can be represented as shown in Fig. VII.20. As before only the $\sigma$-envelopes in the regions of the newly formed

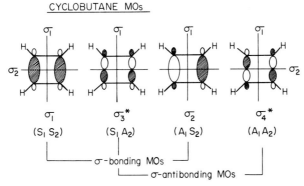

FIGURE VII.20

bonds are shown, although the $\sigma$-orbitals of cyclobutane will be completely delocalized. The symmetries of the orbital combinations of the two ethylenes can now be assigned with respect to $\sigma_1$ and $\sigma_2$. The first combination $(\pi_1 - \pi_2)$ is symmetric with respect to both $\sigma_1$ and $\sigma_2$ and is designated as $(S_1 S_2)$ or more simply $(SS)$. The second combination $(\pi_1 + \pi_2)$ is $(S_1 A_2)$ or $(SA)$, and so on. The cyclobutane $\sigma$-orbitals have corresponding symmetries, as indicated in parentheses in the diagram.

It can be seen from the two orbital diagrams Figs. VII.19 and 20 that the original $\pi$-bonding orbitals of the ethylenic systems combine to give the $(SS)$ and $(SA)$ orbitals of the interacting system, but the final $\sigma$-bonding orbitals of cyclobutane are $(SS)$ and $(AS)$. This is shown more clearly in schematic form (Fig. VII.21), where the two $\pi$-levels are initially shown as degenerate, as are the two $\pi^*$-levels. The $\sigma$-levels are ranked according to their increasing numbers of nodes. The main point is that the bonding orbitals of the reactant system do not correlate directly with the bonding orbitals of the product, thus level-crossings must occur between bonding and antibonding levels during reaction. Although the $\pi_{SS}$ combination decreases in energy along the reaction coordinate, the $\pi_{SA}$ combination increases markedly as the reaction proceeds as it tries to correlate with an antibonding level in the product. The reaction is therefore thermally forbidden, since to conserve orbital symmetry the ground-state reactants are trying to form a doubly excited product and the reaction has a large symmetry-imposed barrier. However, if one of the ethylenic $\pi$-electrons is excited to a $\pi^*$-level, the reaction will then be allowed since the re-

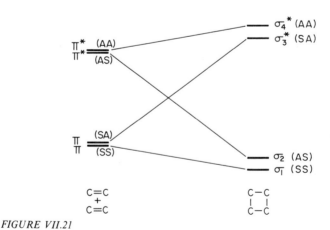

FIGURE VII.21

## VII.5 Intermolecular Cycloadditions

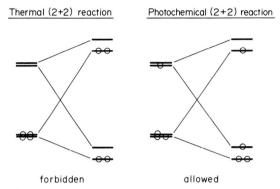

FIGURE VII.22

actant system now correlates directly with a singly excited state of the product, as shown in Fig. VII.22. (This point will be shown more clearly later by means of a state-correlation diagram (Fig. VII.28).)

We next consider the reaction of ethylene and butadiene to form cyclohexene. This prototypical Diels–Alder reaction is called a ($_\pi 4 + _\pi 2$), or simply a (4 + 2) cycloaddition. The transition state for this addition is reasonably considered to have the diene and dienophile in parallel planes with the reacting termini as close together as possible. Thus reactants and products are arranged as shown in Fig. VII.23. Now there is only one element of symmetry that is common to the reactant "molecule" (the interacting diene and dienophile) and the product. This is the plane of symmetry $\sigma_1$ that bisects both the interacting ethylene and butadiene, and also bisects the newly formed cyclohexene double bond.

We now try to relate the six reactant levels, which are all $\pi$-type (four from butadiene, two from ethylene) to the resultant four $\sigma$-type and two $\pi$-type levels of the product cyclohexene. These are depicted in

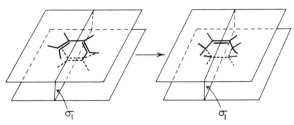

FIGURE VII.23

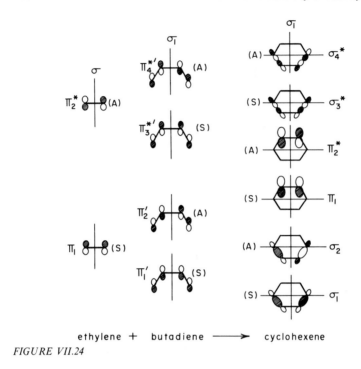

ethylene + butadiene ⟶ cyclohexene

FIGURE VII.24

Fig. VII.24. There is no problem in representing the two sets of reactant MOs, which are shown side-by-side for convenience, and their symmetries with respect to the plane $\sigma$. Similarly there is no problem in describing the $\pi$ and $\pi^*$ orbitals of the product, which are taken to be like the HMO orbitals of ethylene. To deal with the $\sigma$-type orbitals in the product, clearly there must be two $\sigma$-orbitals and two $\sigma^*$-orbitals. These will be delocalized over both newly formed $\sigma$-bonding regions and can only be reasonably represented in those regions as shown. It is also reasonable to rank the more symmetrical $\sigma$ (or $\sigma^*$) orbital as being lower in energy than the less symmetrical one, since it has fewer nodes. The symmetries of the cyclohexene orbitals with respect to $\sigma$ are also shown.

An energy level correlation diagram (Fig. VII.25) can now be constructed. Although the particular ordering and spacings shown for the various energy levels is reasonable, this is not essential to the discussion. Again, it is only the separation into either bonding or antibonding levels that is important.

## VII.5 Intermolecular Cycloadditions

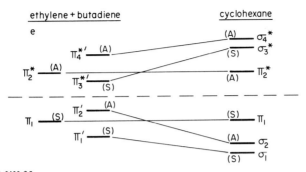

FIGURE VII.25

Again, levels of like symmetry that are closest in energy are connected. Since there are equal numbers of bonding levels of a particular symmetry in both reactants and product, there are no correlations between bonding and antibonding levels. Thus there is no symmetry-imposed barrier and the (4 + 2) reaction is thermally allowed. The most accessible excited state of the reactants would involve promotion of an electron from $\pi_2'$ to $\pi_3^{*\prime}$ of butadiene. This would then give a direct correlation with a much higher and different type $(\sigma \leftarrow \sigma^*)$ of excited state of the product, thus the (4 + 2) cycloaddition is photochemically forbidden. (Again it is not clear why other possible excited state processes should not be allowed. However, it will be shown later by means of state-correlation diagrams that there is a symmetry-imposed barrier to all reasonable photochemical processes and that the reaction is forbidden in this mode.)

To summarize, and compare the above types of intermolecular cycloaddition, it is seen that the $(_\pi 2 + _\pi 2)$ process should be thermally very difficult because orbital symmetry conservation requires a correlation between a bonding level in the reactants and an antibonding level in the product. Thus the energy of the system increases along the reaction coordinate and the reaction is said to be symmetry forbidden. Excitation of an electron in one of the reactants causes the energy change along the reaction coordinate to be decreased, and the reaction becomes more favorable photochemically and is said to be symmetry allowed. On the other hand, the $(_\pi 4 + _\pi 2)$ cycloaddition involves no such correlations of bonding levels in the reactant and antibonding correlations of bonding levels in the reactant and antibonding levels in the product, hence the energy does not increase along the reaction coordinate. Thus the thermal $(_\pi 4 + _\pi 2)$ reaction is said to be symmetry allowed. However, excitation of an electron in the reactants causes the energy change along the reaction coordinate to be increased, and the photochemical ($_\pi 4 + {_\pi}2$) reaction is said to be forbidden.

Similar considerations can be applied to other types of intermolecular cycloaddition reactions, which may involve two, or more, reactant molecules. Using this type of approach, Woodward and Hoffmann have extended the selection rules to cover the general types of intermolecular cycloaddition. These rules are given in Table VII.3. The basis of these rules is exactly as described for the two simple cases above: If orbital symmetry conservation requires one or more correlations between bonding levels in the initial state and antibonding levels in the final state, the reaction will be energetically unfavorable in a thermal sense, but favorable photochemically. If no such correla-

TABLE VII.3.

Selection Rules for Concerted Intermolecular Cycloadditions (and Cycloreversions)

| Type | Allowed processes | | | | | | | |
|---|---|---|---|---|---|---|---|---|
| | Thermal[a] | | | | Photochemical[b] | | | |
| | (numbers of π-electrons involved in each open-chain system) | | | | | | | |
| | K | L | M | N | K | L | M | N |
| $2\pi \to 2\sigma$ | | | 4 | 2 | | | 2 | 2 |
| | | | 6 | 4 | | | 4 | 4 |
| | | | 8 | 2 | | | 6 | 2 |
| $3\pi \to 3\sigma$ | | 2 | 2 | 2 | | 4 | 2 | 2 |
| | | 4 | 4 | 2 | | | | |
| | | 6 | 2 | 2 | | | | |
| $4\pi \to 4\sigma$ | 4 | 2 | 2 | 2 | 2 | 2 | 2 | 2 |

[a] Note that in all the thermally allowed processes the sum of the π-electrons involved in the cycloaddition is $(4n + 2)$.

[b] In all the photochemically allowed processes the sum of the π-electrons involved in the cycloaddition is $(4n)$.

## VII.5 Intermolecular Cycloadditions

tions exist, the reaction will be favorable thermally, but difficult photochemically.

These selection rules are in excellent accord with experiment for the many known cases of thermal and photochemical cycloadditions and cycloreversions.[9] A few general illustrations of the success of these rules are as follows.

The (2 + 2) cycloaddition of two ethylenic systems is a widely observed photochemical reaction, whereas the reaction does not readily take place thermally. The reverse reaction, such as the pyrolysis of cyclobutane, is known, but there is evidence that it is a stepwise process involving a radical intermediate. On the other hand, thermal Diels–Alder or (4 + 2) cycloaddition reactions are extremely widespread and there is overwhelming evidence that these are concerted processes. Photochemical Diels–Alder reactions are much less common and there is less evidence available on whether these are stepwise or concerted. The (4 + 4) cycloaddition is a well-known reaction, but has only been observed photochemically. Thermal [6 + 4] cycloadditions were unknown until formulation of the principle of conservation of orbital symmetry and the selection rules stimulated a search for such processes; they are now well-known reactions. Cycloadditions involving more than two molecules are less well known since entropy factors do not favor such processes, but the thermal reaction of three acetylene molecules to give benzene is an example of a (2 + 2 + 2) cycloaddition. Similar processes are known where two of the three π-systems are incorporated into the same molecule, in a nonconjugated arrangement, such as in (25). Many specific examples of cycloaddition and

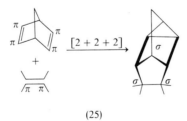

(25)

cycloreversion reactions are given in Woodward and Hoffmann's original monograph on the subject of conservation of orbital symmetry. It is sufficient to say that the success of their selection rules both in explaining known results and in predicting new ones has been truly remarkable.

It has been pointed out that for intramolecular cycloadditions it is necessary to consider state-correlation diagrams to provide a fully satisfactory explanation of the differences in stereochemical preference exhibited by thermal as against photochemical modes of a given reaction. This is also true for intermolecular cycloadditions, since we have already seen that it is not always evident from energy level correlation diagrams why an allowed thermal process should necessarily be photochemically forbidden, or vice versa. Construction of state diagrams for the (2 + 2) and (4 + 2) cycloadditions just considered will illustrate more clearly why this difference occurs between the thermal and photochemical reactions.

### VII.5.1. STATE–CORRELATION DIAGRAMS FOR INTERMOLECULAR CYCLOADDITIONS AND CYCLOREVERSIONS

We proceed in the same way as before by considering the various possible excited states of the reactants and products, and their total state symmetries as well as the symmetry characteristics of the individual electronic components. For the (2 + 2) cycloaddition of two ethylenes, the ground-state configuration of the reactant system and those of its various electronically excited states are shown in Fig. VII.26. The corresponding ground-state and excited state configurations of the product are given in Fig. VII.27.

The symmetries corresponding to the individual levels have already been determined for the energy level correlation diagram in Fig. VII.21.

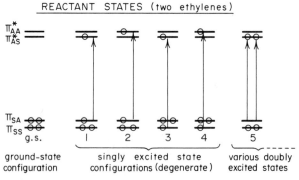

FIGURE VII.26

## VII.5 Intermolecular Cycloadditions

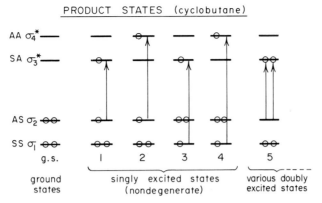

FIGURE VII.27

A table of reactant and product-state symmetries (with respect to elements $\sigma_1$ and $\sigma_2$, respectively) can now be drawn up (Table VII.4), based on the various electronic configurations.

In order to conserve orbital symmetry, both *states* and *levels* of like symmetry correlate, as shown in the state diagram in Fig. VII.28. It can be seen that thermally, the ground-state reaction $(\pi)^2(\pi)^2 \to (\sigma_1)^2(\sigma_2)^2$ has a high symmetry-imposed barrier, since $(SS)^2(SA)^2$ in the reactants correlates with $(SS)^2(SA)^2$ in the products, which is the highly excited state configuration $(\sigma_1)^2(\sigma_3^*)^2$. Therefore, the (2 + 2) reaction is thermally forbidden. Photochemically, one of the degenerate reactant excited states $(SS)^2(SA)^1(AS)^1$ correlates directly with the lowest of the cyclobutane excited states $(SS)^2(AS)^1(SA)^1$ with no barrier due to orbital symmetry conservation. Therefore, the reaction is allowed photochemically. The other excited states [(2)–(4)] only correlate with more highly excited states (not shown in the diagram).

TABLE VII.4

|  | Reactant states |  | Product states |
|---|---|---|---|
| g.s. | $(SS)^2(SA)^2 \to (SS)$ | g.s. | $(SS)^2(AS)^2 \to (SS)$ |
| (1) | $(SS)^2(SA)^1(AS)^1 \to (AA)$ | (1) | $(SS)^2(AS)^1(SA)^1 \to (AA)$ |
| (2) | $(SS)^2(SA)^1(AA)^1 \to (AS)$ | (2) | $(SS)^2(AS)^1(AA)^1 \to (SA)$ |
| (3) | $(SS)^1(SA)^2(AS)^1 \to (AS)$ | (3) | $(SS)^1(AS)^2(SA)^1 \to (AA)$ |
| (4) | $(SS)^1(SA)^2(AA)^1 \to (AA)$ | (4) | $(SS)^1(AS)^2(AA)^2 \to (SS)$ |
| (5) | $(SS)^2(AS)^2 \to (SS)$ | (5) | $(SS)^2(SA)^2 \to (SS)$ |

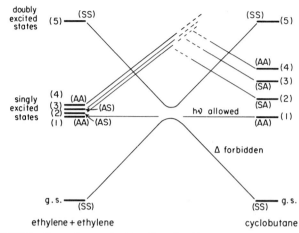

FIGURE VII.28  *State diagram for (2 + 2) cycloaddition.*

Turning to the (4 + 2) cycloaddition, the ground and various singly excited state configurations are shown in Fig. VII.29. The corresponding reactant and product symmetries are given in Table VII.5.

The state-correlation diagram can be constructed as before, conserving orbital symmetry with respect to both states and individual levels and keeping in mind the noncrossing rule. This is shown in Fig. VII.30. Unlike the case of the (2 + 2) reaction, there is now a direct correlation between reactant $[(S)^2(S)^2(A)^2]$ and product $[(S)^2(A)^2(S)^2]$ ground states, and the reaction is thermally allowed. Turning to the various excited state processes, the lowest excited state in the reactants $(S)^2(S)^2(A)^1(S)^1$ correlates directly with a much higher (doubly) excited state of the products $(S)^2(A)^1(S)^2(S)^1$. This process has a large barrier and will be difficult photochemically. It appears at first that either of the two $(S) \rightarrow (S)$ interconversions, which have no symmetry-imposed barrier, would allow the reaction to take place photochemically. However, consideration of the energy level diagrams for (ethylene and butadiene) shows that neither of these processes is feasible, since each necessitates excitation from a bonding orbital in one reactant molecule to an antibonding orbital in the other.† Thus (4 + 2) reaction is still photochemically forbidden.

---

† Alternatively, in the reverse direction, these processes would involve starting from a $(\pi \leftarrow \sigma^*)$ or $(\sigma \leftarrow \pi^*)$ state of cyclobutane to give an ethylene product with one or three electrons and a butadiene with five or three.

## VII.5 Intermolecular Cycloadditions

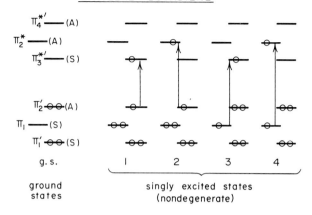

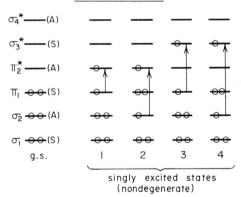

FIGURE VII.29

TABLE VII.5

|       | Reactant states |       | Product states |
|-------|-----------------|-------|----------------|
| (g.s.) | $(S)^2(S)^2(A)^2 \to (S)$ | (g.s.) | $(S)^2(A)^2(S)^2 \to (S)$ |
| (1) | $(S)^2(S)^2(A)^1(S)^1 \to (A)$ | (1) | $(S)^2(A)^2(S)^1(A)^1 \to (A)$ |
| (2) | $(S)^2(S)^2(A)^1(A)^1 \to (S)$ | (2) | $(S)^2(A)^1(S)^2(A)^1 \to (S)$ |
| (3) | $(S)^2(S)^1(A)^2(S)^1 \to (S)$ | (3) | $(S)^2(A)^2(S)^1(S)^1 \to (S)$ |
| (4) | $(S)^2(S)^1(A)^2(A)^1 \to (A)$ | (4) | $(S)^2(A)^1(S)^2(S)^1 \to (A)$ |

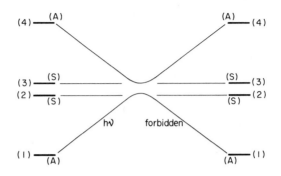

FIGURE VII.30  State diagram for (4 + 2) cycloaddition.

## VII.6  SIGMATROPIC REACTIONS

In the general sense, a sigmatropic change of order $[i,j]$ involves the uncatalyzed migration of a $\sigma$-bond, flanked by one or more $\pi$-systems, to a new position, also flanked by one or more $\pi$-systems, whose termini are $(i-1)$ and $(j-1)$ atoms removed from the original bond termini. In practice, most sigmatropic shifts involve migration of a bond that is flanked by only one $\pi$-system, such as shown in (26). In this shift the

$$\underset{j}{\overset{H_i}{\text{CH}_2{-}\text{CH}{=}\text{CH}_2}} \longrightarrow \underset{j'}{\overset{H_{i'}}{\text{CH}_2{=}\text{CH}{-}\text{CH}_2}}$$

(26)

new termini $i'$ and $j'$ are 0 and 2 atoms, respectively, removed from the original termini $i$ and $j$. Thus the reaction involves a $[1,3]$ shift. Similarly the reaction (27) involves a $[1,5]$ shift. In general, this type of

$$\overset{H}{\text{CH}_2{-}\text{CH}{=}\text{CH}{-}\text{CH}{=}\text{CH}_2} \longrightarrow \overset{H}{\text{CH}_2{=}\text{CH}{-}\text{CH}{=}\text{CH}{-}\text{CH}_2}$$

(27)

## VII.6 Sigmatropic Reactions

migration can be represented as shown in (28). Thus a $[1,j]$ sigmatropic

$$CH_2-(CH=CH)_k-CH=CH_2 \xrightarrow{[1,j]} CH_2=CH-(CH=CH)_k-CH_2$$
with H on the left carbon initially and on the right carbon after migration.

(28)

shift can be thought of as involving migration of a hydrogen atom (or other migrating group) across a π-system containing $(2k + 3)$ electrons. It is clear that where $j = (2k + 3)$, $k$ can be either odd or even.

A less common type of sigmatropic shift is where $i \neq 1$. An example of this type is the degenerate Cope rearrangement[10] of 1,5-hexadiene (29). In this case, each of the new bond termini $i'$ or $j'$ is two atoms

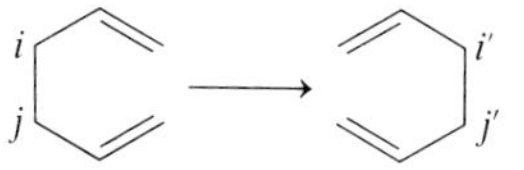

(29)

removed from the old terminus $i$ or $j$. Thus the rearrangement involves a $[3,3]$-sigmatropic shift.

If we consider the simpler $[1,j]$ shifts first, these could take place a priori in either or both of two stereochemically distinct ways, for example, as shown in Fig. VII.31. Where the atom (or group) migrates to the same face of the system the shift is called suprafacial, and where it migrates to the opposite face the shift is called antarafacial. In these cases orbital energy level or state diagrams cannot be used to decide which processes are favored, if any, since there is no element of symmetry that is common to the reactant and product orbitals. However, the transition states for each type of process do possess symmetry

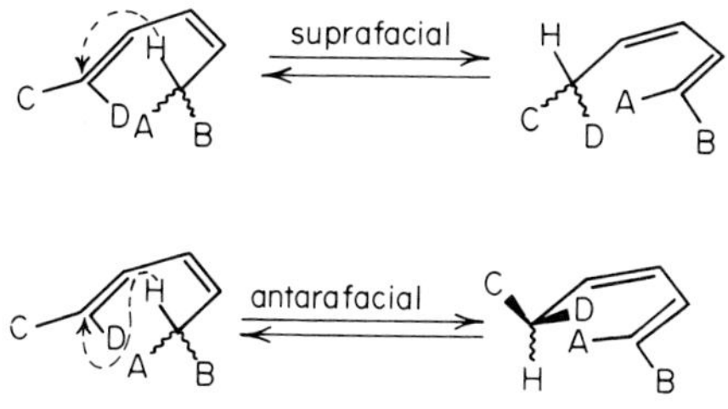

FIGURE VII.31

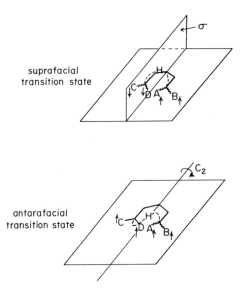

suprafacial transition state

antarafacial transition state

FIGURE VII.32

elements (see Fig. VII.32). In the suprafacial case, the groups $A$, $B$, $C$, and $D$ are all below the plane of the molecule, and at the point where migration is half complete the hydrogen lies in the plane $\sigma$ that bisects the system. (Note that the nature of the saturated groups $A$–$D$ is not important since these do not affect the orbital symmetry characteristics of the $\pi$-system in any significant way.) In the antarafacial case, at the halfway point of the migration, the migrating hydrogen lies in the plane of the molecule, and the groups $A$ and $B$ lie below this plane, with the groups $C$ and $D$ above the plane. (The positions of these groups are only important in indicating the relationships of the migration termini to each other.) Thus the state possesses a $C_2$ axis of symmetry with respect to the $\pi$-system.

The simplest way to treat the orbital symmetry relationships involved in these migrations is to visualize the transition state (VI) as being made

VI

up of a migrating hydrogen atom (with its 1s orbital) and the radical that is left as the C—H bond is broken during migration. Thus in the

## VII.6 Sigmatropic Reactions

[1,5] migration we have a hydrogen migrating from one terminus of a pentadienyl radical to the other. The important MO of the π-framework involved in the migration is considered by Woodward and Hoffmann to be the HOMO. This is the nonbonding allylic-type orbital that possesses a characteristic symmetry which depends on whether $k$ is odd or even in the $(2k + 3)$ electron system of the radical moiety. For example, in the above case we can visualize the process occuring as in (30), where, in order to maintain favorable overlap in the transition-

(30)

state region and avoid a high-energy antibonding situation, it is necessary for the hydrogen to migrate to the same face of the π-system. Therefore, the [1,5]-shift is predicted to take place suprafacially in order to conserve orbital symmetry. For a [1,3] migration, the transition state is again visualized as being made up of a migrating hydrogen atom and the corresponding radical. In this case it is the allyl radical itself, whose HOMO has the symmetry characteristics shown in (31).

(31)

To maintain favorable overlap and conserve symmetry, the migration must now take place in an antarafacial manner.

In general, for any $[1,j]$ shift the nonbonding orbital of the incipient allylic π-orbital involved in the migration has the characteristics shown in Fig. VII.33 (represented in linear form for convenience). Thus for a

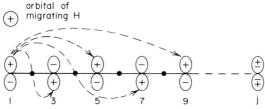

FIGURE VII.33

[1,3] or [1,7] shift, or any migration [1, 2k + 3] where $k$ is even, the transition state has a $C_2$ axis of symmetry and an antarafacial process is preferred.

For a [1,5] or [1,9] shift, or any migration [1, 2k + 3] where $k$ is odd, the transition state has a $\sigma$-plane of symmetry and a suprafacial process is preferred.

These preferences are only valid for the thermal reactions since the principal orbital whose symmetry determines these preferred modes is the HOMO of the ground-state allylic radical. In photochemical processes, the next higher (or LUMO) orbital is considered and the terminal orbital symmetry relationships will be exactly reversed. Thus for [1, 2k + 3] photochemical shifts, even $k$ leads to a suprafacial and odd $k$ to an antarafacial preference. These preferences can again be summarized in a set of selection rules, as shown in Table VII.6.

*TABLE VII.6*

*Selection Rules for [1,j] Sigmatropic Shifts*

| $\sum(1 + j)$ | Thermally allowed | Photochemically allowed |
|---|---|---|
| $4n$ | antara | supra |
| $4n + 2$ | supra | antara |

In considering the agreement between the predictions based on these rules and experiment, it should be pointed out that no shift [1, *j*] is actually symmetry forbidden, either thermally or photochemically. This is because either a suprafacial or an antarafacial process is allowed in every case. However, in practice, only suprafacial shifts are stereochemically reasonable in a number of cases. For shifts involving cyclic systems, for example, antarafacial processes are effectively ruled out, since it is unlikely that sufficient molecular distortion could take place to permit a group to migrate concertedly from one face of such a fairly rigid system to the other. Also in small acyclic systems, where skeletal flexibility is low, antarafacial shifts are unlikely on steric grounds, even though they may be permitted by orbital symmetry.

To illustrate these points,[11] [1,3] thermal shifts are quite rare and have very high activation energies. These would require symmetry-allowed antarafacial processes, which would be sterically very difficult in a three-carbon chain. However, photochemical [1,3] shifts, which can proceed suprafacially, are very common. On the other hand,

## VII.6 Sigmatropic Reactions

[1,5] shifts are very common in thermal reactions, since they are allowed suprafacially. They are less common photochemically and have only been observed in acyclic systems, where an antarafacial process could occur reasonably readily, since an open five-carbon chain has some flexibility.

As far as [1,7] shifts are concerned, cycloheptatriene is an interesting case since the migration termini are necessarily held close together. The rearrangement shown (Fig. VII.34) is not observed thermally since the allowed antarafacial process would be very difficult sterically in this cyclic system. However, the rearrangement is observed photochemically, presumably by an allowed suprafacial process. The thermal [1,7] shift has been observed however in an acyclic case, where the antarafacial mode is less sterically unfavorable.

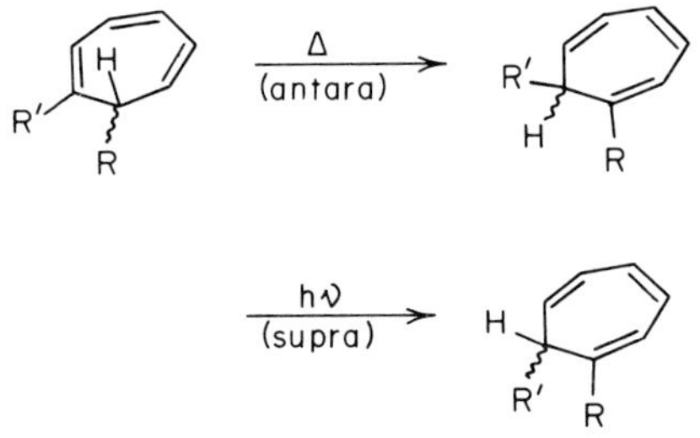

FIGURE VII.34

One very neat demonstration[12] of the steric preference for suprafacial over antarafacial shifts is the case (Fig. VII.35) where the cyclooctatriene, labeled as shown, could rearrange thermally to redistribute the deuterium label. A series of [1,3] sigmatropic shifts would be expected to distribute the deuterium equally to every position, whereas successive [1,5] shifts would only place deuterium at the 3, 4, 7, and 8 positions. Experimentally it was found that the deuterium was statistically distributed between the 3, 4, 7, and 8 positions only,

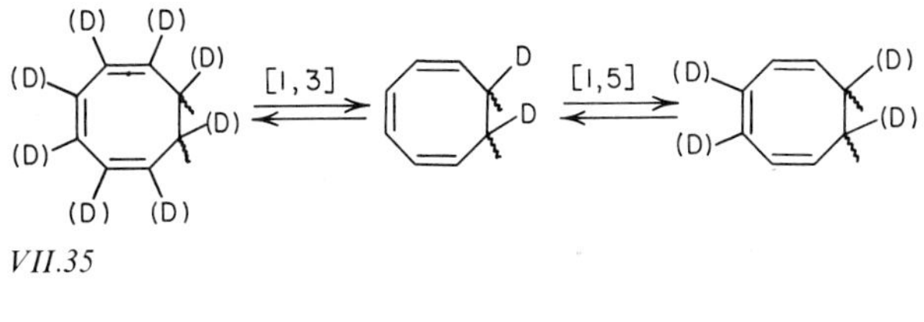

FIGURE VII.35

presumably by symmetry-allowed suprafacial [1,5] shifts, with no antarafacially allowed [1,3] shifts occurring.

For sigmatropic changes of order [i,j], where $i \neq 1$, the migrating bond is flanked by two $\pi$-systems and appropriate topological distinctions must be made with respect to the termini of both systems through which the bond migrates. In such cases there are the four distinct stereochemical possibilities,

supra–supra     antara–supra<br>
supra–antara     antara–antara

although some of these may be indistinguishable chemically or equally allowed by symmetry. It is simplest to consider a specific case, the [3,3] or Cope rearrangement. By analogy with the approach used for [1,j] shifts, this can be viewed as shown (32), where the transition state[†] is

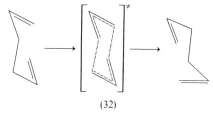

(32)

visualized as two incipient allyl radicals. Suppose that at the initially ruptured $\sigma$-bond, the orbital lobes twist in the same direction as they become parts of the HOMOs of the two allylic $\pi$-systems, as shown (33).

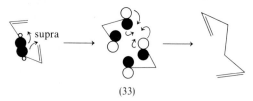

(33)

Then in order to recombine the allylic systems (concertedly) at the other terminus favorable overlap requires the lobes again to twist in the same directions (in either of the ways shown). Thus, if the migrating bond breaks suprafacially, it must also reform suprafacially. Therefore, *supra–supra* is an allowed mode for a thermal [3,3] sigmatropic shift.

[†] Although the representation (32) assumes the reaction proceeds through a chairlike transition state, the alternative boat- or booklike transition state leads to the same conclusion in terms of orbital symmetries. Both types of process are known to occur.

## VII.6 Sigmatropic Reactions

Since there is no a priori reason why initial bond breaking should only occur suprafically, we next consider the analogous antarafacial process as initating the reaction (34). In this case, as the lobes twist in

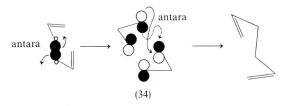

(34)

opposite directions, the symmetry of the two allylic HOMOs is such that recombination at the other termini also requires an antarafacial process to maximize developing overlap. Thus an *antara–antara* mode is equally allowed by symmetry for a thermal [3,3] shift. However, as in the case of [1,*j*] shifts, some allowed processes would be expected to be preferable to others on steric grounds. Using similar arguments about the orbital symmetries of the HOMOs in the incipient allylic radicals, general selection rules can be derived for other [*i*,*j*] sigmatropic processes, both thermal and photochemical. These are listed in Table VII.7.

TABLE VII.7

Selection Rules for [*i*, *j*] Sigmatropic Shifts

| $\sum(i+j)$ | Thermally allowed | Photochemically allowed |
|---|---|---|
| $4n$ | antara–supra<br>supra–antara | supra–supra<br>antara–antara |
| $4n + 2$ | supra–supra<br>antara–antara | antara–supra<br>supra–antara |

Again predictions based on these rules are in excellent agreement with experiment, although the number of known cases is much less than for the more common [1, *j*] shifts. For example, the Cope and Claisen (35)

(35)

rearrangements are well-known examples of [3,3] shifts. Each presumably takes place most readily by the allowed supra–supra pathway, since it would be expected that symmetrical modes such as supra–supra or antara–antara would involve less steric strain than the other possibilities. Corresponding [3,5] sigmatropic shifts would have to take place thermally in either an antara–supra or supra–antara fashion, both of which would be expected to be sterically difficult. Such shifts have only been observed photochemically, where more sterically favorable symmetry-allowed processes are available.

Probably the most striking example of the facility of allowed [3,3] shifts is the degenerate thermal rearrangement of bullvalene,[13] where because of the rigid cyclic structure only supra–supra processes are feasible. This well-known rearrangement is extremely facile as evidenced by the complete NMR equivalence of all 10 protons that occurs when the molecule is heated to even moderate temperatures. Only one example of the very large number of possible[†] degenerate isomerizations of this molecule by means of [3,3] shifts is illustrated (36).

(36)

## VII.7 GENERALIZED SELECTION RULES FOR PERICYCLIC REACTIONS

All the electrocyclic and sigmatropic reactions discussed above can be classified as *pericyclic* reactions, since all major changes in bonding take place concertedly on the perimeter of a closed curve. In this sense they all have the common feature that they can be regarded as concerted cycloaddition processes. Now that the terms suprafacial (*s*) and antarafacial (*a*) have been introduced for sigmatropic changes, these can be applied to the individual components of all cycloaddition processes

---

[†] Although there are 10 carbons (or hydrogens) in the system, leading to 10! distinguishable arrangements, because of its threefold symmetry, there are only 10!/3 (or approximately 1.2 million) nonidentical isomers possible.

## VII.7 Generalized Selection Rules

to arrive at a general rule for all pericyclic reactions. This is stated by Woodward and Hoffmann[2] as:

> A ground-state pericyclic change is symmetry-allowed when the total number of $(4q + 2)_s$ and $(4r)_a$ components is odd (where $q$ and $r$ are integers).

If the above total is even, the reaction is symmetry-forbidden thermally, but in general there will be a corresponding excited state process for which the rule is reversed. Odd-electron systems follow the same pattern as the even-numbered systems containing one more electron.

The above rule can be illustrated by applying it to some of the simple electrocyclic and sigmatropic reactions already considered (thermal reactions only). The conrotatory cyclization of butadiene can be regarded as a $(_\pi 2_s + {}_\pi 2_a)$ cycloaddition, whereas the disrotatory process can be broken down into $(_\pi 2_s + {}_\pi 2_s)$ components (see Fig. VII.36).

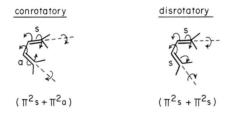

FIGURE VII.36

Thus with $q = 0$, the conrotatory process has one $(4q + 2)_s$ component and no $(4r)_a$ components. The total is odd and the process is therefore symmetry allowed. For the disrotatory process, there are two $(4q + 2)_s$ components and therefore the reaction is symmetry forbidden. Similarly, the cycloadditions of hexatriene can be regarded[†] as shown in Fig. VII.37. Also for the intermolecular cycloadditions in Fig. VII.38 the individual components can be depicted as shown. For the (2 + 2) cycloaddition of two ethylenes it is possible to envisage a different

---

[†] It should be pointed out that these are not necessarily the only ways of breaking down these pericyclic processes into individual components. Other reasonable and equivalent alternatives may be envisaged in many cases. However, the final result in terms of the oddness or evenness of the number of $(4q + 2)_s$ and $(4r)_a$ components does not depend on the particular choice made.

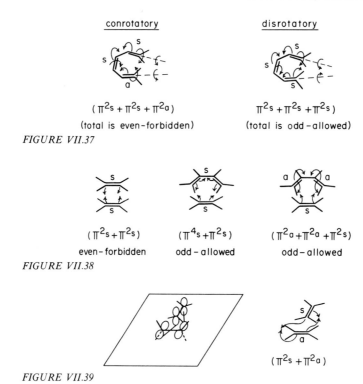

FIGURE VII.37

FIGURE VII.38

FIGURE VII.39

stereochemical approach (Fig. VII.39), whereby one ethylene bonds to opposite faces of the other π-system in a ($_\pi 2_s + {}_\pi 2_a$) process. For such a process the sum of $(4q + 2)_s$ and $(4r)_a$ is odd; therefore, this type of (2 + 2) cycloaddition would be symmetry allowed.

For simple sigmatropic reactions we can have various suprafacial and antarafacial processes, as shown in Fig. VII.40.

To provide complete generalization of the rule for pericyclic reactions, it is necessary to introduce a further convention regarding single atomic orbitals. For example, when $r = 0$ in $4r$, such components must refer to single orbitals. These may be vacant or occupied, and are designated by the symbol $\omega$; thus for a vacant orbital we have $_\omega 0_s$ or $_\omega 0_a$ and for an occupied orbital $_\omega 2_s$ or $_\omega 2_a$. For example, the cycloreversions of a cyclopropyl cation to give an allyl cation can be regarded as shown in Fig. VII.41. Thus the disrotatory processes are allowed, and the conrotatory one is forbidden. [Note that if we had been dealing

### VII.7 Generalized Selection Rules

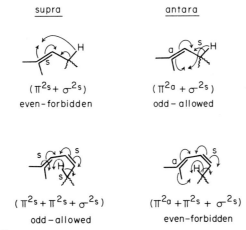

FIGURE VII.40

with the cyclopropyl anion, where the single orbital now becomes $_\omega 2$, the corresponding disrotatory processes would have been $(_\omega 2_s + {_\omega}2_s)$ and $(_\omega 2_a + {_\omega}2_a)$, both of which give even totals for $(4q + 2)_s$ and $(4r)_a$ and are thus symmetry forbidden.]

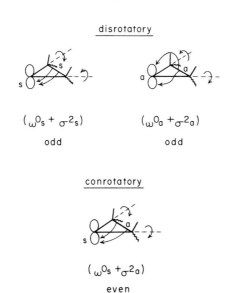

FIGURE VII.41

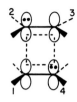 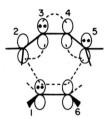

*FIGURE VII.42*

One final point is worth emphasizing: The (2 + 2) and (4 + 2) cycloadditions considered in the previous section are most simply viewed as $(_\pi 2_s + {}_\pi 2_s)$ and $(_\pi 4_s + {}_\pi 2_s)$ processes, of which the former is symmetry forbidden and the latter symmetry allowed according to the generalized selection rule. However, they could also be viewed as highly dipolar processes, involving the electronic arrangements shown in Fig. VII.42. The first of these would then be a $(_\omega 0_s + {}_\omega 2_s + {}_\omega 0_s + {}_\omega 2_s)$ cycloaddition, which is still symmetry forbidden according to the general selection rule; the latter becomes a $(_\omega 2_s + {}_\omega 0_s + {}_\omega 2_a + {}_\omega 0_a + {}_\omega 2_s + {}_\omega 0_s)$ cycloaddition, which is still a symmetry allowed process. This shows that alternative ways of breaking down the same process give the same result. It also illustrates the important point that the polarization of the bonds involved in an electrocyclic process does not alter its orbital symmetry control in any fundamental way. However, a very interesting approach that takes account of the effects of the polarization of pericyclic reactions has been described recently by Epiotis.[14]

The next chapter describes a useful, and in many ways simpler, alternative to the above selection rules for pericyclic reactions.

## PROBLEMS

1. Use the Woodward–Hoffmann rules to decide which of the following reactions are allowed processes according to the principle of conservation of orbital symmetry. Of the allowed processes, which would be facile and which would be very slow reactions? Break each of the given processes down into its individual suprafacial and antarafacial components, and confirm the foregoing predictions.

## Problems

(i), (ii), (iii), (iv), (vi), (vii), (viii), (ix), (x) [reaction schemes]

2. Construct an energy level correlation diagram for the degenerate (Cope) rearrangement of 1,5-hexadiene (assume a booklike geometry for this process). Consider both the thermal and photochemical processes.

3. Construct an energy level correlation diagram for the cycloaddition of ethylene and 1,3,5-hexatriene ($\pi^2 + \pi^6$). Use this to predict which process ($\Delta$ or $h\nu$) should be most facile.

4. Construct state-correlation diagrams for the thermal and photochemical cycloadditions of two butadienes to give 1,5-cyclooctadiene. Which process is preferred?

## REFERENCES

1. R. B. Woodward and R. Hoffmann, *J. Am. Chem. Soc.* **87**, 395, 2045, 2046, 2511 (1965).
2. R. B. Woodward and R. Hoffmann, "The Conservation of Orbital Symmetry." Verlag Chemie, Weinheim, 1970. R. Hoffmann and R. B. Woodward, *Acc. Chem. Res.* **1**, 17 (1968).
3. E. Vogel, *Justus Liebigs Ann. Chem.* **615**, 14 (1958); R. Criegee and K. Noll, *Justus Liebigs Ann. Chem.* **627**, 1 (1959).
4. E. Vogel, W. Grimm, and E. Dinne, *Tetrahedron Lett.* p. 391 (1965); E. N. Marvell, G. Caple, and B. Schatz, *Tetrahedron Lett.* p. 385 (1965); D. S. Glass, J. W. H. Watthey, and S. Winstein, *Tetrahedron Lett.* p. 377 (1965).
5. R. B. Woodward and R. Hoffmann, *J. Am. Chem. Soc.* **87**, 395 (1965).
6. P. von R. Schleyer, J. M. Su, M. Saunders, and J. C. Rosenfeld, *J. Am. Chem. Soc.* **91**, 5174 (1969).
7. J. J. Bloomfield, J. S. McConaghy, Jr., and A. G. Hortmann, *Tetrahedron Lett.* p. 3723 (1969).
8. C. A. Coulson, "Valence," p. 65. Oxford Univ. Press, London and New York, 1952.
9. Reference 2, p. 73.
10. S. J. Rhoads, *in* "Molecular Rearrangements" (P. de Mayo, ed.), Vol. 1, Ch. 11. Wiley (Interscience), New York, 1963.
11. Reference 2, p. 120.
12. W. R. Roth, *Justus Liebigs Ann. Chem.* **671**, 25 (1964).
13. W. von E. Doering and W. R. Roth, *Tetrahedron* **19**, 715 (1963); W. von E. Doering and W. R. Roth, *Angew. Chem., Int. Ed. Engl.* **2**, 115 (1963); W. von E. Doering and J. F. M. Oth, *Angew. Chem., Int. Ed. Engl.* **6**, 414 (1967).
14. N. D. Epiotis, *J. Am. Chem. Soc.* **94**, 1924, 1935, 1941, 1946 (1972).

## SUPPLEMENTARY READING

Fukui, K., *Acc. Chem. Res.* **4**, 57 (1971).
Gilchrist, T. L., and Storr, R. C., "Organic Reactions and Orbital Symmetry." Cambridge Univ. Press, London and New York, 1972.
*Woodward, R. B., and Hoffmann, R., "The Conservation of Orbital Symmetry." Verlag Chemie, Weinheim, 1970.

# VIII | THE MÖBIUS–HÜCKEL CONCEPT

## VIII.1 HÜCKEL SYSTEMS

From simple HMO calculations we have seen that systems containing cyclic arrays of $(4n + 2)$ $\pi$-electrons possess special electronic stability and that this stability arises primarily from their possession of closed-shell electron configurations. Such systems obey Hückel's rule and are said to be aromatic. For other systems, with cyclic arrays of $4n$ $\pi$-electrons, there is found to be a corresponding instability due to the possession of open-shell configurations. Such systems do not obey Hückel's rule and are said to be antiaromatic. As will be shown later, this special stability or instability associated with the possession of either $(4n + 2)$ or $(4n)$ electrons is true of cyclic transition states in some pericyclic reactions as well as all ground-state molecules of the type considered so far.

However, Hückel's rule, whether applied to ground or transition states, is only appropriate for considering certain types of cyclic system. These are systems composed of cyclic arrays of orbitals in which there are always zero, or an even number of sign inversions resulting from plus–minus orbital interactions. This is necessarily the case for the simple ground-state systems we have considered so far, since no matter how the array of basis orbitals is initially set up, there must always be

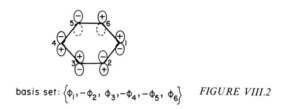

basis set: $\{\phi_1, \phi_2, \phi_3, \phi_4, \phi_5, \phi_6\}$    FIGURE VIII.1

0 or 2 or 4 or 6 ... plus–minus overlaps. Therefore, any MO whatever formed from such a basis set must also contain similar numbers of nodes or plus–minus interactions.

For example, the basis set orbitals for benzene are most conveniently set up as shown in Fig. VIII.1. This set clearly contains no plus–minus overlaps. Suppose instead we had chosen a different basis set (Fig. VIII.2). This would not have changed the final calculated results, but the basis set now contains four plus–minus overlaps. We could start with any similar combination of these AOs whatever as basis set, and in each case there would be an even number of plus–minus overlaps.

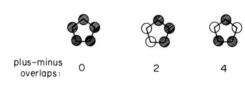

basis set: $\{\phi_1, -\phi_2, \phi_3, -\phi_4, -\phi_5, \phi_6\}$    FIGURE VIII.2

This is clear from a consideration of any one basis set such as the first. Changing the sign of any one orbital always introduces (or removes) two plus–minus overlaps. This is equally true of odd-membered cyclic systems such as pentadienyl, where several possible basis sets are shown in planar projection in Fig. VIII.3.

The foregoing property, which is completely independent of how the total orbital array is specified, is only characteristic of certain systems. These have been designated by Zimmerman[1] as Hückel sys-

plus–minus overlaps:    0    2    4

FIGURE VIII.3

VIII.2  Möbius Systems

tems since they have simple MO solutions of the normal Hückel type and fit the rule that $(4n + 2)$ electrons lead to stability, whereas $(4n)$ leads to instability.

## VIII.2  MÖBIUS SYSTEMS

On the other hand, many systems consist of, or can be thought of, as monocyclic arrays of orbitals in which there is one or some other odd number of plus–minus overlaps.[†] Such systems are called Möbius systems because of their analogy with Möbius strips.

Heilbronner[2] has pointed out that large cyclic polyenes, such as annulenes with twenty-ring carbons or more, could exist as planar conjugated systems with continuous parallel $\pi$-orbital overlap, or as twisted systems with gradually changing overlap characteristics. These possibilities are shown schematically in Fig. VIII.4.

Again, for a Möbius system, every change of sign of a given basis orbital will introduce (or remove) two plus–minus overlaps so that

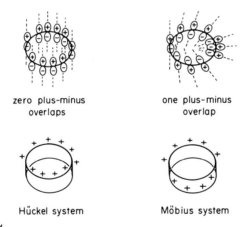

zero plus-minus overlaps    one plus-minus overlap

Hückel system    Möbius system

FIGURE VIII.4

[†] Although, as stated by Zimmerman,[1] this requirement of an odd number of plus–minus overlaps is characteristic of pericyclic Möbius systems, the statement should be modified for nonpericyclic systems. In the case of barrelene, for example, the orbital array *can* be represented with an even number of plus–minus overlaps. However, it can also be represented as having an odd number of such overlaps, which is not the case for any Hückel system.

there will always be an odd number of such overlaps for arrays of this type.

It turns out for all Möbius-type systems (and not just large monocyclic polyenes) that, contrary to Hückel systems, the possession of $4n$ electrons leads to special stability and $(4n + 2)$ leads to instability. This interesting relationship between the two types of systems is summarized in Table VIII.1.

TABLE VIII.1

Comparison of the Stability Requirements of Hückel and Möbius Systems

| Number of electrons in cyclic array | Hückel systems[a] | Möbius systems[b] |
|---|---|---|
| $4n$ | antiaromatic | aromatic |
| $4n + 2$ | aromatic | antiaromatic |

[a] Even number of plus–minus overlaps in orbital array.
[b] Odd number of plus–minus overlaps in orbital array.

In considering possible Möbius systems, we are not restricted to a consideration of basis sets consisting of only $2p_z$-type AOs. Other types of AO may be considered, and provided that no matter how these basis functions are initially described, there is one or an odd-number of plus–minus overlaps, the system is Möbius. Other types of AO may also be included in arrays designated as Hückel, as illustrated by the examples in Fig. VIII.5. (As pointed out by Zimmerman[1] such hypothetical arrays are intended to be illustrative only.) Note from the foregoing that oddness or evenness of the number of plus–minus overlaps can be forced on the system by the particular *types* of orbitals involved, as well as by the geometry of the framework (as in twisted annulenes), but not by the particular choice of sign orientations used for the basis set. It should also be reemphasized that this choice of sign orientations for the basis set orbitals has no effect on the final MO results; these are independent of the particular sign conventions chosen for the basis set.

Heilbronner[2] has worked out the HMO energies for Möbius systems of the twisted monocyclic polyene type using normal Coulomb integral values $H_{ii} = \alpha$, but modified bond integrals $H_{ij} = \beta \cos \omega_{ij}$, where $\omega_{ij}$ is the average angle of twist between successive $\pi$-centers. The energy levels for such Möbius systems were found to be expressible

## VIII.2 Möbius Systems

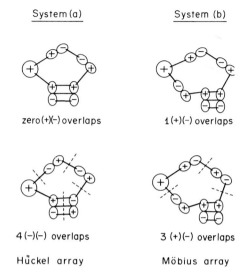

FIGURE VIII.5

by

$$\varepsilon_j^m = \alpha + 2\beta^m \cos(2j+1)\pi/n \qquad (j = 1, 2, 3, \ldots, n)$$

where $\beta^m = \beta \cos(\pi/n)$ and $\pi/n$ is the angle between two consecutive $2p_z$ AOs on a Möbius ring of $n$ atoms. It will be recalled that for Hückel systems the corresponding energy levels for monocyclic polyenes were given by

$$\varepsilon_j = \alpha + 2\beta \cos(2j\pi/n)$$

Thus the pattern of the roots for Möbius systems is quite different, and so is the spacing unit of energy:

**Möbius roots**

$$x_j^m = -2\cos(2j+1)\pi/n \quad \text{and} \quad \varepsilon_j^m = \alpha - x_j\beta^m$$

**Hückel roots**

$$x_j = -2\cos 2j\pi/n \quad \text{and} \quad \varepsilon_j = \alpha - x_j\beta$$

The energy unit $\beta^m$ is clearly smaller than $\beta$, and takes account of the diminished overlap due to twisting. Note that if Hückel systems were to be twisted without forming Möbius rings, decreased possibilities for overlap would also yield a smaller value of $\beta$.

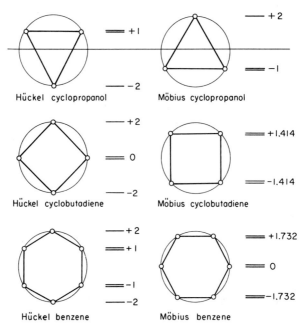

FIGURE VIII.6  *Energy level patterns for simple Hückel and Möbius systems.*

It is now interesting to compare the energy level patterns for several typical small ring systems of both the Hückel and Möbius types. This is not meant to imply that such small ring $\pi$-systems can necessarily exist in Möbius form but the comparisons shown in Fig. VIII.6 demonstrate a very useful mnemonic device for rapidly obtaining the energies of various delocalized systems on a relative scale. This device is useful since it shows energy level patterns as a function of the geometric arrangement of the basis orbitals, not only for cyclic polyenes (as shown) but also for other systems whose orbitals form cyclic arrays. It can readily be seen that such patterns lead to the kind of relationships presented in Table VIII.2, where $(4n + 2)$ electrons confer aromatic stability on Hückel systems, but $4n$ electrons are required for analogous stabilization in Möbius systems. For each type these are the numbers of electrons, respectively, that lead to closed-shell electronic configurations.

We now consider some specific examples of Möbius systems, beginning with ground-state molecules.

## VIII.2.1 TWIST TRIMETHYLENEMETHANE

In normal trimethylenemethane (I), the basis orbitals as usually set up contain no sign inversions. If the outer $sp^2$ centers are now each twisted through 90°, the following orbital arrangement results (II). In

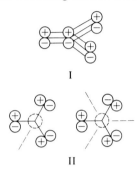

twist trimethylenemethane

this system, the central AO can be ignored since it is now orthogonal to, and hence cannot lead to any net overlap with, the other three. No matter how the orbitals are represented, this cyclic outer array must contain an odd number of sign inversions and is hence a Möbius system. The energy level pattern of the MOs formed from these orbitals will be as shown for Möbius cyclopropenyl in Fig. VIII.7. The radius of the circle will not be $\beta$, however, as for Hückel systems, because of the reduced overlap possibilities, and can be assigned a value of $\varepsilon\beta$, where $\varepsilon < 1$. Thus a twist trimethylenemethane would have a closed-shell configuration if it contained $4n$ electrons, as expected for a Möbius system.

## VIII.2.2 CYCLOPROPANE

This molecule has been considered to possess unusual bonding arising from the types of orbitals shown (III) in order to explain its anomalous properties in comparison with other cycloalkanes. As in

FIGURE VIII.7

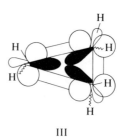

III

the Walsh model,[3] each carbon is taken to be sp² hybridized. Two of the sp² hybrids on each carbon are directed towards the hydrogens, with C—H bonding arising from sp²-ls overlaps. The remaining sp²-orbital on each carbon is directed toward the center of the ring, and overlaps with the other two sp² orbitals, all lying in the plane of the ring. The remaining orbital on each carbon is taken to be a pure p-orbital. These three p-orbitals, also lying in the plane of the ring, can then overlap as shown. Since the two arrays of orbitals involved in the cyclopropane ring system (neglecting C—H bonding) are geometrically orthogonal to each other, they can be treated approximately as independent systems. Each is a cyclic three-membered array; the sp²-set is a Hückel array since there are no sign inversions (as shown in Fig. VIII.8) and the p-set is a Möbius array analogous to twist trimethylenethane. The energy level pattern of each system can be represented using the circle mnemonic device, recalling that the radii of the two circles will be different since overlaps and hence $\beta_{ij}$ values will be greater for the sp²-orbitals than the p-orbitals. Also the circles will be centered at different heights on an energy scale, since the sp²-orbitals will be initially lower in energy than the p-orbitals. This model, and its approximate energy level distribution, shows qualitatively that six electrons

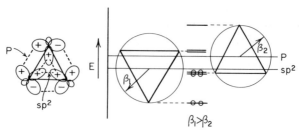

FIGURE VIII.8

## VIII.2 Möbius Systems

are necessary to give a stabilized closed-shell arrangement. This corresponds to the delocalized $\sigma$-framework of the Walsh model.

It is interesting that the MO system formed from the $sp^2$-orbitals is predicted to contain two electrons, and that formed from the Möbius-type overlap of the p-orbitals should contain four. Thus the most accessible electrons energetically should be those of the somewhat unusual "$\pi$-system" formed from the p-orbitals. These are also more numerous and more exposed to external reagents than those of the interior $sp^2$ system. This is in good accord with the known properties of cyclopropane systems, which show evidence of delocalization and high reactivity towards typical electrophilic reagents.[4] Also, the ring-strain in cyclopropane, estimated from heats of combustion, is significantly lower than would be expected in comparison with other small ring systems.[5] Evidently cyclopropane is electronically more stable than a simple localized structure would predict.

### VIII.2.3   BARRELENE

If it is assumed that the strongest bonding interactions in this system involve the three "isolated" ethylenic systems, as seems reasonable, then barrelene can be treated as an array of $\pi$ MOs (IV) rather than

IV

AOs and it thus becomes a Möbius system.[†] Because of their symmetry characteristics, the set of three bonding MOs of the ethylenic systems can only interact with each other, as can the set of three antibonding MOs. Thus we can consider the mixing of these MOs in two separate sets (V) using the Möbius circle device (see Fig. VIII.9). Since the energies of isolated, noninteracting ethylenic MOs are initially at $(\alpha + \beta)$ and $(\alpha - \beta)$, respectively, the circles representing the BMO and ABMO

---

[†] Strictly speaking it is not necessary to do this, since changing the sign of any one of the AOs, as represented above, results in an odd number of plus–minus overlaps; hence barrelene really is a Möbius system.

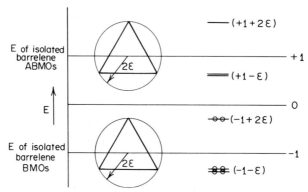

FIGURE VIII.9

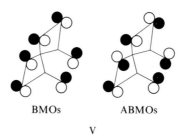

V

interaction patterns are centered at $\beta$ and $-\beta$, respectively (relative to $\varepsilon = \alpha$). Also, the radii of the two circles, which give the strength of the bonding or antibonding interactions, will be smaller than $2\beta$ because the transannular orbital interactions in a barrelene system will have bond integrals $\beta'$ which will be significantly smaller than the normal value $\beta$ based on parallel $2p_z$ overlap. If we let the ratio $\beta'/\beta = \varepsilon$, the radius of each circle will be $2\varepsilon$, in terms of the roots, or $2\varepsilon\beta$ in terms of MO energies. Thus the roots (and energies) of the interacting $\pi$-system are predicted to be

$$-1 - \varepsilon, \qquad \varepsilon_1 = \alpha + \beta + \varepsilon\beta = \varepsilon_2$$
$$-1 + 2\varepsilon, \qquad \varepsilon_3 = \alpha + \beta - 2\varepsilon\beta$$
$$+1 - \varepsilon, \qquad \varepsilon_4 = \alpha - \beta + \varepsilon\beta = \varepsilon_5$$
$$+1 + 2\varepsilon, \qquad \varepsilon_6 = \alpha - \beta - 2\varepsilon\beta$$

The total $\pi$-energy of the system obtained from the occupancy of the various levels is

$$E_\pi = 4(\alpha + \beta + \varepsilon\beta) + 2(\alpha + \beta - 2\varepsilon\beta) = 6\alpha + 6\beta$$

This is the same energy that the three ethylenic systems would have in the absence of any transannular interaction, and the value of $DE$ for barrelene is therefore zero.

Using a value of $\varepsilon\beta = \beta'$ for transannular bond integral terms and a value of $\beta$ for normal ethylene-type terms, a regular HMO calculation on barrelene gives exactly the same result as above for the individual energy levels and hence for $E_\pi$ and $DE$. Thus although the barrelene MOs are not the same as those for three isolated ethylenes and barrelene is therefore a delocalized $\pi$-system, there is no gain in energy from this delocalization. This is a similar situation to that of cyclobutadiene, where the MOs are also delocalized yet the system is no more stable than two isolated ethylenic double bonds. Hence barrelene is also antiaromatic in the same sense that cyclobutadiene is.

This emphasizes the point made earlier (see Table VIII.1) since for

Möbius barrelene, $DE = 0$ for a $(4n + 2)$ system (six electrons)

and for

Hückel cyclobutadiene, $DE = 0$ for a $(4n)$ system (four electrons)

The Möbius–Hückel differentation can be useful for assessing interactions in more complex ground-state systems and a number of interesting examples have been given by Zimmerman,[1] who gives the following general rule:

Any three orbitals (either AOs or MOs) will interact to give two levels of lower net energy and one of high energy if the product of the three overlaps is negative. Where this product is positive, two levels of higher, and one of lower, net energy are formed.

## VIII.3 APPLICATION OF THE MÖBIUS–HÜCKEL DIFFERENTIATION TO CONCERTED REACTIONS

The Möbius–Hückel approach can be easily applied[1] to predict whether concerted reactions of the types previously discussed will be allowed or forbidden. This approach does not require explicit consideration of orbital symmetry relationships. All that is needed is to decide whether the cyclic array of orbitals formed at the transition state stage of a reaction is a Hückel or a Möbius system. Then the number of electrons involved (either $4n$ or $4n + 2$) indicates whether the transition state will be stabilized or destabilized. Reactions whose transition states resemble Hückel arrays and that possess $4n + 2$ (or $4n + 1$) electrons will be thermally allowed. Those whose transition states resemble Möbius arrays require $4n$ (or $4n - 1$) electrons for stabilization, and for the reaction to be allowed thermally. The following examples illustrate the simplicity and utility of this approach.

## VIII.3.1  (2 + 2) CYCLOADDITION

The ($_\pi 2 + _\pi 2$) reaction of two ethylenes to give cyclobutane can be represented as shown in Fig. VIII.10. No matter how the orbital signs in the starting ethylenes are initially arranged, it is clear that the transition-state orbital arrays must contain zero or an even number of plus–minus overlaps. Therefore, the transition state in a concerted ($_\pi 2 + _\pi 2$) reaction is a Hückel system and its energy level pattern can be obtained from the circle device (and will resemble the pattern in cyclobutadiene). This system will thus contain two degenerate NBMOs. The energy levels of the initial and final states can be represented reasonably as shown in Fig. VIII.11. Connecting energy levels of corresponding energy illustrates how the electrons in the system can proceed most smoothly from reactant through transition state to the

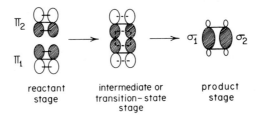

FIGURE VIII.10

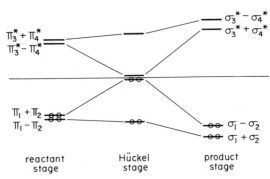

FIGURE VIII.11

## VIII.3  Möbius–Hückel Differentiation

product stage. From this diagram it is clear that the system must undergo an increase in energy at the transition state; hence the reaction is unfavorable thermally. In other words, since the transition state is a Hückel system containing $4n$ electrons, it is destabilized and the reaction must overcome an energy barrier at this stage.

### VIII.3.2  (4 + 2) CYCLOADDITION

The $(_\pi 4 + _\pi 2)$ reaction of butadiene and ethylene is represented (Fig. VIII.12) in a similar manner to that in Fig. VIII.10. Again, however the orbital signs are chosen in the reactants, the transition-state array of orbitals must contain zero or an even number of plus–minus overlaps. Therefore, the $(_\pi 4 + _\pi 2)$ reaction also involves a Hückel transition state. The difference in this case is that the system contains six electrons, and at the Hückel stage of the reaction, its energy level pattern will resemble that of benzene. Thus it will be a stabilized transition state and the reaction will be thermally allowed. The energy level diagram (Fig. VIII.13) also shows that there will be no significant barrier to reaction

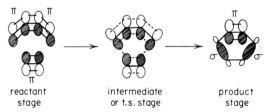

FIGURE VIII.12

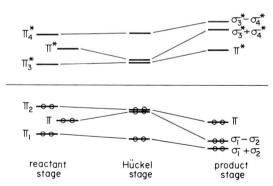

FIGURE VIII.13

in proceeding from reactants to products, as was the case for the ($_\pi 2 +\ _\pi 2$) reaction. The relationships are general for the above types of intramolecular cycloaddition and lead to the same selection rules as described previously, that is, those cyclizations (or cycloreversions) that involve ($4n + 2$) electrons will be thermally allowed and those that involve ($4n$) electrons will be thermally forbidden.

### VIII.3.3 CYCLOADDITION OF BUTADIENE

This reaction can take place in either a conrotatory or a disrotatory fashion (see Fig. VIII.14). In the conrotatory mode, no matter how the basis orbitals are represented there will always be one or an odd number of plus-minus overlaps in the system during ring closure. Thus the transition state for the conrotatory process is of the Möbius type. Conversely for the disrotatory process, the transition-state orbital arrays will always involve an even number of plus–minus overlaps; therefore, for this process the transition state is of the Hückel type.

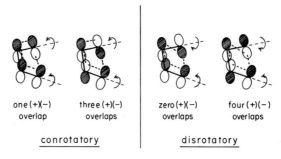

one (+)(−) overlap    three (+)(−) overlaps    zero (+)(−) overlaps    four (+)(−) overlaps

conrotatory    |    disrotatory

FIGURE VIII.14

The two processes can be represented and compared on the same energy level diagram, making use of the circle device to obtain the energy level patterns for the conrotatory and disrotatory transition states (Fig. VIII.15). The reactant (butadiene) levels are placed at the center, and the product (cyclobutene) levels at the two extremes of the diagram.

It can be seen from the diagram that the reaction will be easier energetically when a Möbius transition state is involved (i.e., there are $4n$ electrons involved in the reaction) since there is no large energy barrier, unlike the case of the disrotatory process. Thus the conrotatory

## VIII.3 Möbius–Hückel Differentiation

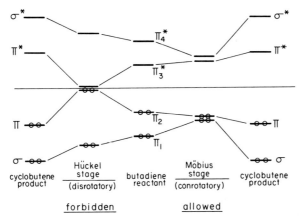

FIGURE VIII.15

mode is preferred (allowed) and the disrotatory mode is difficult (forbidden).

These relationships are typical of thermal intramolecular cycloadditions, since all conrotatory processes have Möbius transition states and are favored when the reacting system contains $4n$ electrons; all disrotatory processes, on the other hand, have Hückel transition states and are favored when there are $(4n + 2)$ electrons present. It can be seen that these simple ideas lead to the same selection rules as derived by Woodward and Hoffmann. To illustrate the generality of the above treatment, one more intramolecular cycloaddition will be considered briefly.

### VIII.3.4 CYCLOADDITION OF HEXATRIENE

Here the transition states are like distorted benzene systems, one being of the Hückel type (disrotatory) and the other of the Möbius type (conrotatory). (See Fig. VIII.16.)

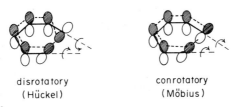

FIGURE VIII.16

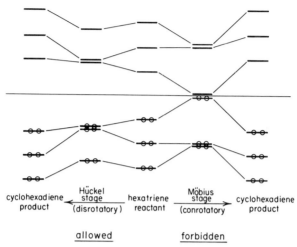

FIGURE VIII.17

Again it is seen (Fig. VIII.17) that one of the processes has an energy barrier arising from a destabilized transition state. For this $(4n + 2)$ electron system it is the Möbius transition state and hence the conrotatory process is forbidden thermally. The disrotatory process, involving a stabilized Hückel transition state, has no such barrier and is more favorable energetically; thus the thermal disrotatory process is allowed.

### VIII.3.5 SIGMATROPIC REACTIONS

One advantage of the Möbius–Hückel approach is that it is possible to predict the preferred mode of a reaction without recourse to the symmetry characteristics of the system, and indeed even where the system possesses no symmetry (other than trivial $C_s$ symmetry). The approach is thus independent of the types or combinations of orbitals involved during reaction. It is also usually very simple using this approach to determine whether a reaction will be allowed or forbidden, without constructing any correlation diagrams at all. Thus the Möbius–Hückel differentiation is particularly appropriate for sigmatropic reactions, where there is generally no symmetry element common to the reacting system and where correlation diagrams cannot be constructed.

### VIII.3 Möbius–Hückel Differentiation

suprafacial migration
(Hückel)

antarafacial migration
(Möbius)

FIGURE VIII.18

It is evident that for a sigmatropic process, any single antarafacial component will introduce one or an odd number of plus–minus overlaps into the orbital array at the transition state. Thus for sigmatropic shifts of the type [1,j] antarafacial processes always lead to Möbius transition states, whereas suprafacial shifts lead to Hückel transition states. If we consider [1,7] shifts, as shown in Fig. VIII.18, the fact that this system possesses eight electrons that are involved in the pericyclic process means that the Möbius array would be stabilized and the Hückel array would be destabilized. Thus the antarafacial process has a more stabilized transition state and will be preferred over the suprafacial. This is in accord with the Woodward–Hoffmann rules. Similar considerations of other [1,j] shifts lead to the analogous rules given in Table VIII.2. These and corresponding rules that can be developed for

TABLE VIII.2

Selection Rules for [1, j] Sigmatropic Shifts Based on the Möbius–Hückel Differentiation

|   | No. of electrons involved | Favored T.S. | Preferred mode |
|---|---|---|---|
| [1,3] | 4 | Möbius | antara |
| [1,5] | 6 | Hückel | supra |
| [1,7] | 8 | Möbius | antara |
| [1,9] | 10 | Hückel | supra |
| [1, 2k + 3] | k even | Hückel | supra |
|  | k odd | Möbius | antara |

electrocyclic reactions on the basis of the Möbius–Hückel differentiation have been referred to as the Dewar–Zimmerman rules.[1] These can be summarized in the statement:

Concerted reactions that involve Hückel-type transition states are thermally preferred if the system contains $(4n + 2)$ electrons, but photochemically preferred if it contains $(4n)$ electrons. Reactions that involve Möbius-type transition states are thermally preferred if the system contains $(4n)$ electrons and photochemically preferred where it contains $(4n + 2)$ electrons.[†]

Turning to sigmatropic shifts of the type $[i,j]$, where there are the four possibilities shown (as in the accompanying tabulation), then if

|  | T.S. type |
| --- | --- |
| supra–supra | Hückel |
| supra–antara | Möbius |
| antara–supra | Möbius |
| antara–antara | Hückel |

each antarafacial component introduces one or an odd number of sign changes, only those processes that involve one such component will lead to Möbius transition states. Any others will involve Hückel transition states. This is illustrated in Fig. VIII.19 for the case of the Cope ([3,3]) rearrangement.

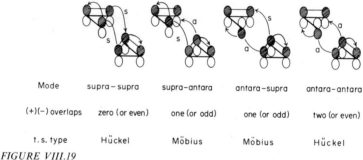

| Mode | supra–supra | supra–antara | antara–supra | antara–antara |
| --- | --- | --- | --- | --- |
| (+)(−) overlaps | zero (or even) | one (or odd) | one (or odd) | two (or even) |
| t.s. type | Hückel | Möbius | Möbius | Hückel |

FIGURE VIII.19

Therefore, shifts $[i,j]$, where $(i + j) = 4n + 2$ will take place preferably in a supra–supra or antara–antara fashion in the thermal reaction. This includes the Cope rearrangement. Shifts $[i,j]$, where $(i + j) = 4n$ will prefer to take place thermally in an antara–supra fashion, since these result in a Möbius transition state.

[†] As before, odd-electron systems show the same preference as the even-electron system containing one more electron.

## VIII.3 Möbius–Hückel Differentiation

That the Dewar–Zimmerman rules are fully consistent with the generalized selection rule given earlier for all pericyclic reactions, is shown in Table VIII.3. The number of $(4q + 2)_s$ components in any reaction must be either odd or even, as must the number of $(4r)_a$ components. Thus there are only the four basic possibilities shown. Similarly, if each antarafacial component generates an odd number of plus–minus overlaps and each suprafacial component gives an even number of such overlaps, the types of transition states corresponding to each overall process can easily be determined. Also the net number of electrons involved [in terms of whether it is $(4n)$ or $(4n + 2)$] can easily be arrived at from the oddness or evenness of the various components.[†]

In many ways the Möbius–Hückel approach is easier to apply to concerted reactions, since all it involves is a visualization of the stereochemistry of the pericyclic process and a counting of the number of electrons involved in the reaction.

### VIII.3.6 THE SOURCE OF STABILITY IN MÖBIUS SYSTEMS

An important question is: How can Möbius systems be more stable in some cases than the corresponding Hückel systems, when every Möbius system must contain at least one plus–minus overlap in its basis set? Thus every MO in a Möbius system must contain at least one node, which represents an antibonding, energy-raising interaction. The lowest energy MO in Hückel systems generally contains no such nodes.

There are two reasons for this greater stabilization in some Möbius systems:

(i) It turns out from calculations that Möbius-type MOs have the property of having very small values of the LCAO–MO coefficients near the site of the plus–minus overlaps. Thus the nodal interaction is only weakly antibonding. This is particularly so for the lower energy MOs.

---

[†] Note that an odd number of $(4q + 2)$ components must involve $(4n + 2)$ electrons, whereas an even number involves $4n$ electrons; whereas odd or even numbers of $(4r)$ components always involve $(4n)$ electrons.

TABLE VIII.3

Consistency of Generalized Selection Rules for Pericyclic Reactions

| No. of $(4q+2)_s$ components[a] | No. of $(4r)_a$ components[b] | Sum of components | Total number of electrons | Number of plus–minus overlaps developed | Type of T.S. | Allowed processes |
|---|---|---|---|---|---|---|
| even | even | even | $4n$ | even + even | Hückel | $h\nu$ |
| even | odd | odd | $4n$ | even + odd | Möbius | $\Delta$ |
| odd | even | odd | $4n+2$ | even + even | Hückel | $\Delta$ |
| odd | odd | even | $4n+2$ | even + odd | Möbius | $h\nu$ |

[a] The terms in this column always generate an even number of plus–minus overlaps.
[b] Only the odd terms in this column generate an odd number if plus–minus overlaps.

# Problems

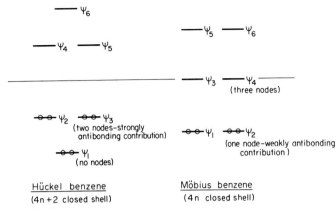

FIGURE VIII.20

(ii) The lowest energy MOs in Möbius systems come in pairs, whereas it is the next higher and successive Hückel MOs that come in pairs.

If we compare the nodal interactions in Hückel and Möbius benzene, this latter point becomes evident. (Fig. VIII.20). If we compare the bonding MOs for the two closed-shell systems, then although $\psi_1$ in the Hückel case contains no energy-raising nodes in its wave function, the next two MOs $\psi_2$ and $\psi_3$ each contain two strongly antibonding interactions. Thus although the two lowest energy electrons experience no antibonding interactions in their wave function, the next four electrons each experience two such interactions. On the other hand, $\psi_1$ and $\psi_2$ in the Möbius system each contain only one node. Hence all four electrons in the closed-shell Möbius system only experience one weakly antibonding interaction. Also, with more than one node in the system, as in the case with $\psi_2$ and $\psi_3$ in the Hückel system, the MO coefficients cannot as easily minimize the antibonding interaction as would be the case where the wave function has only one node.

## PROBLEMS

1. Use the Möbius–Hückel approach to predict which of the following processes is allowed or forbidden. Of the allowed processes, designate any that would be energetically difficult.

(i) [reaction scheme: two dienes + Δ → cyclooctatetraene]

(ii) [reaction scheme with Me, Me, H substituents → diene products with Me and H]

(iii) [reaction scheme: D, Me, H cyclohexadiene hv → Me, H, D product]

(iv) [reaction scheme: bicyclic with H → benzene, Δ]

(v) [reaction scheme: bicyclic hv → cyclooctatetraene]

(vi) [norbornene + alkene Δ → cage product]

(vii) [D, Me, H cycloheptatriene Δ → (D), Me, H, (D) product]

(viii) [bicyclic–Me Δ → Me-bicyclic]

(ix) [t-Bu, H, t-Bu, H cyclooctatetraene hv → t-Bu, H, t-Bu, H bicyclic]

(x) [bicyclic H,H Δ → cycloheptatriene H,H]

2. Carry out a Hückel calculation on Möbius trimethylenemethane. Ignore the central carbon, and use a bond integral of $0.75\beta$. Calculate $E_\pi$ for this system. Compare the $\pi$-bonding energy per electron with the value for Hückel trimethylenemethane.

3. Obtain the energy level distributions for the planar and Möbius cyclooctatetraene systems using the polygon rule. Calculate and compare $E_\pi$ for each system using either a scaled diagram or the Heilbronner formula.

## REFERENCES

1. H. E. Zimmerman, *Acc. Chem. Res.* **4**, 272 (1971).
2. E. Heilbronner, *Tetrahedron Lett.* p. 1923 (1964).
3. A. D. Walsh, *Trans. Faraday Soc.* **45**, 179 (1949).
4. M. Saunders, P. Vogel, E. L. Hagen, and J. Rosenfeld, *Acc. Chem. Res.* **6**, 53 (1973); R. T. Morrison and R. N. Boyd, "Organic Chemistry," 2nd Ed., p. 277. Allyn & Bacon, Boston, Massachusetts, 1966.
5. J. D. Roberts and M. C. Caserio, "Basic Principles of Organic Chemistry," pp. 112–113. Benjamin, New York, 1964.

## SUPPLEMENTARY READING

Heilbronner, E., *Tetrahedron Lett* p. 1923 (1964).
*Zimmerman, H. E., *Acc. Chem. Res.* **4**, 272 (1971).

# IX | SYMMETRY, TOPOLOGY, AND AROMATICITY

## IX.1 AROMATICITY FOR PERICYCLIC AND OTHER TOPOLOGIES

Hückel's rule has already been discussed and found applicable to simple monocyclic systems with $n = 0, 1, 2, \ldots, 7$. Even when this type of system is structurally modified in various ways, as illustrated in Fig. IX.1, those systems containing $(4n + 2)$ electrons still retain aromatic character. Although these structural modifications may include saturated or unsaturated transannular fragments, incorporation of orthogonal $\pi$-systems, interruptions of the cyclic systems by saturated groups, or other forms of perturbation, all these molecules

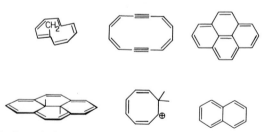

FIGURE IX.1  Perturbed pericyclic systems.

## IX.1 Aromaticity and Topology

retain a more or less intact pericyclic $\pi$-system. For all such ground-state systems and for cyclic transition states in thermal, concerted reactions, the possession of $(4n + 2)$ $\pi$-electrons in a pericyclic array confers aromatic (or pseudoaromatic) character.

Goldstein and Hoffmann[1] have recently examined the factors that lead to stabilization in various systems of interacting orbitals, particularly with respect to the topology of these interactions. They point out that there are three basic factors to be considered in assessing the stabilization of any system of interacting orbitals:

(a) The symmetry properties of the component orbitals; that is, whether they are of $p$- or $d$-type symmetry, which for example would be important in a comparison of the stabilities of structurally analogous borazine and phosphonitrilic systems.

(b) The magnitude of the orbital overlap, which for example may be of the normal parallel $p_z$-type, or of a less effective type as in homo-conjugated systems, pericyclic transition states, or twisted systems.

(c) The topology of the orbital interactions; so far only pericyclic topologies have been considered, but it is clear that Möbius systems differ in their modes of stabilization from Hückel systems.

Goldstein and Hoffmann have paid particular attention to the third factor and have given a very interesting and detailed treatment of the conditions for aromatic stabilization in pericyclic and other topologies. A summary of this treatment follows.

The fundamental building blocks of stabilized conjugated systems can be thought of as *intact conjugated polyene segments*, which are called *ribbons*. Such ribbons may be linked together directly, as shown for benzene (I); they may be linked by single bonds to form longer

ribbon

I

ribbons as shown for octatetraene (II); or they may be connected by

ribbons

II

insufficiently insulating centers, as in the homotropylium ion (III).

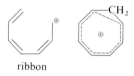

III

Four possible topologies are considered for the linking together and interactions of one or several such segments or ribbons. These are shown schematically in Fig. IX.2. There are other modes of ribbon annelation, even for three ribbons, but the ideas developed from these simple cases can be extended to other systems. From the foregoing it is clear that the only possible topology for a single ribbon is pericyclic. For two ribbons, a spirocyclic topology is also possible, and for three or more ribbons both longicyclic and laticyclic topologies are possible, in addition to pericyclic.

### IX.1.1 MODES OF RIBBON INTERACTION

It is necessary to place certain arbitrary restraints on the modes of ribbon interaction in order to simplify the topological arguments. These restraints are:

(i) Significant interactions between ribbons must only occur at the termini.

(ii) The twisting of any ribbon must be less than 90°. This excludes Möbius ribbons, but does not necessarily exclude Möbius-type systems.

(iii) The two termini of any one ribbon must remain indistinguishable with respect to their connectivities and also with respect to the type of connections made (i.e., σ- or π-type).

Because of these restrictions, systems such as those shown in IV and V would be included, but VI would be excluded. Restriction (i) implies

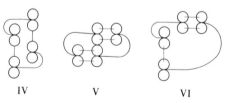

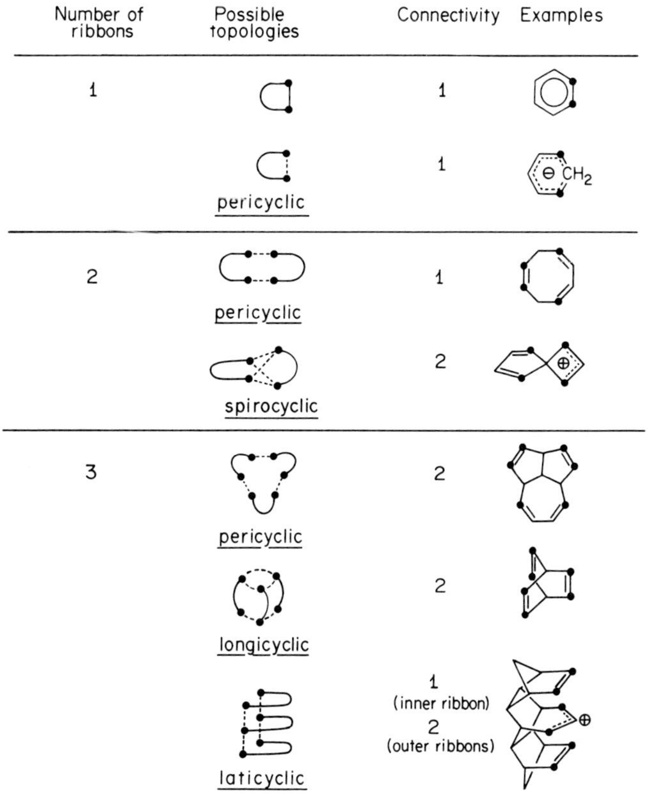

IX.1 *Aromaticity and Topology* 317

| Number of ribbons | Possible topologies | Connectivity | Examples |
|---|---|---|---|
| 1 | pericyclic | 1 | |
| | pericyclic | 1 | ⊖CH₂ |
| 2 | pericyclic | 1 | |
| | spirocyclic | 2 | ⊕ |
| 3 | pericyclic | 2 | |
| | longicyclic | 2 | |
| | laticyclic | 1 (inner ribbon) 2 (outer ribbons) | ⊕ |

FIGURE IX.2  The symbol · — · is used where the connection of two ribbons is made directly through a single bond, and the symbol ···· where the connection is made through an insulating center such as a methylene group. Connectivity is defined by the number of linkages or interactions at each ribbon terminus.

that the effectiveness of interaction between any two ribbons depends only on the electron density and phase relationships at the terminal positions. It will be recalled that the phase of the wave function at reacting termini is of vital importance in considering electrocyclic and sigmatropic reactions.

Next, the pseudosymmetry of any ribbon orbital is defined relative to the phase relationships at its termini as being either pseudo-p ($\psi$p) or pseudo-d ($\psi$d). By pseudosymmetry is meant the symmetry characteristic of a particular orbital, exclusively with respect to the

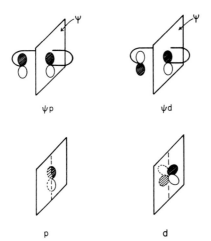

ψp   ψd

p   d   FIGURE IX.3

phases of the terminal lobes of the orbital. The remainder of the orbital (and the molecule itself) is ignored in assigning a pseudoplane of symmetry ($\psi$) to the system, as shown in Fig. IX.3. Thus any π-type MO is either $\psi$p or $\psi$d, by analogy with the symmetry characteristics of p- or d-orbitals with respect to such a pseudoplane. The well-known pattern of orbitals for simple polyene ribbons with up to six π-centers is illustrated below. This shows that for any ribbon the orbitals alternate between $\psi$p and $\psi$d with increasing energy, starting from the lowest energy MO, which is always $\psi$p.

If we now consider the HOMO and LUMO of conjugated polyene systems corresponding to these ribbons (Fig. IX.4), it is evident that their symmetry characteristics depend on the number ($n$) of π-centers present and on the electron occupancy of the polyene, as measured by the net charge $Z$. Thus each polyene segment can be defined by the symbol $n^Z$. For example, ethylene is $2°$, the allyl cation is $3^+$, and butadiene is $4°$.

Only four HOMO–LUMO patterns are possible for any linear polyene fragment or ribbon, since the HOMO must be singly or doubly occupied, and must be of pseudosymmetry $\psi$p or $\psi$d; similarly the LUMO must be vacant and also either $\psi$p or $\psi$d. These four patterns are shown in Table IX.1, with specific examples of each type.

The single π-centers corresponding to an isolated vacant ($1^+$), singly occupied ($1°$), or doubly occupied ($1^-$) $p_z$-orbital have been included for the sake of completeness. Note that each of the four

## IX.1 Aromaticity and Topology

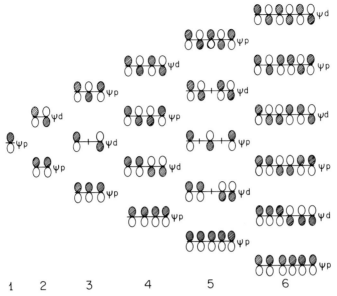

FIGURE IX.4  *Orbital pattern for simple polyene ribbons. Note that for any ribbon, the molecular orbitals alternate between $\psi p$ and $\psi d$ with increasing energy.*

TABLE IX.1

*HOMO–LUMO Patterns of Simple Polyene Ribbons*

| LUMO<br>HOMO<br>Mode: | —— $\psi d$<br>—•— $\psi p$<br>1 | —— $\psi d$<br>—••— $\psi p$<br>2 | —— $\psi p$<br>—•— $\psi d$<br>3 | —— $\psi p$<br>—••— $\psi d$<br>0 |
|---|---|---|---|---|
| Examples: | (1°) | (1⁻)<br>2° | | (1⁺) |
| | | 3⁺ | 3° | 3⁻<br>4° |
| | 5° | 5⁻ | | 5⁺ |
| | | 6°<br>7⁺ | 7° | 7⁻ |

patterns has a homologation factor of four electrons, that is, 2° (ethylene) has the same HOMO–LUMO pattern as 6° (hexatriene), 3⁻ (the allyl anion) has the same pattern as 7⁻ (the heptatrienyl anion), and so forth. This leads to the following useful definition:

The *mode of a ribbon* is a convenient quantity that describes the HOMO–LUMO pattern of any ribbon and is designated as $(n - z)$ *modulo* 4.[†]

For example, ethylene has a HOMO–LUMO pattern

$$\text{———} \;\psi d$$
$$\text{–••–} \;\psi p$$

and is also characterized as the ribbon 2°. Thus $(2 - 0)$ *modulo* $4 = 2$, and ethylene is an example of a mode 2 ribbon. Hexatriene (6°) has the same pattern, and its mode is given by $(6 - 0)$ *modulo* 4, which is also 2. The allyl cation ($3^+$) has a mode $= (3 - (+1))$ *modulo* $4 = 2$, whereas the allyl anion ($3^-$) has a mode $= (3 - (-1))$ *modulo* $4 = 0$. Considering larger systems, the heptatrienyl radical (7°) has mode $= (7 - 0)$ *modulo* $4 = 3$. Examination of the terminal orbital symmetry relationships for this radical reveal them to be exactly the same as for the HOMO and LUMO of the simpler allyl radical (3°). Therefore, the HOMO–LUMO pattern of any ribbon is unambiguously defined by its mode. Note that neutral molecules and ionic species generally have even modes, whereas radicals have odd modes.

These modes are additive, so that when two ribbons of mode $\mu_1$ and $\mu_2$ interact (even weakly) in an acyclic way to give an extended ribbon, this new composite is of mode $(\mu_1 + \mu_2)$ *modulo* 4.[‡] This is independent of the relative energies of the interacting orbitals. Conversely, the partitioning of any ribbon into any number of shorter ribbons, which can still interact, leaves the orbital symmetry relationships of the ribbons unchanged in a modal sense.

Normally when two levels interact to produce one level of higher energy and one of lower energy, it is the higher level that acquires an extra node. However, in these ribbon interactions the newly developed node is buried somewhere in the center of the ribbon system and the new terminal symmetry relationships are not affected. However, $\psi p$ are not always lower than $\psi d$ levels as a result of these interactions. The various possible ribbon extensions are illustrated in Fig. IX.5. The numbers of electrons shown occupying these interacting levels have simply been chosen to demonstrate the mode additivity relationships. It is noted that two interacting orbitals of like symmetry generate $\psi p$

---

[†] Modulo 4 simply means subtract the highest multiple of 4 from the result of performing the arithmetical operation indicated, which in this case is $(n - z)$.

[‡] The proof of this relationship is given by Goldstein and Hoffmann.[1]

## IX.1 Aromaticity and Topology

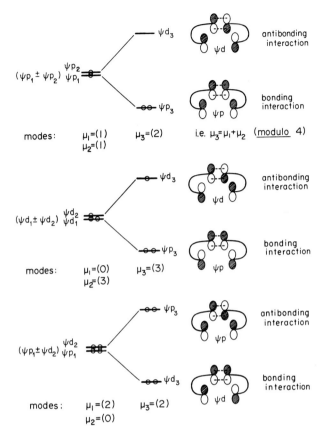

FIGURE IX.5

below $\psi d$ in the extended ribbon, and interacting orbitals of unlike symmetry generate $\psi d$ below $\psi p$.

We now consider the various topologies of interacting ribbons, restricting the discussion to even modes for the sake of simplicity. This means that only molecules and ions will be considered.

### IX.1.2 PERICYCLIC SYSTEMS

This is the only closed topology available to a single ribbon. The terminal orbital interactions in a pericyclic system may be of two types (Fig. IX.6), depending on the pseudosymmetry of the orbital

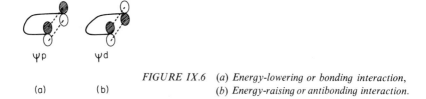

FIGURE IX.6  (a) Energy-lowering or bonding interaction, (b) Energy-raising or antibonding interaction.

involved. With this in mind we can consider the terminal interactions and resulting energy level patterns for mode 0 and mode 2 ribbons (Fig. IX.7).

*Thus stabilization is only available to a mode 2 ribbon in a pericyclic interaction, and not to a mode 0 ribbon.* This will also be true of any number of ribbons interacting in a pericyclic fashion, provided their modal sum is two. It can be noted from Table IX.1 that all mode 2 ribbons contain exactly $(4n + 2)$ electrons.

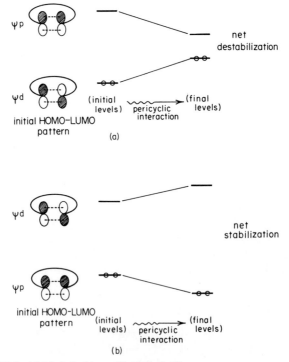

FIGURE IX.7  (a) Mode 0 ribbons, (b) Mode 2 ribbons.

## IX.1 Aromaticity and Topology

The above statement about mode 2 ribbons corresponds to Hückel's rule when the interactions between ribbons (or ribbon termini) are as great as the interactions within the ribbons. It also corresponds to the requirement for a stabilized pericyclic transition state in either a fully suprafacial sigmatropic reaction or a disrotatory electrocyclic process. The statement is also applicable to interrupted conjugation of the homoconjugative type. These four types of pericyclic interaction are qualitatively the same, the only distinctions between them being quantitative in the sense that different energy levels and interaction energies may be involved. Whatever the system, net stabilization always results from pericyclic interactions involving mode 2 ribbons, although the magnitude of this stabilization may vary. Several illustrative examples of stabilized pericyclic systems are given in Fig. IX.8.

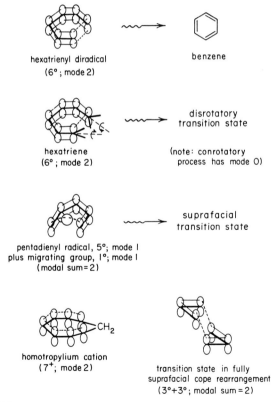

FIGURE IX.8

Other types of pericyclic interaction may be envisaged as involving ring closure of two ribbons, rather than the simple ribbon extensions considered earlier. In these cases there will be two interactions of the types discussed previously, one at each terminus. However, the situation is different from ribbon extension in that $\psi p$ orbitals can now only interact effectively at their termini with other $\psi p$ orbitals; similarly $\psi d$ can only interact effectively with $\psi d$. This is illustrated for the three basic possibilities of two-ribbon pericyclic interaction (Fig. IX.9). In case (c), each bonding interaction at one terminus is accompanied by a corresponding antibonding interaction at the other terminus. Thus there is no net interaction.

*Thus, only orbitals of the same pseudosymmetry can interact in a two-ribbon intermolecular pericyclic topology.* This result can be used to draw interaction diagrams for various kinds of ring systems to see which pericyclic interactions of the two-ribbon type lead to stability,

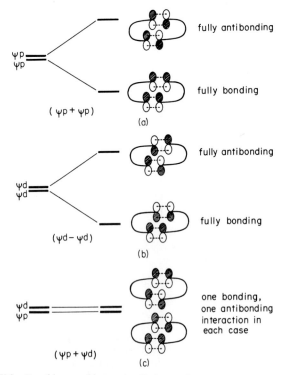

FIGURE IX.9  Possible two-ribbon pericyclic interactions.

## IX.1 Aromaticity and Topology

and what the associated modal properties are in each case. Various possible combinations of mode 0 and mode 2 ribbons are illustrated in Fig. IX.10(a) and (b) with specific examples of molecules containing such ribbons.

Because only orbitals of the same pseudosymmetry can interact, only the modal interactions of type (0,2) can lead to stabilization in

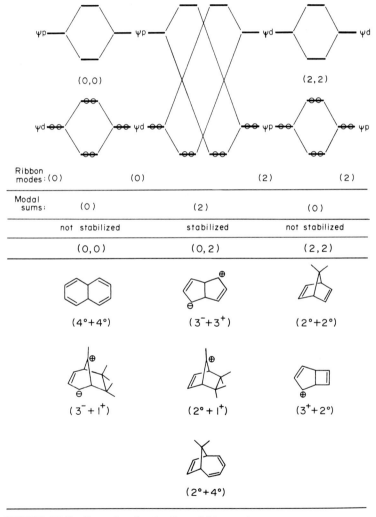

FIGURE IX.10 Examples of pericyclic modal interactions.

these pericyclic systems. This can be seen from the diagram, as can the fact that (0,0) and (2,2) interactions give offsetting energy-lowering and energy-raising contributions, and thus lead to no net stabilization.

### IX.1.3  SPIROCYCLIC SYSTEMS

Although various types of terminal orbital overlaps are possible in spirocyclic systems, a simple example can be used to show that the only effective interaction is between two $\psi d$ orbitals (Fig. IX.11). From this a similar diagram to that for various pericyclic interactions can be constructed for spirocyclic systems (Fig. IX.12).

Again it is shown that net stabilization can only result from (0,2) modal interactions, and not from either (0,0) or (2,2) interactions. Some realistic but as yet unknown examples of spirocyclic systems that are expected to be stabilized are shown by VII, VIII, and IX. Note that

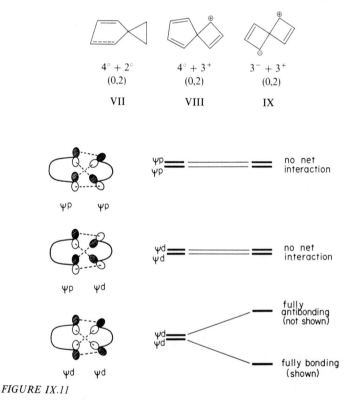

FIGURE IX.11

## IX.1  Aromaticity and Topology

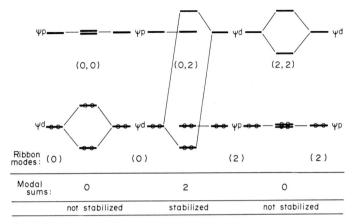

FIGURE IX.12  *Examples of spirocyclic modal interactions.*

for even these spirocyclic systems, exactly $(4n + 2)$ electrons are again required for stabilization.

### IX.1.4  LONGICYCLIC SYSTEMS

The same types of argument can be applied to longicyclic systems. Since at least three ribbons are involved in these cases, interactions between two of them are considered first, then the resulting system is allowed to interact with the third ribbon. The results of such interactions are independent of how the interactions might be permuted; longicyclic interactions are both associative and commutative. This can be verified by considering representative systems. Thus in a longicyclic system containing a mode 0 ribbon and two mode 2 ribbons, the following combinations are equivalent:

$$(0,2) + 2 = 0 + (2,2) = (2,0) + 2$$

All three combinations have modal sum equal to 2, and are designated as (0,2,2) systems. The interaction diagrams for longicyclic systems are more complex than the cases already discussed, but modal analysis shows that there are two sets of stabilizing interactions of three ribbons and two sets that are nonstabilized (see Table IX.2). It is very interesting to note that although stabilization again comes only from (0,2) interactions, longicyclic systems may require either $(4n)$ or $(4n + 2)$ electrons for stabilization, in contrast to pericyclic and spirocyclic systems, which

TABLE IX.2

|  | Modal sum | No. of electrons |
|---|---|---|
| Stabilized systems: (0,2,2) | 0 | $4n$ |
| (0,0,2) | 2 | $4n + 2$ |
| Nonstabilized systems: (0,0,0) | 0 | $4n$ |
| (2,2,2) | 2 | $4n + 2$ |

always require $(4n + 2)$. Also, either $(4n)$ or $(4n + 2)$ electrons can lead to destabilization in a longicyclic, unlike pericyclics and spirocyclics for which $4n$ electrons are necessary for a nonstabilized system.

Goldstein and Hoffmann have given some interesting examples of both stabilized and nonstabilized longicyclics, and although some are as yet unknown, there is good experimental evidence in support of the above predictions about stabilities.

For example, barrelene X is a $(2°2°2°)$ longicyclic, thus its ribbon

$2° + 2° + 2°$
$(2,2,2)$

X

modes are $(2,2,2)$. It is therefore predicted to be nonstabilized, in agreement with both the results of simple MO calculations and with experimental evidence.[2] The hydrogenation of barrelene is more exothermic than that of most other alkenes, suggesting that its ethylenic systems are of very high energy. The norbornadienyl anion XI is a $(2°2°1^-)$ system, which means that its modes are also $(2,2,2)$. All reasonable attempts to generate this anion have failed, whereas there is ample evidence from reaction kinetics that the norbornadienyl cation XI $(2,2,0)$ is a stabilized reaction intermediate.[3a,b] Similarly there is evidence that in the $(3,2,2)$ bicyclic system it is the anion XIII which is stabilized, whereas the cation XIV is not.[4a,b] In this system the anion is $(2°,2°,3^-)$ with modal interactions of the stabilized $(2,2,0)$ type, but the cation is $(2°,2°,3^+)$ with modal interactions of the nonstabilized $(2,2,2)$ type.

## IX.1 Aromaticity and Topology

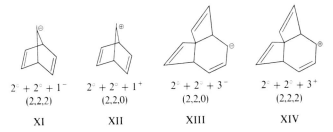

| XI | XII | XIII | XIV |
|---|---|---|---|
| 2° + 2° + 1⁻ (2,2,2) | 2° + 2° + 1⁺ (2,2,0) | 2° + 2° + 3⁻ (2,2,0) | 2° + 2° + 3⁺ (2,2,2) |

Similar supporting evidence comes from rate comparisons of the following systems.[5] (See Fig. IX.13.) From this type of result, Goldstein and Hoffman suggest that there is more stabilization available in a longicyclic than in a pericyclic system. In these cases there are clearly more stabilizing (2,0) interactions available to the longicyclic. Similarly, from the rates of allylic proton exchange[6] in the bicyclics (Fig. IX.14) it appears that the (2,2,0) longicyclic can again achieve more stabilization than the analogous (2,0) pericyclic system.

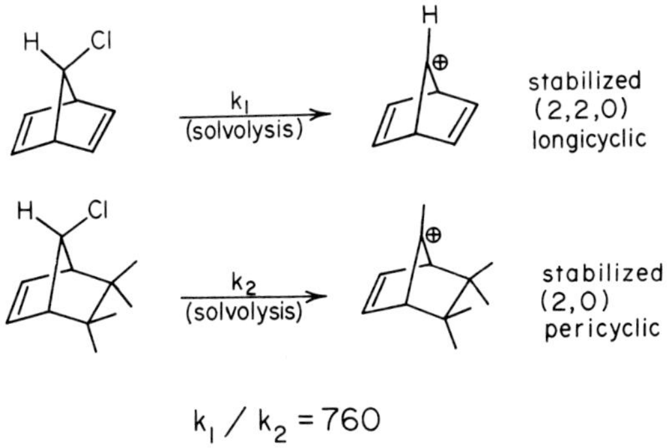

$k_1 / k_2 = 760$

FIGURE IX.13

$k_1 / k_2 = 750$

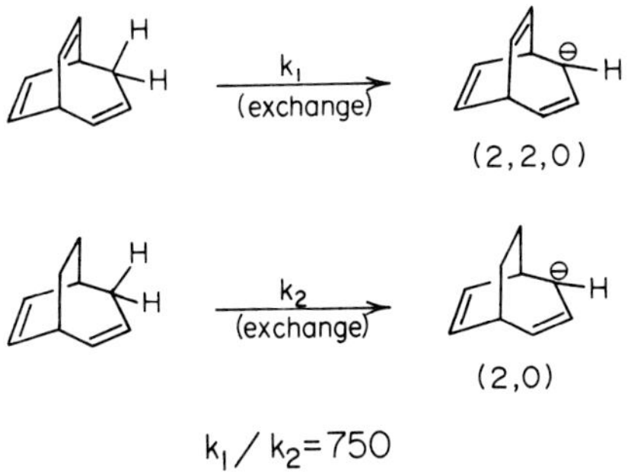

FIGURE IX.14

## IX.1.5 LATICYCLIC SYSTEMS

The same approach can be used for laticyclics, again considering the various terminal interactions of three or more ribbons with $\psi p$ or $\psi d$ pseudosymmetries. However, although these interactions are permutationally invariant, interactions of exterior ribbons with each other are probably much weaker than their interactions with interior or central ribbons. Therefore, it is necessary to define explicitly that (2,0,2) or preferably $(2 + 0 + 2)$ means that the central ribbon is mode 0. Thus in modal terms $(2 + 0 + 2)$ is not equivalent to $(2 + 2 + 0)$ in laticyclic interactions. Thus for a given set of three ribbons there are various possibilities of modal interaction, depending on the geometrical arrangement of the ribbons. This leads to a hierarchy of stabilities for laticyclics, which depends on the possible numbers of stabilizing (2,0) and destabilizing (0,0) or (2,2) interactions (see Table IX.3). Again

TABLE IX.3

|  |  | Modal sum | Total number of electrons |
|---|---|---|---|
| Stabilized by two interactions | $(0 + 2 + 0)$ | 2 | $4n + 2$ |
|  | $(2 + 0 + 2)$ | 0 | $4n$ |
| Stabilized by one interaction | $(0 + 0 + 2)$ | 2 | $4n + 2$ |
|  | $(2 + 2 + 0)$ | 0 | $4n$ |
| Not stabilized | $(0 + 0 + 0)$ | 0 | $4n$ |
|  | $(2 + 2 + 2)$ | 2 | $4n + 2$ |

it is noted that either $(4n)$ or $(4n + 2)$ electrons can lead to stability, or indeed, to instability. Thus there is no simple rule such as Hückels rule for these topologies.

There is little evidence as yet concerning the relative stabilities of laticyclic systems, but the order of stabilities predicted for typical laticyclic arrangements is shown in Fig. IX.15.

## IX.1.6 GENERALIZATIONS

From similar considerations, the general conclusion may be drawn that for three-ribbon systems of various topologies, the expected order

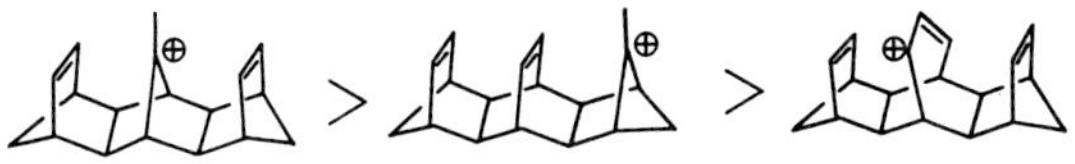

## IX.1 Aromaticity and Topology

Type: $(2^\circ + 1^+ + 2^\circ)$   $(2^\circ + 2^\circ + 1^+)$   $(2^\circ + 3^+ + 2^\circ)$

Modal characteristics: $(2+0+2)$   $(2+2+0)$   $(2+2+2)$

FIGURE IX.15

of stability is

(i) symmetrical but nonhomogeneous† laticyclics,
(ii) isoconjugate longicyclics,
(iii) isomeric unsymmetrical laticyclics,
(iv) homogeneous laticyclics, and
(v) homogeneous longicyclics.

Examples of these five types are given in Fig. IX.16. Further examples of these types of system and their stabilities will be discussed in the next section on HMO calculations on nonplanar polyene systems.

The above discussion can be extended to other topologies since all annelations implicitly involve combinations of simpler pericyclic interactions. For any topology, however large and complex, a hierarchy of stabilities can be set up simply by counting the number of $(0 + 2)$ stabilizing interactions and subtracting the number of those $(0 + 0)$ or $(2 + 2)$ interactions that are not stabilizing. Extension of the above

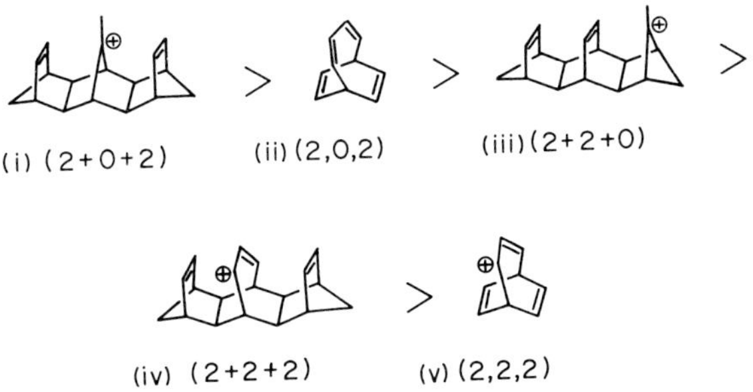

(i) $(2+0+2)$   (ii) $(2,0,2)$   (iii) $(2+2+0)$

(iv) $(2+2+2)$   (v) $(2,2,2)$

FIGURE IX.16

† In a modal sense; thus $(2 + 0 + 2)$ is nonhomogeneous whereas $(2 + 2 + 2)$ is homogeneous, where both are symmetrical laticyclics.

ideas to four-ribbon longicyclics and laticyclics is reasonably straightforward, but care must be taken in considering spirocyclics and the various possible $\psi$p- and $\psi$d-type interactions in these systems.

It is very interesting that although the total number of electrons involved in these systems does not give any general indication of stability, one version of Hückel's rule has survived in the sense that stabilization can only come from individual (0 + 2) modal interactions, and each such interaction always involves $(4n + 2)$ electrons.

This leads to a useful generalization that is due to Goldstein. It is clear that stabilization in a bicyclic topology is not associated with any magic number of electrons, such as $(4n + 2)$ in a pericyclic system. However, in order to achieve stabilization, particularly for longicyclic topologies, the system must consist of component electronic systems that are nonhomogeneous. In other words, if one component of the bicyclic system contains $(4n + 2)$ electrons, at least one other component must contain $(4n)$ electrons. This is also true of laticyclic systems, although the degree of stabilization will depend on whether the nonhomogeneous components are symmetrically or unsymmetrically arranged. For longicyclic systems this requirement of nonhomogeneity can be thought of as the analog of Hückel's rule for pericyclic systems.

### IX.1.7 AROMATICITY AND BICYCLOAROMATICITY

The concept of aromaticity is a well-known one and is frequently used by organic chemists in describing pericyclic systems, although precisely what constitutes this special quality is not always clearly defined. One useful way of defining aromaticity[7] is based on comparisons of pericyclic and analogous acyclic π-systems. For example, if we compare benzene and hexatriene, these two structures contain equivalent numbers of π-electrons (and formal double bonds) and can be said to be isoconjugate. Therefore we could reasonably define aromaticity in terms of the special stabilization that arises when the acyclic topology of hexatriene is closed to form the pericyclic topology of benzene. Since benzene has an HMO energy $E_\pi = 6\alpha + 8\beta$, and hexatriene has $E_\pi = 6\alpha + 6.99\beta$, the pericyclic is clearly strongly stabilized with respect to its isoconjugate acyclic, and the difference $\Delta E_\pi$ could be defined as constituting aromatic stabilization.

The same kind of approach is illustrated more generally in the Table IX.4 where some simple cyclic and acyclic π-systems are compared.

## IX.1 Aromaticity and Topology

Table IX.4

| Pericyclic | $E_\pi$ | Isoconjugate acyclic | $E_\pi$ | $\Delta E_\pi$ |
|---|---|---|---|---|
| cyclopropenyl cation | $2\alpha + 4\beta$ | allyl cation | $2\alpha + 2.83\beta$ | $+1.17\beta$ |
| cyclopropenyl anion | $4\alpha + 2\beta$ | allyl anion | $4\alpha + 2.83\beta$ | $-0.83\beta$ |
| cyclobutadiene | $4\alpha + 4\beta$ | butadiene | $4\alpha + 4.48\beta$ | $-0.48\beta$ |
| cyclopentadienyl cation | $4\alpha + 5.24\beta$ | pentadienyl cation | $4\alpha + 5.46\beta$ | $-0.22\beta$ |
| cyclopentadienyl anion | $6\alpha + 6.48\beta$ | pentadienyl anion | $6\alpha + 5.46\beta$ | $+1.02\beta$ |
| benzene | $6\alpha + 8\beta$ | hexatriene | $6\alpha + 6.99\beta$ | $+1.01\beta$ |
| tropylium cation | $6\alpha + 9\beta$ | heptatrienyl cation | $6\alpha + 8.05\beta$ | $+0.95\beta$ |
| tropylium anion | $8\alpha + 8.10\beta$ | heptatrienyl anion | $8\alpha + 8.05\beta$ | $+0.05\beta$ |
| cyclooctatetraene | $8\alpha + 9.6\beta$ | octatetraene | $8\alpha + 9.52\beta$ | $0.08\beta$ |

By this definition, systems with significant positive $\Delta E_\pi$ values (comparable in magnitude with that of benzene) could be said to be aromatic, that is, the cyclopropenyl cation, cyclopentadienyl anion, and tropylium cation. Pericyclic systems whose acyclic analog is significantly more stabilized than they are could then be defined to be antiaromatic, that is, the cyclopropenyl anion, cyclobutadiene, and the cyclopentadienyl cation. Other systems such as the tropylium anion or cyclooctatetraene, where there is little difference in $E_\pi$ between the cyclic and acyclic forms, could be said to be nonaromatic.

This kind of approach can be extended to bicyclic systems to define a property called bicycloaromaticity.[7] In this case the choice of reference system is less clear, but Goldstein has chosen the isoconjugate bishomo pericyclic system. The situation is also complicated by the fact that numerical values of $E_\pi$ for bicyclic and bishomo systems will depend on the particular choice of bond integrals used (see next section). Nonetheless, a qualitative approach can be adopted since it is clear from the foregoing discussion of ribbon interactions which bicyclic systems should be stabilized and which should be nonstabilized. Similarly, it is clear from the simple pericyclic analogs which bishomo systems should be stabilized and which should not, although the numerical value of such stabilizations may not be specified. In this case only if a certain bicyclic system is stabilized and its isoconjugate bishomo pericyclic is not, the system will be defined to be bicycloaromatic. The qualitative comparisons shown in Table IX.5 illustrate the general idea involved. The limited experimental evidence available, as described previously, is in reasonable accord with the definition of special stabilization for bicyclic systems.

TABLE IX.5

| Bicyclic system | Isoconjugate bishomo pericyclic | Bicycloaromaticity |
|---|---|---|
| norbornadienyl$^\oplus$ (stabilized) | bishomocyclopentadienyl$^\oplus$ (nonstabilized) | yes |
| norbornadienyl$^\ominus$ (nonstabilized) | bishomocyclopentadienyl$^\ominus$ (stabilized) | no |
| barrelene (nonstabilized) | bishomocyclohexatriene (stabilized) | no |
| bicyclo[3.2.2]nonatrienyl$^\oplus$ (nonstabilized) | bishomotropylium$^\oplus$ (stabilized) | no |
| bicyclo[3.2.2]nonatrienyl$^\ominus$ (stabilized) | bishomotropylium$^\ominus$ (nonstabilized) | yes |

## IX.2  HMO CALCULATIONS ON NONPLANAR SYSTEMS

Since nonplanar polyene systems of various topologies have just been discussed and compared in a qualitative way, it is of some interest to obtain a more quantitative idea of the stabilities and other properties

## IX.2 HMO Calculations on Nonplanar Systems

of such systems. This can be done by carrying out HMO calculations on some typical nonplanar systems. Although the method of calculation for this type of system is inherently no different from that used for the planar pericyclic systems treated in the first part of this book, there are enough points of difference in the calculations themselves to make it of interest to follow these through for two nonplanar $\pi$-systems to illustrate potential difficulties and sources of error.

The two systems that will be considered are simple representatives of the longicyclic and laticyclic topologies, namely, the norbornadienyl system XV and the isoconjugate laticyclic XVI. In this way the $(2°1^+2°)$

XV        XVI

cation can be compared with the $(2°1^-2°)$ anion in each case, and also the results for the longicyclic topology can be compared with those for the corresponding laticyclic system.

### IX.2.1  THE NORBORNADIENYL SYSTEM

As before, the first step in the HMO treatment is to draw the basic $\pi$-structure and label each position (Fig. IX.17). Also, to simplify the calculation, the symmetry properties of the system will be used. Norbornadienyl can clearly be placed in the $C_{2v}$ group, with the symmetry elements as shown. However, before carrying out any symmetry operations or selecting the symmetry orbitals, it is necessary

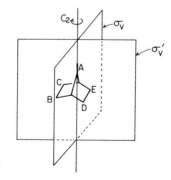

FIGURE IX.17

to specify the orbitals of the basis set clearly and unambiguously. The reason for this will become obvious. One basis set array is as shown

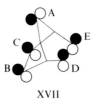

XVII

(XVII). This can be designated as $\{+\phi_A + \phi_B + \phi_C + \phi_D + \phi_E\}$. Once this particular sign convention is chosen it must be retained throughout the calculation. Any Coulomb integrals of the type $H_{AB}$ will be positive $\beta$ terms, as will integrals of the type $H_{BD}$, since as set up the orbital interactions involved in these integrals are favorable. However, any Coulomb integrals of the type $H_{AD}$ will be negative $\beta$ terms, since as defined the interactions involved will be unfavorable or antibonding.

Next the transformation table is set up as before, noting that certain symmetry operations on nonplanar systems may transform a particular orbital into the negative of itself or the negative of another orbital. This is shown in Table IX.6 and IX.7. Since the symmetry operations may transform $\phi_A$ into either $\phi_A$ or $-\phi_A$, the numbers of positions unchanged in label at the bottom of the table may be positive or negative. The numbers of independent symmetry orbitals belonging to each type of irreducible representation are

$$n_{\Gamma_{A_1}} = \frac{5 - 1 - 1 + 1}{4} = 1, \qquad n_{\Gamma_{A_2}} = \frac{5 - 1 + 1 - 1}{4} = 1$$

$$n_{\Gamma_{A_1}} = \frac{5 + 1 - 1 - 1}{4} = 1, \qquad n_{\Gamma_{A_2}} = \frac{5 + 1 + 1 + 1}{4} = 2$$

TABLE IX.6

Transformation table

| | E | $C_2$ | $\sigma_v$ | $\sigma_v'$ |
|---|---|---|---|---|
| | $\phi_A$ | $-\phi_A$ | $-\phi_A$ | $\phi_A$ |
| | $\phi_B$ | $\phi_E$ | $\phi_D$ | $\phi_C$ |
| | $\phi_C$ | $\phi_D$ | $\phi_E$ | $\phi_B$ |
| | $\phi_D$ | $\phi_C$ | $\phi_B$ | $\phi_E$ |
| | $\phi_E$ | $\phi_B$ | $\phi_C$ | $\phi_D$ |
| | 5 | $-1$ | $-1$ | 1 |

TABLE IX.7

Character table

| | E | $C_2$ | $\sigma_v$ | $\sigma_v'$ |
|---|---|---|---|---|
| $A_1$ | 1 | 1 | 1 | 1 |
| $A_2$ | 1 | 1 | $-1$ | $-1$ |
| $B_1$ | 1 | $-1$ | 1 | $-1$ |
| $B_2$ | 1 | $-1$ | $-1$ | 1 |

## IX.2 HMO Calculations on Nonplanar Systems

Proceeding as before by taking the appropriate dot products between the rows of the transformation table (Table IX.6) and character table (Table IX.7), the symmetry orbitals chosen are:

$$S_1 = \tfrac{1}{2}(\phi_B + \phi_C + \phi_D + \phi_E)$$
$$S_2 = \tfrac{1}{2}(\phi_B - \phi_C - \phi_D + \phi_E)$$
$$S_3 = \tfrac{1}{2}(\phi_B - \phi_C + \phi_D - \phi_E)$$
$$S_4 = \phi_A$$
$$S_5 = \tfrac{1}{2}(\phi_B + \phi_C - \phi_D - \phi_E)$$

Particular care should be taken, especially with more complex nonplanar systems, to check that these symmetry orbitals do transform under each group operation with the correct character, before beginning the calculations. This is clearly the case for the simple system shown in Fig. IX.18.

Before writing down the simplified secular determinant, it is first necessary to assign an appropriate value (as well as sign) to each Coulomb integral, since there are now several different types of nearest-neighbor overlap. As in the simpler systems considered previously $H_{BC} = H_{DE} = +\beta$ since these refer to normal ethylene type $2p_z$–$2p_z$ interactions. However, the transannular interactions between $\phi_A$ and $\phi_B$ or between $\phi_B$ and $\phi_D$ will be of reduced effectiveness. For simplicity it will be assumed that each of these will be only half as effective as the normal parallel p-orbital interaction. Cross-ring interactions such as between $\phi_B$ and $\phi_E$ will be assumed to be zero since

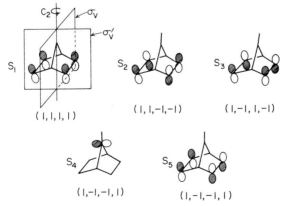

FIGURE IX.18

these do not involve centers directly joined by one insulating center. Thus

$$H_{BC} = H_{DE} = \beta, \qquad H_{AB} = H_{AC} = \beta/2$$
$$H_{BD} = H_{CE} = \beta/2, \qquad H_{AD} = H_{AE} = -\beta/2$$

The simplified secular determinant is then

$$\begin{vmatrix} H_{11} - \varepsilon & & & & \\ & H_{22} - \varepsilon & & & \\ & & H_{33} - \varepsilon & & \\ & & & H_{44} - \varepsilon & H_{45} \\ & & & H_{54} & H_{55} - \varepsilon \end{vmatrix} = 0$$

The individual terms can easily be evaluated as before, giving

$$H_{11} = \alpha + 3\beta/2, \qquad H_{44} = \alpha$$
$$H_{22} = \alpha - 3\beta/2, \qquad H_{55} = \alpha + \beta/2$$
$$H_{33} = \alpha - \beta/2, \qquad H_{45} = \beta$$

Expressed in terms of $X$, the determinant becomes

$$\begin{vmatrix} X + 1.5 & & & & \\ & X - 1.5 & & & \\ & & X - 0.5 & & \\ & & & X & 1 \\ & & & 1 & X + 0.5 \end{vmatrix} = 0$$

This yields the following equations and roots:

$$P_{A_1} = x + 1.5 = 0, \qquad x_1 = -1.5$$
$$P_{A_2} = x - 1.5 = 0, \qquad x_5 = +1.5$$
$$P_{B_1} = x - 0.5 = 0, \qquad x_3 = +0.5$$
$$P_{B_2} = x^2 + 0.5x - 1 \qquad x_2 = -1.281$$
$$= 0, \qquad x_4 = +0.781$$

The energy level diagram for the cation and anion is depicted schematically in Fig. IX.19. Hence for the cation the energy terms are given by

$$E_\pi^+ = 4\alpha + 5.562\beta, \qquad B_\pi = 5.562\beta$$
$$DE^+ = 1.562\beta, \qquad DEPE^+ = 0.391\beta$$

## IX.2 HMO Calculations on Nonplanar Systems

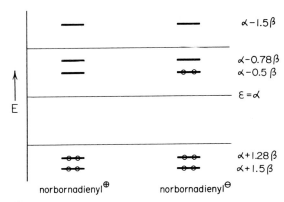

FIGURE IX.19

and for the anion the corresponding quantities are

$$E_\pi^- = 6\alpha + 4.562\beta, \qquad B_\pi^- = 4.562\beta$$
$$DE^- = 0.562\beta, \qquad DEPE^- = 0.094\beta$$

Thus the cation is predicted to be much more stable than the anion both on the basis of $\pi$-bonding energy and delocalization energy. This is clear from the fact that the cation has all four electrons in bonding levels, whereas the anion has two of its electrons in an antibonding level. These results are in excellent agreement with the qualitative Goldstein and Hoffmann approach, since the cation $(2°1^+2°)$ is modally a (2,0,2) system and therefore stabilized, whereas the anion $(2°1^-2°)$ is modally a nonstabilized (2,2,2) system. It is also worth noting that in the Möbius–Hückel sense, the norbornadienyl system is Möbius. Therefore, it should be a stable system when it contains $(4n)$ but not $(4n + 2)$ electrons, as supported by the calculated HMO results on the cation (four electrons) and anion (six electrons).

The MO coefficients for norbornadienyl are easy to obtain from the simplified secular equations. In the first three cases (where $x_i = -1.5$, $+1.5$, $+1.5$, $+0.5$) the solution of the secular equation is trivial since $\psi_i$ has the same coefficients as the appropriate symmetry orbital. For the roots $x_2 = -1.281$ and $x_4 = +0.781$ substitution into the equations,

$$C_1 x + C_2 = 0, \qquad C_1 + C_2(x + 0.5) = 0$$

followed by normalization, gives the other two sets of coefficients. The five sets of MO coefficients are as tabulated in Table IX.8, and the

*TABLE IX.8*

Norbornadienyl MOs

| $\psi_i$ | $x_i$ | $C_A$ | $C_B$ | $C_C$ | $C_D$ | $C_E$ | $\varepsilon_i$ |
|---|---|---|---|---|---|---|---|
| $\psi_1$ | $-1.5$ | 0 | 0.5 | 0.5 | 0.5 | 0.5 | $\alpha + 1.5\beta$ |
| $\psi_2$ | $-1.281$ | 0.615 | 0.394 | 0.394 | $-0.394$ | $-0.394$ | $\alpha + 1.281\beta$ |
| $\psi_3$ | $+0.5$ | 0 | 0.5 | $-0.5$ | 0.5 | $-0.5$ | $\alpha - 0.5\beta$ |
| $\psi_4$ | $+0.781$ | 0.788 | $-0.308$ | $-0.308$ | 0.308 | 0.308 | $\alpha - 0.781\beta$ |
| $\psi_5$ | $+1.5$ | 0 | 0.5 | $-0.5$ | $-0.5$ | 0.5 | $\alpha - 1.5\beta$ |

MOs themselves are depicted schematically in Fig. IX.20. Note that the increase in number of nodal regions is not as simply related to the energy as with planar systems. It is of interest to compare the calculated electron densities in the cation and anion. These are given in Table IX.9.

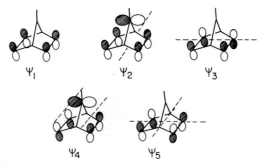

*FIGURE IX.20*

*TABLE IX.9*

| Norbornadienyl$^+$ | Norbornadienyl$^-$ |
|---|---|
| $q_A = 0.756$; $\xi_A = +0.244$ | $q_A = 0.756$; $\xi_A = +0.244$ |
| $q_B = q_C = q_D = q_E = 0.809$; $\xi_B = 0.189$ | $q_B = 1.309$; $\xi_B = -0.309$ |
| $\rho_{AB} = 0.485$ | $\rho_{AB} = 0.485$ |
| $\rho_{BC} = 0.810$ | $\rho_{BC} = 0.310$ |
| $\rho_{BD} = 0.190$ | $\rho_{BD} = 0.689$ |
| (+0.24) 0.485 (+0.19) 0.810 0.190 | (+0.24) 0.485 (−0.31) 0.310 0.689 |

## IX.2 HMO Calculations on Nonplanar Systems

Although both systems are delocalized, as expected the ionic charge is more evenly distributed over the bicyclic framework in the cation than in the anion. In the anionic system, the bond order is greatly reduced in the "double-bond" regions in comparison with ethylene.

### IX.2.2 THE $(2° + 1 + 2°)$ LATICYCLIC SYSTEM

The approach used is very similar to that given previously, except that the interacting orbitals in the laticyclic case XVIII are best represented in planar projection form XIX. $C_{2v}$ symmetry is used in this

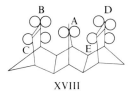

XVIII

case also, and the sign conventions adopted for the AO basis set are as shown. Bond integrals of the type $H_{AB}$ will be set equal to $3\beta/4$, since

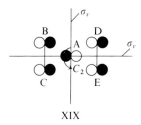

XIX

inspection of models shows that interactions of this type will be much more effective than any of the transannular interactions in norbornadienyl systems, (which were set equal to $\beta/2$), and will probably be nearly as effective as those of the $H_{bc}$ type, which are normally set at $\beta$. Thus the bond integral parameters used for this calculation are

$$H_{AB} = H_{AC} = H_{AD} = H_{AE} = 3\beta/4$$
$$H_{BC} = H_{DE} = \beta$$
$$H_{BD}^\dagger = H_{CE}^\dagger = H_{BE} = H_{CD} = 0$$

---

† The bond integrals representing these interactions are probably not zero, since overlap between AOs on these centers is significant. However, they have been set equal to zero for the sake of simplicity since their inclusion would not materially affect the final results in terms of a comparison of the cation and anion.

TABLE IX.10

| | E | $C_2$ | $\sigma_v$ | $\sigma_v$ |
|---|---|---|---|---|
| | $\phi_A$ | $-\phi_A$ | $\phi_A$ | $-\phi_A$ |
| | $\phi_B$ | $-\phi_E$ | $\phi_C$ | $-\phi_D$ |
| | $\phi_C$ | $-\phi_D$ | $\phi_B$ | $-\phi_E$ |
| | $\phi_D$ | $-\phi_C$ | $\phi_E$ | $-\phi_B$ |
| | $\phi_E$ | $-\phi_B$ | $\phi_D$ | $-\phi_D$ |
| | 5 | $-1$ | 1 | $-1$ |

The transformation table under the $C_{2v}$ operations is given in Table IX.10 and the numbers of symmetry orbitals of each type are thus

$$N_{A_1} = 1, \quad N_{A_2} = 1, \quad N_{B_1} = 2, \quad N_{B_2} = 1$$

These symmetry orbitals are as follows:

$$S_1 = \tfrac{1}{2}(\phi_B + \phi_C - \phi_D - \phi_E), \quad S_4 = \tfrac{1}{2}(\phi_B + \phi_C + \phi_D + \phi_E)$$
$$S_2 = \tfrac{1}{2}(\phi_B - \phi_C + \phi_D - \phi_E), \quad S_5 = \tfrac{1}{2}(\phi_B - \phi_C - \phi_D + \phi_E)$$
$$S_3 = \phi_A,$$

The simplified secular determinant is then

$$\begin{vmatrix} H_{11} - \varepsilon & & & & \\ & H_{22} - \varepsilon & & & \\ & & H_{33} - \varepsilon & H_{34} & \\ & & H_{43} & H_{44} - \varepsilon & \\ & & & & H_{55} - \varepsilon \end{vmatrix} = 0$$

Evaluation of the matrix elements in terms of $\alpha$ and $\beta$ gives

$$H_{11} = \alpha + \beta$$
$$H_{22} = \alpha - \beta$$
$$H_{33} = \alpha, \quad H_{34} = 3\beta/2$$
$$H_{44} = \alpha + \beta$$
$$H_{55} = \alpha - \beta$$

which with the usual substitutions gives

$$\begin{vmatrix} x+1 & & & & \\ & x-1 & & & \\ & & x & 1.5 & \\ & & 1.5 & x+1 & \\ & & & & x-1 \end{vmatrix} = 0$$

## IX.2 HMO Calculations on Nonplanar Systems

From solution of this determinant the roots and energy values are

$$x_2 = -1 \quad \text{(from } \Gamma_{A_1}\text{)}, \quad x_3 = +1 \quad \text{(from } \Gamma_{A_2}\text{)}$$
$$x_1 = -2.081 \quad \text{and} \quad x_5 = +1.081 \quad \text{(from } \Gamma_{B_1}\text{)},$$
$$x_4 = +1 \quad \text{(from } \Gamma_{B_2}\text{)}$$

Thus the energy level diagram for the cation and anion is as shown in Fig. IX.21. From the given eigenvalues and occupancies, the energy terms of interest are, for the cation,

$$E_\pi^+ = 4\alpha + 6.162\beta, \qquad B_\pi = 6.162\beta$$
$$DE^+ = 2.162\beta, \qquad DEPE^+ = 0.541\beta$$

and for the anion,

$$E_\pi^- = 6\alpha + 4.162\beta, \qquad B_\pi = 4.162\beta$$
$$DE^- = 0.162\beta, \qquad DEPE^- = 0.027\beta$$

Therefore, for the laticyclic systems also, the cation is significantly stabilized, while the anion is not stabilized.† This is in further agreement

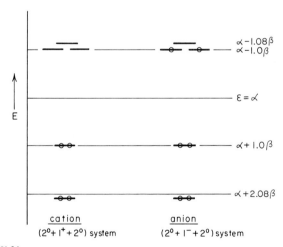

FIGURE IX.21

---

† The fact that the anion turns out to have a diradical ground-state arrangement is not significant, since $\psi_3$, $\psi_4$, and $\psi_5$ are all very similar in energy; small changes in the $\beta$ terms could easily result in $\psi_3$ being raised in energy above $\psi_4$ and $\psi_5$. In addition, the degeneracy of these two MOs would be removed if bond integrals of the $H_{BD}$ type were included.

with the Goldstein–Hoffman approach,[1] since the cation is modally a $(2 + 0 + 2)$ system, while the anion is $(2 + 2 + 2)$. It is also interesting that, as suggested by these authors, the stabilization available to the $(2°1^+2°)$ laticyclic system is calculated to be significantly greater than that found for the analogous longicyclic system, the norbornadienyl cation. This calculated result is partly dependent on the fact that the transannular bond integral terms in the laticyclic case were (reasonably) assigned higher values than the analogous terms in the longicyclic system. However, there are more nonzero bond integral terms in the latter, so that the result is probably still valid on a comparative basis.

The coefficients are easy to calculate from the simplified secular equations and are as given in Table IX.11.

TABLE IX.11

| $\psi_i$ | $x_i$ | $C_A$ | $C_B$ | $C_C$ | $C_D$ | $C_E$ | $\varepsilon_i$ |
|---|---|---|---|---|---|---|---|
| $\psi_1$ | −2.081 | 0.585 | 0.406 | 0.406 | 0.406 | 0.406 | $\alpha + 2.081\beta$ |
| $\psi_2$ | −1.000 |  | 0.500 | 0.500 | −0.500 | −0.500 | $\alpha + \beta$ |
| $\psi_3$ | +1.000 |  | −0.500 | −0.500 | 0.500 | −0.500 | $\alpha - \beta$ |
| $\psi_4$ | +1.000 |  | 0.500 | −0.500 | −0.500 | 0.500 | $\alpha - \beta$ |
| $\psi_5$ | +1.081 | 0.811 | −0.292 | −0.292 | −0.292 | −0.292 | $\alpha - 1.081\beta$ |

Again, the electron distributions in the cation and anion are interesting, and are shown in Table IX.12. Once more, the stabilized (delocalized) cation has a more even charge distribution than the nonstabilized anion, and the bond orders in the "double bond" regions are greatly reduced in the anion.

Similar calculations can easily be carried out on more complex nonplanar systems providing care is taken with the sign conventions and

TABLE IX.12

| Cation | Anion |
|---|---|
| $q_A = 0.685$, $\xi_A = +0.315$ | $q_A = 0.685$, $\xi_A = +0.315$ |
| $q_B = 0.830$, $\xi_B = +0.170$ | $q_B = 1.330$, $\xi_B = -0.330$ |
| $\rho_{BC} = 0.830$ | $\rho_{BC} = 0.330$ |
| $\rho_{AB} = 0.475$ | $\rho_{AB} = 0.475$ |

symmetry operations. The next section describes a simple and rapid approach to the evaluation of the energy level patterns, stabilities, and molecular orbitals of these and other systems without doing any numerical calculations at all.

## IX.3  ORBITAL INTERACTION DIAGRAMS

The fundamental building blocks of organic $\pi$-systems are the $sp^2$-hybridized carbons that form the $\pi$-framework of the molecules. Similarly, the basic components of the molecular orbitals that describe the behavior of $\pi$-electrons in these systems are the $p_z$-type orbitals on each center. It is quite simple, starting from these basic components, to arrive at representations of the MOs in more complex systems by allowing simpler orbital systems to interact using the following rules that are based on quantum mechanical ideas, and particularly on perturbation theory. In addition, it is possible to rank the MOs thus obtained in approximate order of increasing energy, and to make reasonable estimates of which delocalized systems would be expected to be stabilized and which would be nonstabilized relative to simpler localized systems such as ethylene. This approach, which involves the use of orbital interaction diagrams, is merely qualitative and pictorial and the MOs obtained only give the phases of the wave function in each nuclear region; that is to say, only the signs and not the magnitudes of the MO coefficients are obtained. Similarly, any stabilization energies estimated from these diagrams are only roughly qualitative. Nonetheless, it is possible to obtain very useful insights into the nature of typical $\pi$-systems, their MOs, energy level distributions, and stabilities without doing any calculations at all.[8]

### General Rules for Orbital Interactions

1. When two orbitals whose energy levels are degenerate or nearly degenerate interact, they combine strongly to yield a new level of significantly lower energy and a corresponding level of significantly higher energy. Any new node developed as a result of this interaction is taken by the higher energy or antibonding combination.

2. When two levels of unlike energy interact, they can also combine to yield a new level of lower energy and one of higher energy in which the lower energy orbital has mixed into the higher orbital in an anti-

bonding manner, and the higher orbital has mixed into the lower in a bonding manner.

3. Only levels that are reasonably close to each other in energy interact strongly; the more nearly degenerate they are, the stronger the interaction and splitting to give the new levels.

4. Generally the new higher energy level is somewhat more destabilized than the new lower energy level is stabilized, as a result of the interaction.

5. Only levels of like symmetry (or pseudosymmetry)[†] can interact strongly. Orbitals of different symmetry types that have no net overlap will have zero interaction.

6. If any one level ($A$) interacts with two or more other levels ($B, C, \ldots$) of different energy, the interactions are additive in a pairwise manner and the results of these interactions are permutationally invariant. However, if one of the levels ($B$) is of the same energy as another ($A$), this interaction must be taken into account first.

7. Although it is fairly clear from the above, it is important to emphasize that any $m$ levels interacting with any other $n$ levels must generate exactly $(m + n)$ new levels.

We begin by constructing an orbital interaction diagram (Fig. IX.22) for a very simple π-system, that of ethylene, which can be thought of as

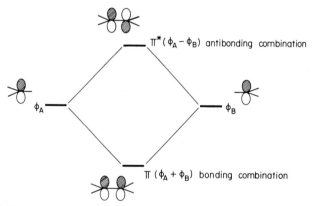

FIGURE IX.22

[†] It will be assumed that terminal orbital interactions are generally of controlling importance in producing the new levels.

## IX.3 Orbital Interaction Diagrams

arising from a combination of two $p_z$-type orbitals on adjacent centers. This gives the familiar pattern of $\pi$ and $\pi^*$ orbitals and levels of the ethylene system. Suppose we now allow two ethylenes to combine in a linear fashion to form butadiene. The two $\pi$-levels of the interacting systems are degenerate, as are the two $\pi^*$ levels, therefore we would expect the strongest pairwise interactions to involve orbitals of each type. In addition, the $\pi$-levels developed are of different symmetry to the $\pi^*$ levels; thus any $\pi$–$\pi^*$ interactions will be zero. (It will be recalled that only MOs of the same symmetry class as the molecule are acceptable wave functions. Interactions between $\pi$ and $\pi^*$ levels would lead to unacceptable functions and hence are not allowed.) The orbital interaction diagram is as shown in Fig. IX.23. Thus $\pi_1$ and $\pi_2$ combine in a bonding fashion to give $(\pi_1 + \pi_2)$, or from the previous diagram, a $(\phi_A + \phi_B + \phi_C + \phi_D)$ combination. They also interact in an antibonding fashion to give $(\pi_1 - \pi_2)$ or $(\phi_A + \phi_B - \phi_C - \phi_D)$. Similarly $\pi_1^*$ and $\pi_2^*$ interact to give a bonding combination $(\pi_1^* - \pi_2^*)$ or $(\phi_A - \phi_B - \phi_C + \phi_D)$ and an antibonding combination $(\pi_1^* + \pi_2^*)$ or $(\phi_A - \phi_B + \phi_C - \phi_D)$. These signs of the wave function correspond exactly to those of the calculated MO coefficients of butadiene, although naturally the numerical weights are not given. Also the ordering of the four resulting levels is in good correspondence with the calculated eigenvalues. It can be deduced from a comparison of the electronic

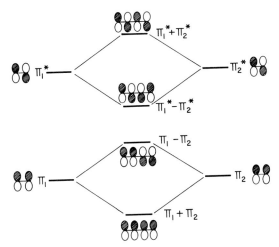

FIGURE IX.23

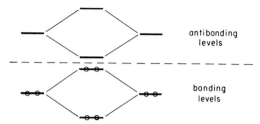

**FIGURE IX.24**

energies in the interaction that butadiene would not be greatly stabilized relative to two ethylenes as a result of the formation of these new levels. While two of the electrons are decreased in energy (see Fig. IX.24), the other two are increased in energy to roughly the same extent. Thus, qualitatively, one would conclude that butadiene is not a particularly stabilized system in the delocalized sense, which is in agreement with the calculated *DE* value.

The same kind of interaction diagram can be constructed for the allyl system (Fig. IX.25) by combining the $\pi$-orbitals of ethylene with the $p_z$ orbital of an isolated sp² carbon. In this case, although it is clear that three new levels will result, the new orbitals are not quite as easy to depict. First, $\phi_C$ will mix into $\pi$ in a bonding manner, producing the orbital $(\pi + \phi_C)$ or from the designation in the earlier diagram, $(\phi_A + \phi_B + \phi_C)$. Similarly $\phi_C$ will mix into $\pi^*$ in an antibonding manner giving $(\pi^* + \phi_C)$ or a $(\phi_A - \phi_B + \phi_C)$ combination. In addition, $\pi^*$ will mix into $\phi_C$ in a bonding manner and $\pi$ will mix into $\phi_C$ in an anti-

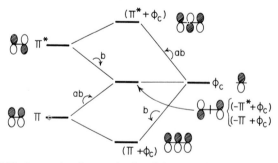

**FIGURE IX.25** *Interaction diagram for the allyl system showing three types of mixing, b = bonding; ab = antibonding.*

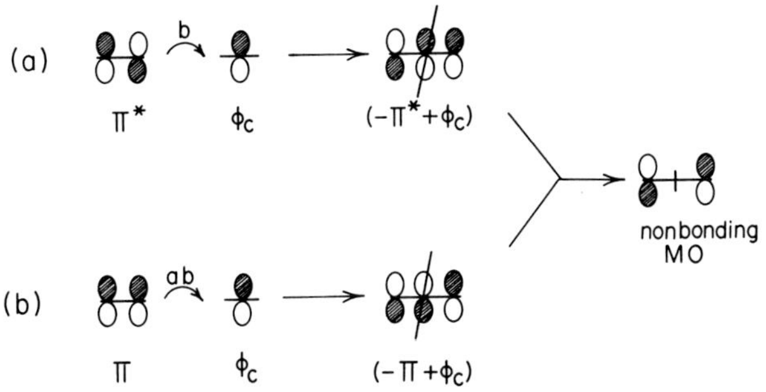

## IX.3 Orbital Interaction Diagrams

FIGURE IX.26

bonding manner. If we consider the first of these interactions this will yield the result in Fig. IX.26(a). Similarly the second interaction will yield the result shown in Fig. IX.26(b). The combination or superposition of these two will give an MO where the terminal lobes are reinforced since they are of similar phase, while the central lobe will be diminished since the wave function is of opposite sign in the two contributing forms. If we assume the two contributions are equal (or approximately so), the resulting MO will be of the same energy as $\phi_c$, that is, it will be a nonbonding MO, and will have zero value of the wave function at position 2. This set of orbitals is again in agreement with those calculated for the allyl system, both with respect to the signs of the wave function at various positions and also the ordering of their energy levels. If we again compare the energy of an allyl system (the cation) with respect to that of ethylene and an isolated carbon, it is clear from Fig. IX.27 that allyl should be a stabilized system, since one of the three interacting levels remains at the same energy, while the most stable level is significantly lowered in energy. Thus for the cation, the two $\pi$-electrons of the ethylene system would be significantly lowered in

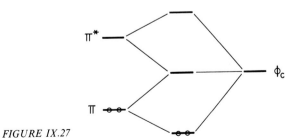

FIGURE IX.27

**350**   IX  *Symmetry, Topology, and Aromaticity*

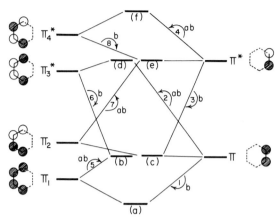

**FIGURE IX.28**

energy on formation of this cation. (The same conclusion would hold for the radical or anion, although the net stabilization would be less on a per electron basis.) This is again in agreement with the calculated HMO results.

More complicated acyclic systems can be generated by allowing end-on interactions of simpler polyene systems and constructing appropriate interaction diagrams. These could be used to illustrate why odd-membered linear polyenes have intrinsically more stable $\pi$-systems than the corresponding even-membered neighbors in the same way that ethylene and allyl are compared above.

It is now of interest to allow combinations of simple systems in a cyclic fashion and consider the orbital interactions that result. This will be done for the combination of the ethylene and butadiene $\pi$-systems, interacting at their termini to give the benzene system. (See Fig. IX.28) The interacting orbital systems are shown in vertical rather than horizontal projection. The individual orbital interactions are numbered for clarity and the final levels designated (a)–(f). (Note that the degeneracy of some of the levels does not come directly from the orbital interaction treatment, although it is clear from the treatment that some of the resulting levels will be very similar to each other in energy.) Considering the individual interactions one at a time, where the terminal symmetry characteristics are appropriate, we obtain:

1, which involves mixing of $\pi$ into $\pi_1$ in a bonding way, yielding a benzene orbital (a);

## IX.3  Orbital Interaction Diagrams

(1)

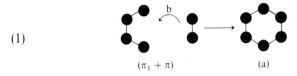

$(\pi_1 + \pi)$        (a)

2, which involves mixing of $\pi$ into $\pi_3^*$ in an antibonding way, yielding (d);

(2)

$(\pi_3^* - \pi)$        (d)

5 and 6, which involve mixing of $\pi_1$ into $\pi$ (antibonding) and $\pi_3^*$ into $\pi$ (bonding), respectively. Thus the resulting orbital is (b). Note that these are the only possible interactions involving these three orbitals $(\pi_1, \pi_3^*, \text{ and } \pi)$.

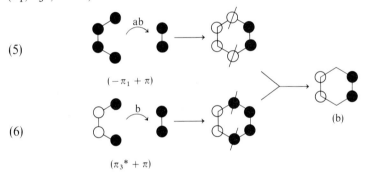

(5)

$(-\pi_1 + \pi)$

(6)

$(\pi_3^* + \pi)$        (b)

Next we consider:

3, which involves mixing of $\pi^*$ into $\pi_2$ in a bonding way, yielding (c);

(3)

$(\pi_2 + \pi^*)$        (c)

4, which involves mixing of $\pi^*$ into $\pi_4^*$ in an antibonding way, yielding (f);

(4)

$(\pi_4^* - \pi^*)$        (f)

7 and 8, which involve mixing of $\pi_2$ and $\pi_4^*$ into $\pi^*$ in an antibonding and bonding way, respectively, yielding (e).

(7)

(8)

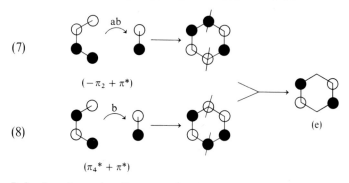

$(-\pi_2 + \pi^*)$

$(\pi_4^* + \pi^*)$

It is clear from the diagrams that levels (b) and (c) will be very similar in energy, as will levels (d) and (e). Thus the six benzene MOs derived from the interaction treatment and their approximate ordering in energy will be as shown in Fig. IX.29. It will be seen that these are in very good qualitative agreement with the phases of the MO wave functions calculated from the benzene MO coefficients, as is the ranking

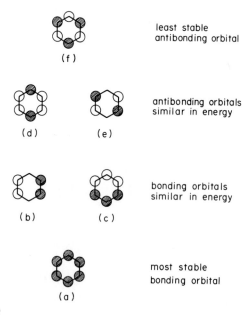

FIGURE IX.29

## IX.3 Orbital Interaction Diagrams

of the various MOs on an energy scale. (Again the relative magnitudes of the coefficients are not obtained.) It is also clear from the interaction diagram that benzene will be a stabilized $\pi$-system, relative to those of ethylene and butadiene, since of the three (occupied) bonding levels, two remain approximately unchanged in energy while the third is significantly lowered by the interactions shown.

It is very interesting to extend this approach to nonpericyclic topologies and obtain the results of orbital interactions in some of the systems already considered in the previous two sections. Beginning with a very simple case, the norbornenyl system (XX), we obtain the

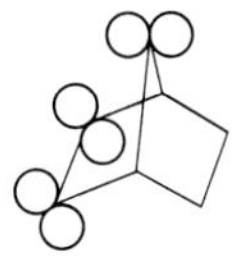

XX

interaction diagram (Fig. IX.30). Here the p-orbital cannot interact with $\pi^*$ since there is no net overlap. It can, however, mix into the lower energy $\pi$ level in a bonding way. Similarly, $\pi$ can mix into the p-level in an antibonding way. The resulting norbornenyl MOs are therefore of the type and relative energy shown. If we consider the norbornenyl cation first, this should be a stabilized system, since the only occupied level is significantly lowered in energy. However, if we were to consider the anion, this would be a nonstabilized system, since two of the four

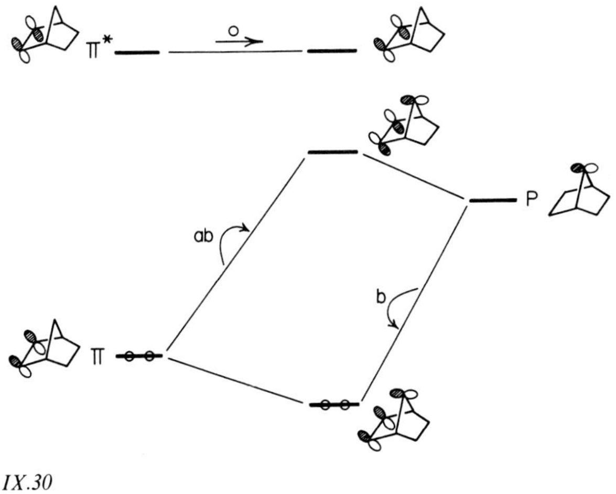

FIGURE IX.30

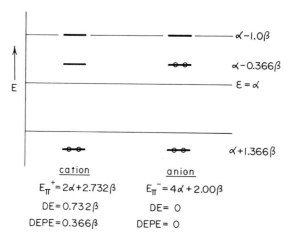

FIGURE IX.31 *Calculated energy levels for norbornenyl systems.*

electrons are lowered and the other two comparably raised in energy. These conclusions are in accord with our previous expectations since the cation is modally a (2,0) system, whereas the anion is (2,2). These qualitative results are also in good agreement with HMO calculations[†] on this system, which are given in Fig. IX.31.

It is of interest to compare the foregoing results with those for the norbornadienyl cation and anion. It is simpler first to consider norbornadiene, for which the orbital interaction diagram is shown in Fig. IX.32. Thus norbornadiene itself would not be expected to be a stabilized system. [Also it is a (2,2) system and calculated HMO results show that it has a *DE* value of zero.]

A further interaction diagram (Fig. IX.33) can now be constructed by allowing a p-orbital at the bridgehead position to interact with the norbornadiene orbitals, yielding norbornadienyl. The only effective interaction will be between $\pi_2$ and p, as shown. From the diagram it is seen that the norbornadienyl cation [a (2,0,2) system] will be stabilized since two of the norbornadiene electrons remain at the same energy level and two are significantly lowered by delocalization. Again, the anion [a (2,2,2) system] will be nonstabilized since the two additional electrons in this system would be raised in energy on delocalization

---

[†] These were obtained using a reduced bond integral value of $\beta/2$ for the transannular interaction.

## IX.3 Orbital Interaction Diagrams

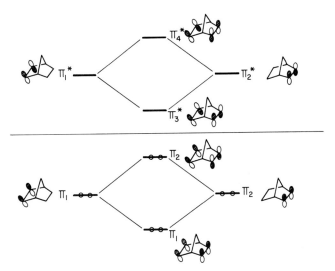

FIGURE IX.32

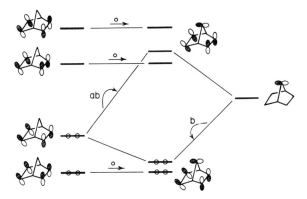

FIGURE IX.33

to form the norbornadienyl MOs, offsetting the gain in energy of those in the lowest lying MO.

The calculated HMO energies correspond very well with the above qualitative picture, as shown by the energy level diagram (Fig. IX.34) and associated quantities such as $E_\pi$ and $DE$ that were calculated previously.

Using the result of the interactions that give norbornadiene, a further ethylenic interaction can be superimposed on this moiety to yield

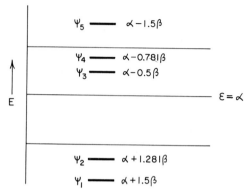

FIGURE IX.34  *HMO energy level diagram for the norbornadienyl system.*

the barrelene system (Fig. IX.35) from which it is seen that this six electron Möbius system is nonstabilized, as concluded previously. Also the energy level pattern is very similar to that obtained either by calculation or by the circle mnemonic device. However, as in the case of benzene, the degeneracies of some of the pairs of levels do not come directly out of this qualitative approach. However, it is clear that some of the levels will be very close in energy and in fact the patterns of the orbital coefficients for each level do correspond with those calculated by the HMO method.

Finally, the approach can be extended to laticyclic systems; for example, the $(2° + 1 + 2°)$ laticyclic system already treated by the

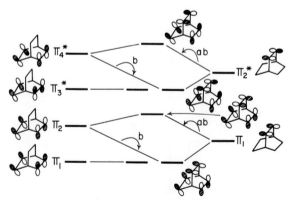

FIGURE IX.35

## IX.4 MO Following

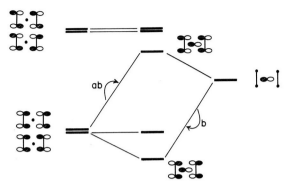

FIGURE IX.36  Orbital interaction diagram for the (2 + 1 + 2) laticyclic system.

Goldstein–Hoffmann approach[1] and by an HMO treatment, has an interaction diagram as shown in Fig. IX.36. The resulting MOs and levels are very similar to those already obtained. (In this case, either zero or very weak interactions are assumed to occur between the separated ethylenic moieties.) Clearly, the cation will be stabilized (as shown), but not the anion, as found earlier.

This kind of approach can be applied to other more complicated systems, provided the simple rules are observed and care is taken to allow only orbitals of appropriate symmetry to interact. This often gives very useful insights into problems of chemical stability and reactivity. Similar treatments can also be developed for delocalized $\sigma$-orbital systems. It should not be concluded, however, that use of orbital interaction diagrams always leads to the prediction that it is the cationic species which is stabilized and the anion is not. This point could be illustrated by constructing an interaction diagram for the [3.2.2]-bicyclononyl system, which would show that the anion was clearly stabilized relative to the cation.

### IX.4  MO FOLLOWING

This qualitative MO approach to chemical reactions was developed recently by Zimmerman,[9] who calls it "the molecular orbital counterpart of electron pushing." It involves following the *form* of the MOs of the system that change during reaction. As pointed out by Zimmerman, the forms of MOs have already been used to define their symmetry characteristics and enable the construction of correlation diagrams for electrocyclic reactions. Also the forms of the HOMO and LUMO have been used in treating sigmatropic processes. However, for such treatments to be useful, one of two conditions is necessary: either that certain symmetry be maintained during the reaction or that the reacting system comprise a cyclic array of orbitals. This is true whether the orbital symmetry conservation or the Möbius–Hückel approach is being used.

Where no suitable molecular symmetry exists or where the reaction does not involve a simple cyclic array of orbitals, reactions have been treated by perturbation approaches. Zimmerman's approach offers a simple, qualitative MO treatment that is also based on perturbation theory, and which is equally capable of dealing with reactions that do not meet the above conditions, as well as those that do and can be subjected to the approaches described previously.

In addition to allowing predictions as to whether a given reaction will be allowed or forbidden, MO following often gives an insight into the factors that make a reaction energetically favorable or unfavorable, and also which geometric changes of the system will tend to lower the barrier to a forbidden reaction.

The method involves two basic and complementary approaches, which we briefly describe. The first is to assess the electronic characteristics of the system at half-reaction, and from the form of the MOs at this point to develop correlations of reactant–transition state–product MOs. This approach is based on a presumed continuity of nodal character as reaction proceeds.[10] The second approach is to use perturbation theory in a simple qualitative way to assess the changes in each reactant MO as the system proceeds to half-reaction (or transition state) and on to products.

The treatment involves several basic assumptions, all of which are quantum mechanically reasonable. These assumptions are as follows:

1. The energies of MOs increase with increasing number of nodes.
2. The nodal character of MOs changes in a continuous manner, and in a linear or totally cyclic array of basis orbitals the number of nodes does not change.
3. In a linear system the nodes tend to remain symmetrically disposed about the center of the array.
4. In linear systems, levels corresponding to MOs that have the same nodal parity (odd or even numbers of nodes) tend not to cross.

## IX.4 MO Following

5. Appreciable interaction between MOs as new bonds are formed during reaction only occurs when the two bonding portions have matching nodal character. Such interactions also lead to noncrossing of MO levels.

Note that in the above statements only nodes *between* atomic or hybrid orbitals are taken into account. Nodes that occur *within* these component orbitals are ignored.

The method can best be illustrated by using specific examples, beginning with reactions that lack suitable symmetry for the other approaches to be applied. One interesting case considered by Zimmerman is the known 1,2-shift of a hydrogen or alkyl group in a carbene[11] to give an ethylene (XXI). This process can be represented

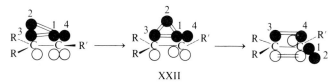

XXI

in terms of the gross changes in the orbitals involved as in (XXII).

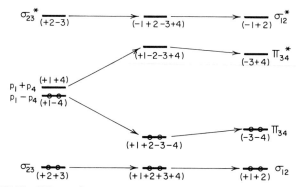

XXII

Based on these orbitals, and their changes in interaction as reaction proceeds, the following energy level correlation diagram (Fig. IX.37)

FIGURE IX.37  *MO transformations in 1,2-migration in carbenes.*

can be set up. It should be noted that for the reactant, the orbitals at the carbene center are shown arbitrarily as pure p-orbitals, although it is known that there may be added s-character. However, it is not necessary to admix any s-character at this point. Further, although these orbitals (and their levels) could be designated simply as $p_1$ and $p_4$, it is preferable to use linear combinations $(p_1 + p_4)$ and $(p_1 - p_4)$ since it is evident that some of the MOs at half-reaction acquire both $p_1$ and $p_4$ character. Any linear combination of these two orbitals is acceptable for the starting orbitals. The starting orbitals are easy to designate, since there will be both a $\sigma$ and a $\sigma^*$ orbital involving 2 and 3, and also the linear combinations of p-orbitals at the carbene center as previously described. Similarly, in the product there will be an ethylenic $\pi$ and $\pi^*$ orbital involving 3 and 4, as well as a $\sigma$ and $\sigma^*$ orbital involving 1 and 2. The form of the MOs at half-reaction is obtained from their analogy to butadienoid orbitals, since the four interacting orbitals form a linear array of our partially delocalized orbitals. Hence nodal properties can be assigned to the half-reaction orbitals. It should be noted that no weighting of any of the basis orbitals is implied and the notations given under each level simply give the nodal characteristics. For example, the notation $(+1 - 2 - 3 + 4)$ only implies that orbitals 2 and 3 in the half-reaction MO have opposite sign from 1 and 4, and that there are nodes between orbitals 1 and 2, and between 3 and 4.

The form of the localized reactant and product MOs is based on the nature of the orbitals from which they are formed. Thus $\sigma_{23}$ is designated as $(2 + 3)$ and $\sigma_{23}^*$ as $(+2 - 3)$. Similarly for the products $\sigma_{12}$ is $(+1 + 2)$ and $\sigma_{12}^*$ is $(-1 + 2)$. (Note that multiplying any set of MO coefficients by $-1$ changes nothing physically.)

The form of each MO from reactant to product state can now be followed with a view to determining what new character each acquires during reaction and in what direction the energy is likely to change. Thus the electronic changes can be followed to examine if the overall reaction should be favorable or unfavorable energetically. For example, if we follow $\sigma_{23}$, which has the form $(+2 + 3)$ (i.e., no node between orbitals 2 and 3), then this should correlate directly with MO1 of the half-reacted species but not with MO2, if nodal character persists. Thus $\sigma_{23}$ acquires some $+1 + 4$ character at half-reaction, and this should occur with no marked increase in energy. Similarly, we can follow the next higher occupied reactant orbital, which is one of the two degenerate linear combinations, namely $(p_1 - p_4)$, which has $(+1 - 4)$ character, or a node between orbitals 1 and 4. As this acquires

## IX.4 MO Following

some 2 and 3 character, it will correlate directly with second half-reaction MO $(+1 + 2 - 3 - 4)$ if nodal character persists, and not with MO3, which is $(+1 - 2 - 3 + 4)$. Because this starting orbital is essentially nonbonding and correlates with the second lowest butadienoid MO at half-reaction, this change should occur with a lowering of energy. Thus of the four electrons occupying the reactant MOs, two are involved in energy-lowering interactions and two are not expected to be significantly changed energetically. Thus the overall migration should be allowed. The form of the antibonding MOs could also be followed in a similar manner, again assuming the persistence of nodal character.

Another (and complementary) way to decide which orbitals mix into which and to obtain information about the changes in MO energies during reaction, is to use a simple first-order perturbation approach. Each reactant MO incorporates some of the character of other reactant MOs as reaction takes place; and to decide which character it acquires, Zimmerman gives the following simple rule based on perturbation theory. Each reactant MO $(k)$ is added to the MO in question $(o)$ using a mixing factor $Q_k$ given by the expression

$$Q_k = cS_{ko}^{new}/(E_k - E_o)$$

The numerator of this expression is proportional to the overlap $(S_{ko})$ of the original $(o)$ MO under consideration with the MO $(k)$ being added. However, only newly bonding (or interacting) orbitals are included in a consideration of the inter-MO overlap. This new overlap $S_{ko}^{new}$ may be positive (i.e., $++$ or $--$) or negative (i.e., $+-$ or $-+$). The denominator is positive if a higher energy MO is being mixed into the original MO, and negative if a lower energy MO is being admixed. Thus if both numerator and denominator have the same sign, the MO being mixed in is added to the original MO; if numerator and denominator are of different sign, the MO being mixed in is subtracted. (Thus the proportionality constant $c$ is positive.)

This may be restated by saying that a given MO will acquire some of the character of another MO as reaction proceeds, if the change in overlap between the two MOs, considering all sites of increased bonding, is nonzero. Also, any MO character acquired by another MO is added in so that the low-energy MO obtains a bonding contribution and the high-energy MO an antibonding contribution.

This rule can be applied to the perturbation of the reactant MOs in the carbene rearrangement already considered. For example, the MO $\sigma_{23}$ can only interact effectively with $(p_1 + p_4)$. The $+2$ part of this MO overlaps positively with $p_1$ [giving a (+ +) contribution] and the +3 part overlaps with $p_4$ [again giving a (+ +) contribution]. Thus $S_{ko}^{new}$ is positive. Also $\sigma_{23}$ is lower in energy than $(p_1 + p_4)$ so that $(E_k - E_o)$ is positive. Thus $Q_k$ for the mixing in of $(p_1 + p_4)$ into $\sigma_{23}$ is positive, resulting in $\sigma_{23} + (p_1 + p_4)$ character at half reaction, or $(+1 + 2 + 3 + 4)$. The MO $\sigma_{23}$ cannot interact effectively with either $(p_1 - p_4)$ or $\sigma_{23}{}^*$ in this way, since each positive overlap term has a corresponding negative overlap term. Thus $S_{ko}^{new} = 0$ and no admixing of these MOs results.

If we turn to the MO $(p_1 - p_4)$, this can only interact with the antibonding orbital $\sigma_{23}{}^*$. The $+p_1$ part of the MO overlaps positively (+ +) with the +2 part of MO $\sigma_{23}{}^*$ and the $-p_4$ part overlaps with the $-3$ part of $\sigma_{23}{}^*$, giving a (− −) or positive overlap contribution. Thus $S_{ko}^{new}$ is positive. In this case $\sigma_{23}{}^*$ is higher in energy than $(p_1 - p_4)$, so that $E_k - E_o$ is again positive. Thus $\sigma_{23}{}^*$ mixes into $(p_1 - p_4)$ positively, yielding the combination $(+1 + 2 - 3 - 4)$ at half-reaction. Attempted admixing of either $\sigma_{23}$ or $(p_1 + p_4)$ into the MO $(p_1 - p_4)$ again gives corresponding positive and negative overlap contributions and no admixing of these MOs results.

Instead of this qualitative description, we could have used equation (1) explicitly and obtained all possible interaction terms I, which can be designated as $I(MO_o:MO_k)$. Thus for the starting MO $\sigma_{23}$ of the carbene system we have $I(\sigma_{23}:(p_1 - p_4))$, which gives

$$Q_k = \frac{cS_{ko}^{new}}{E_k - E_o} = \frac{c(++) + (+-)}{(+)} = 0$$

Thus $I(\sigma_{23}:(p_1 - p_4))$ leads to no interaction. $I(\sigma_{23}:(p_1 + p_4))$, which gives

$$Q_k = \frac{c((++) + (++))}{(+)} > 0$$

leading to $(\sigma_{23} + (p_1 + p_4))$ or $(+1 + 2 + 3 + 4)$. $I(\sigma_{23}:\sigma_{23}{}^*)$, which gives

$$Q_k = \frac{c((++) + (+-))}{+} = 0$$

or no interaction. Similarly for the MO $(p_1 - p_4)$ we obtain

$I(p_1 - p_4:\sigma_{23}) = 0$ since $Q_k = 0$
$I(p_1 - p_4:p_1 + p_4) = 0$ since $Q_k = 0$
$I(p_1 - p_4:\sigma_{23}{}^*) = (+1 + 2 - 3 - 4)$ since $Q_k > 0$

## IX.4 MO Following

For the MO $(p_1 + p_4)$ we have only one nonzero interaction

$$I(p_1 + p_4 : \sigma_{23}) = (+1 - 2 - 3 + 4) \quad \text{since} \quad Q_k < 0$$

and also for $\sigma_{23}{}^*$ only one nonzero interaction, namely,

$$I(\sigma_{23}{}^* : p_1 - p_4) = (-1 + 2 - 3 + 4) \quad \text{since} \quad Q_k < 0$$

This treatment[†] leads directly to the form of the half-reaction MOs shown in Fig. IX.37, and the correlation diagram is completed by connecting levels of like nodal parity between reactant or product MOs and those at half-reaction.

An interesting result obtained from the correlations in Fig. IX.37 is that the carbenic orbital $(p_1 + p_4)$ cannot be the orbital that becomes part of the $sp^2$ hybrid in the product. If this were to be the case, the orbital would have to be lowered in energy from a nonbonding MO in the reactant to a bonding MO in the product; the MO would then contain an electron pair at the onset of reaction but would become antibonding (see Fig. IX.37) as reaction proceeded. Such processes are forbidden. This forbiddenness is understandable since $(p_1 + p_4)$ is a vertically oriented orbital (in the plane of the paper in XXII), with its positive lobe directed upwards. In the product this would put the $sp^2$ hybrid and the $R'$ group in the plane of the paper and would thus generate a twisted ethylenic bond.

If the $(p_1 - p_4)$ orbital, which is perpendicular to the plane of the paper, were to acquire some $S$ character, it would be lowered in energy as it approached an $sp^2$ configuration, but would still correlate directly with a bonding orbital in the product without passing through an antibonding level at half-reaction. In this case the $sp^2$ hybrid and the group $R'$ would remain horizontal to the plane of the paper during the allowed reaction.

Zimmerman has used the above approach to predict which processes are allowed in a number of other interesting cases such as carbene insertions, 1,2-migrations, sigmatropic rearrangements, and cheletropic reactions.[9] Its main advantages are that one can follow the changes in electron density distribution, the motion of nodes, and the changing characteristics of entire MOs as reactions proceed, and that it does not require cyclic arrays of orbitals at half-reaction or the preservation of molecular symmetry to be applicable. Rather than simply giving

---

[†] Note that in any interaction such as $I(\sigma_{23}{}^* : p_1 - p_4)$, this is equivalent to the sum of $I(\sigma_{23}{}^* : p_1)$ and $I(\sigma_{23}{}^* : -p_4)$ since only newly developed overlaps are considered.

predictions of allowedness or forbiddenness, the approach also gives insight into the factors controlling the energy changes during reaction that arise from electronic delocalization as bonds are made and broken.

## PROBLEMS

1. Determine the modal properties of the polyene ribbons in each of the following systems and decide which systems will be stabilized, and which will be destabilized:

I

V

II

VI

III

VII

IV

VIII

IX

Rank these systems in approximate order of increasing $\pi$-electron stabilization.

2. Calculate $E_\pi$ and $DE$ for the barrelene system (X). Use $\beta$ for the

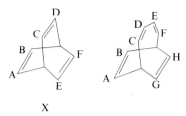

X

$H_{AB}$-type integrals and $\beta/2$ for the $H_{AC}$ type. Set transannular integrals such as $H_{AD} = 0$.

3. Calculate $E_\pi$ and $DE$ for the [4.2.2]-bicyclodecatetraene (above) system. Use values of $\beta$ for normal $H_{AB}$-type bond integrals, $\beta/2$ for the $H_{AC}$ type, and $3\beta/4$ for the $H_{AG}$ type. Set each $H_{AF}$-type integral $= 0$.

4. Calculate $E_\pi$, $DE$, and $DEPE$ for the [3.2.2]-bicyclononadienyl cation (XI). Determine the coefficients of the occupied MOs only, and use these to obtain the electron densities and bond orders. Use $H_{AB}$ etc. $= \beta$, $H_{AC}$ etc. $= \beta/2$, $H_{AF}$ etc. $= \beta/2$, $H_{AE}$ etc. $= 0$.

XI

5. Construct an orbital interaction diagram for the [3.2.2]-bicyclononadienyl system, and compare the resulting energy level distribution with that obtained in Problem 4.
6. Construct an orbital interaction diagram, starting from two butadienes, to obtain the energy level pattern and form of the MOs for planar cyclooctatetraene.

## REFERENCES

1. M. J. Goldstein and R. Hoffman, *J. Am. Chem. Soc.* **93**, 6193 (1971).
2. H. E. Zimmerman and R. M. Pauffer, *J. Am. Chem. Soc.* **82**, 1514 (1960); H. E. Zimmerman, G. L. Grunewald, R. M. Pauffer, and M. A. Sherwin, *J. Am. Chem. Soc.* **91**, 2330 (1969).
3a. S. Winstein and C. Ordronneau, *J. Am. Chem. Soc.* **82**, 2084 (1960).

3b. P. R. Story and M. Saunders, *J. Am. Chem. Soc.* **82**, 6199 (1960); P. R. Story, L. C. Snyder, D. C. Douglas, E. W. Anderson, and R. L. Kornegay, *J. Am. Chem. Soc.* **85**, 3630 (1963).
4a. J. B. Grutzner and S. Winstein, *J. Am. Chem. Soc.* **90**, 6562 (1968).
4b. S. W. Staley and D. W. Reichard, *J. Am. Chem. Soc.* **91**, 3998 (1969).
5. Reference 3a.
6. Reference 4b.
7. M. J. Goldstein, *J. Am. Chem. Soc.* **89**, 6357 (1967).
8. W. L. Jorgensen and L. Salem, *The Organic Chemist's Book of Orbitals*, Academic Press, New York, 1973.
9. H. E. Zimmerman, *Accounts Chem. Res.* **5**, 393 (1972).
10. J. A. Altmann, I. G. Csizmadia and K. Yates, *J. Am. Chem. Soc.* **96**, 4196 (1974); **97**, 5217 (1975).
11. W. Kirmse, "Carbene Chemistry," p. 52. Academic Press, New York, 1964.

## SUPPLEMENTARY READING

* Goldstein, M., and Hoffmann, R., *J. Am. Chem. Soc.* **92**, 6193 (1971).
* Zimmerman, H., *Acc. Chem. Res.* **5**, 393 (1972).

# *INDEX*

## A

Allyl cation, $\pi$-energy and reactivity of, 230
Allyl system
  bond orders for, 71
  $B_\pi$, values of, 48
  *DE* values of, 50
  electron densities for, 76
  $E_\pi$, values of, 48
  HMO treatment of, 46
  molecular orbitals of, 64
$\alpha$, *see* Coulomb integral
Alternant hydrocarbons,
  definition of, 79
  dipole moments of, 83
  electron densities of, 83
  energy levels of, 80–83
Aniline
  HMO charge distribution in, 221
  localization energies for, 227
Annulenes, 144–150
Anthracene,
  *DE* value of, 15
Aromaticity
  and bicycloaromaticity, 332
  criteria for, 135, 142
  for pericyclic and other topologies, 314–334
Azulene
  *DE* value of, 151
  dipole moment of, 84, 207
  HMO predictions of reactivity in, 219

## B

Barrelene
  as longicyclic system, 328
  *DE* value of, 301
  energy level distribution in, 299
  $\pi$-energy of, 300
  orbital interaction diagram for, 356
*BEPE*, *see* $\pi$-bonding energy per electron
$\beta$, *see* Bond integral
Benzaldehyde, HMO charge distribution in, 221
Benzene
  bond orders of, 123
  delocalization energy of, 118
  electron densities for, 123
  energy levels of, 117
  localization energy for, 228
  molecular orbitals of, 122
  resonance energy of, 118–120
Benzyl cation, $\pi$-energy and reactivity of, 230
Bicycloaromaticity, 332
  definition of, 334
Bicyclodecapentaene, [6.2.0], *DE* value of, 151
Bicyclotetradecaheptane, 151
Bond integral
  experimental estimates of, 213–218
  in EHMO method, 191
  in MO theory, 22, 38

**367**

Bond order
  consistency with bond integral, 189
  correlation with bond length, 209
  correlation with infrared frequency, 210
  definition of, 70
$B_\pi$, see $\pi$-bonding energy
Bullvalene, degenerate rearrangement of, 284
Butadiene
  bond orders for, 72
  DE value of, 55
  electron densities for, 76
  HMO treatment of, 51
  molecular orbitals of, 69
  $\pi$-energy of, 54

## C

Carbene, 1,2 shifts in, 359
Character table, 101
  for $C_{2v}$ group, 110
Characteristic equation, see Secular polynomial
Charge density, definition of, 77
Charge-transfer spectra, and simple HMO results, 213
Claisen rearrangement, 283
Coefficients, 42, 43
  calculation of, 60–63
  generalized method, 66
  including overlap, 179
Cofactors
  definition of, 52
  method of, 52–54
Cope rearrangement, 282, 283
  application of Möbius–Hückel approach to, 308
Core potential, 159
Correlation diagrams, 250–256
Coulomb integral
  in MO theory, 22, 37
  in valence bond theory, 13
Cyclobutadiene
  bond order for, 73
  DE value of, 57
  HMO treatment of, 56
  $\pi$-energy of, 56

Cyclobutadienyl, energy levels for, 137
Cycloheptatrienyl, energy levels for, 139
Cyclooctatetraenyl, energy levels for, 141
Cyclopentadienyl, energy levels for, 130, 138
Cyclopropane, energy level distribution in, 297
Cyclopropenyl, energy levels for, 136

## D

DE, see Delocalization energy
Delocalization energy
  and bond order, 71
  and resonance stabilization, 118
  and resonance stabilization energy, 133
  and solvolytic reactivity, 230
  definition of, 50
  of polyenes
    branched chain, 134
    cyclic and acyclic, 131
    linear, 132
  of typical hydrocarbons, 120
  per electron, 50
DEPE, see Delocalization energy, per electron
Dewar–Zimmerman rules, 307, see also Möbius–Hückel concept
  consistency with generalized selection rules, 309, 310
Diels–Alder reaction
  and simple HMO results, 231–233
  and Woodward–Hoffmann rules, 271
  orbital symmetry relationships in, 267
Dipole moment
  and simple HMO results, 207
  of alternant and nonalternant hydrocarbons, 83

## E

EHMO method, see Extended Hückel method
Electrocyclic reactions, definition of, 241

# Index

Electron density
 definition of, 75
 distribution of, 65
$E_\pi$, see $\pi$-electronic energy
Ethylene
 bond order for, 71
 $B_\pi$, value of, 46
 EHMO energy levels for, 198
 EHMO treatment of, 192
 electron density for, 76
 $E_\pi$, values of, 48
 HMO treatment of, 44
 molecular orbitals of, 62
  in EHMO treatment, 196
 $\pi$-system for, 31
Exchange integral, in valence bond theory, 13
Extended Hückel method, 190–201

## F

Formaldehyde
 dipole moment of, 209
 HMO charge distribution for, 208
Free valence index, definition of, 74
Frontier electron approach, 222
Fulvene, dipole moment, 207

## H

Hamiltonian operators, 8–10
Hermitian operator, 4
Heteroatoms
 bond integrals for, 161
 coulomb integrals for, 157
 table of parameters for, 162
 treatment of by $\omega$-technique, 188
 treatment of in HMO theory, 156–163
Hückel MO method, 27–44
 application to simple $\pi$-systems, 44
 approximations in, 37–39
 basic assumptions in, 27–29
 basic procedure in calculations, 43, 44
 extensions and improvements of, 156–204

Hückel's rule, 134–142
 and annulenes, 143–150
 and polycyclic hydrocarbons, 150
 applicability of, 143, 291
Hückel systems, see also Möbius–Hückel concept, definition of, 291
 energy level patterns for, 296
 stability requirements of, 294
Hydrogen molecule, 11
 MO wave function for, 24

## I

Intermolecular cycloadditions, 263
 selection rules for, 270
Intramolecular cycloadditions, 242
 selection rules for, 248
Ionization potentials, and simple HMO results, 215

## L

Laticyclic systems, 317, 330
 simple HMO treatment of, 341–344
LCAO method, 19–21
Localization energy, definition of, 227
Longicyclic systems, 317, 327–329

## M

Matrix elements, 100
Möbius–Hückel concept, 291
 application to concepted reactions, 301–309
Möbius systems, see also Möbius–Hückel concept, definition of, 293
 energy level patterns for, 296
 stability,
  requirements, 294
  source of, 309–311
Mode, definition of, 320
MO following approach, 357–364
Molecular orbital theory, 18–25

## N

Naphthalene
  DE value of, 150
  reactivity at $\alpha$ and $\beta$ positions, 221
NDO approximation, *see* Neglect of differential overlap
Neglect of differential overlap, 39, 40
Nodal character, continuity of, 358
Nodes, 64, 69, 122, 311
Nonalternant hydrocarbons, definition of, 83
Noncrossing rule, 261
Nonplanar systems, HMO calculations on, 334–345
Norbornadienyl
  anion, stability of, 328
  cation, stability of, 328
  DE values of, 338, 339
  energy levels for, 339
  molecular orbitals of, 340
  orbital interaction diagram for, 355
  $\pi$-energies of, 338, 339
  simple HMO treatment of, 335
Norbornenyl
  anion, stability of, 354
  cation, stability of, 354
  energy level distribution for, 353
Normalization, 4, 193
  including overlap, 178

## O

$\omega$-technique, 180
  and charge distribution in allyl cation, 184
  and delocalization energy, 185
  and dipole moment, 208
  and heteroatoms, 188
  definition of, 181
Orbital interaction diagrams, 345–357
  general rules for constructing, 345, 346
Orbital symmetry, conservation of, 239
Overlap, inclusion of in HMO calculations, 170–180

Overlap integral
  in MO theory, 23, 39
  in valence bond theory, 14
  typical values for, 200

## P

Pauli principle, 16, 20
Pentadienyl, energy levels of, 130
Pericyclic systems, 321–326
Phenanthrene, DE value of, 150
$\pi$-bonding energy
  definition of, 46
  per electron, 48
$\pi$-electronic energy
  definition of, 46
  including overlap, 173–177
  of cyclic polyenes, 127
  of linear polyenes, 127
$\pi \rightarrow \pi^*$ transition energy, and simple HMO results, 212
Point group, 90
Polygon rule, 128
Pseudosymmetry, 317
Pyrene, DE value of, 150

## Q

Quantum mechanics, basic postulates, 2–8

## R

Reactivity
  and simple HMO results, 218–234
  delocalization approaches, 229
  frontier electron approach, 222
  initial-state calculations, 219
  perturbation approach, 358
  perturbation methods, 223–226
  product stability approach, 231
  superdelocalizability index, 225
Redox potentials, and simple HMO results, 214
Resonance integral, *see* Bond integral

# Index

Resonance stabilization energy, vertical and nonvertical, 119
Ribbons, definition of, 315

## S

Schrödinger equation, 7
Secular determinant, 33, 34
 generalized Hückel form, 40
Secular equations, 33, 34
Secular polynomial, 33
Selection rules
 for intermolecular cycloadditions, 270
 for intramolecular cycloadditions, 248
 for sigmatropic shifts, 280, 283
 generalized,
  based on Möbius–Hückel approach, 310
 for pericyclic reactions, 284–288
$S_i$, see Symmetry orbital
$S_{ij}$, see Overlap integral
Sigmatropic reactions, 276
 definition of, 242
 Möbius–Hückel approach to, 306–309
 selection rules,
  for $[i,j]$ shifts, 283
  for $[1,j]$ shifts, 280
Signed minors, see Cofactors
Slater atomic orbitals, 190
 expressions for, 193
Slater exponent, 193
Spirocyclic systems, 317, 326, 327
State-correlation diagrams, 256–263, 272
Superdelocalizability index, 225
Symmetry element, 250
Symmetry operations, 89, 90
Symmetry orbitals, 99, 102
 and secular determinants, 100, 105
 for norbornadienyl, 337

## T

Transformation table, 94, 102
Trimethylenemethane
 $DE$ value of, 59
 HMO treatment of, 58
 $\pi$-energy of, 59
 twist form, 297

## U

Urea
 $DE$ value of, 168
 energy levels of, 167
 HMO treatment of, 165
 molecular orbitals of, 169

## V

Valence-bond theory, 10–18
Variation method, 30, 34

## W

Walsh model, for cyclopropane, 298
Wheland intermediate, HMO treatment of, 226
Wolfsberg–Helmholtz approximation, 191
Woodward–Hoffmann rules, 233, 240, 245, 249

## Z

ZDO approximation, see Neglect of differential overlap
Zero differential overlap, see Neglect of differential overlap